Software Test Attacks to Break Mobile and Embedded Devices

Chapman & Hall/CRC Innovations in Software Engineering and Software Development

Series Editor
Richard LeBlanc
Chair, Department of Computer Science and Software Engineering, Seattle University

AIMS AND SCOPE

This series covers all aspects of software engineering and software development. Books in the series will be innovative reference books, research monographs, and textbooks at the undergraduate and graduate level. Coverage will include traditional subject matter, cutting-edge research, and current industry practice, such as agile software development methods and service-oriented architectures. We also welcome proposals for books that capture the latest results on the domains and conditions in which practices are most effective.

PUBLISHED TITLES

Software Development: An Open Source Approach
Allen Tucker, Ralph Morelli, and Chamindra de Silva

Building Enterprise Systems with ODP: An Introduction to Open Distributed Processing
Peter F. Linington, Zoran Milosevic, Akira Tanaka, and Antonio Vallecillo

Software Engineering: The Current Practice
Václav Rajlich

Fundamentals of Dependable Computing for Software Engineers
John Knight

Introduction to Combinatorial Testing
D. Richard Kuhn, Raghu N. Kacker, and Yu Lei

Software Test Attacks to Break Mobile and Embedded Devices
Jon Duncan Hagar

Software Designers in Action: A Human-Centric Look at Design Work
André van der Hoek and Marian Petre

CHAPMAN & HALL/CRC INNOVATIONS IN
SOFTWARE ENGINEERING AND SOFTWARE DEVELOPMENT

Software Test Attacks to Break Mobile and Embedded Devices

Jon Duncan Hagar

CRC Press
Taylor & Francis Group
Boca Raton London New York

CRC Press is an imprint of the
Taylor & Francis Group an **informa** business

A CHAPMAN & HALL BOOK

CRC Press
Taylor & Francis Group
6000 Broken Sound Parkway NW, Suite 300
Boca Raton, FL 33487-2742

© 2014 by Taylor & Francis Group, LLC
CRC Press is an imprint of Taylor & Francis Group, an Informa business

No claim to original U.S. Government works

Version Date: 20130611

International Standard Book Number-13: 978-1-4665-7530-1 (Paperback)

Library of Congress Cataloging-in-Publication Data

Hagar, Jon Duncan.
   Software test attacks to break mobile and embedded devices / Jon Duncan Hagar.
      pages cm. -- (Chapman & Hall/CRC innovations in software engineering and software
      development series)
   Includes bibliographical references and index.
   ISBN 978-1-4665-7530-1 (pbk.)
   1. Penetration testing (Computer security) 2. Mobile computing--Security measures. 3. Embedded
computer systems--Security measures. I. Title.

QA76.9.A25H343 2013
005.8--dc23   2013021984

Visit the Taylor & Francis Web site at
http://www.taylorandfrancis.com

and the CRC Press Web site at
http://www.crcpress.com

Contents

Foreword by Dorothy Graham

IF YOU HAVE A MOBILE PHONE, a car, or even a washing machine, you are affected by software in mobile or embedded systems. If your workplace has a website, it will increasingly be viewed not only on computers but on phones, tablets, and other mobile devices. If you are testing software, your future may include more and more testing on such devices.

Much of traditional testing wisdom can be applied to testing in this new and growing space, but traditional wisdom is not enough. Special considerations and concerns for testing in embedded systems are known to some specialists but too few general testers. Many are now finding that they need to test on mobile devices and are encountering unexpected problems and wondering how they should be testing in this new space.

This book fills a need in two major areas: first, for traditional testers when moving into the mobile and embedded area, it bridges the gap between IT and mobile and embedded system testing, showing how to apply some traditional approaches and outlining new approaches. Second, for those working with mobile and embedded systems who may not have an extensive background in testing, this book brings testing ideas, techniques, and solutions that are immediately applicable to testing these devices.

Using the framework of attacks popularized by James Whittaker's books, Jon Duncan Hagar describes those that are relevant here and extends the approach with new attacks specifically for mobile and embedded systems. He provides detailed information and guidance on how to test more effectively and efficiently in the mobile and embedded world. Some of the attacks may be familiar but are more important for these devices. Some attacks may need to be applied in different ways. Some of the attacks may be completely new to you. In any case, Jon shows you what to test and how to test, giving ideas that you can use to do better testing of mobile devices now and save yourself serious trouble later on.

Jon's extensive experience (much of it in the embedded world), his thorough research, and his deep knowledge give this book a solid foundation and provide helpful guidance and steps to take in applying testing attacks to mobile and embedded devices.

Foreword by Lisa Crispin

I WAS SO PLEASED WHEN I HEARD THAT JON WAS WRITING A BOOK about embedded software testing. First, Jon is a great teacher. I learn new ideas and techniques every time I participate in one of his workshops. Second, I know next to nothing about testing embedded software, and as Jon points out in his introduction, this is no longer an area we can avoid. Products that run only in a web browser or on a desktop are increasingly rare. And so I find myself working on a team whose product must work well on mobile devices.

As I started reading this book, I was intrigued by the idea of patterns for test attacks. This is also an area where I lack experience, and it is clearly a critical skill that most, if not all, software testers must acquire and hone.

One reason I have avoided testing embedded software and mobile devices for so long is that, honestly, the idea of doing that kind of testing seems scary to me. It appears to involve endless permutations of testing scenarios. I imagine the complex hardware and software required. The different configurations, operating systems and versions of operating systems, amount of memory, and other variable features of mobile devices appear to be boundless. Mobile device testing experts I know test on 70 or more phones. Creative approaches such as crowd sourcing may be required, and that is another area that is new for me.

Maybe you share my initial hesitation. Maybe you think, "I've avoided learning how to do testing attacks on mobile and embedded devices all this time, and I can probably get by without knowing how for a few more years." But look at all the mobile devices around you. It is time to put these tools in your testing toolbox and aspire to be not just a great but a magnificent tester.

Fear not. As I said, Jon is a wonderful teacher. He starts with the basics and guides you step by step. He explains patterns you can use and techniques ranging from simple mind mapping to sophisticated test labs.

Every tester who wants to keep current needs to read this book, and you can read with confidence knowing you are being guided by the best in this business. As Jon says in the introduction, start reading and have some fun. And do not just read. Try the techniques. Do the exercises. Experiment with the techniques. Learn by doing, with this book as your guide.

Preface

THIS BOOK is written for testers who are working in the expanding world of "smart" devices driven by software. To be successful with the attacks in this book, a tester should have other reference materials from the test world, as well as materials from hardware, software, and systems. No single book is a complete work on software or testing. Testers also need to know about test planning, design, implementation, attacks, as well as documentation, because good testers can use all of these skills in continuous learning, testing, critical thinking, and exploration to advance their careers. Testers must use and take advantage of certain skills such as critical thinking. Testers must have a lot of skill in many testing concepts, approaches, types, and techniques. The diversity of skill with these concepts is needed because testing any complex software system is intractable, and so testers are faced with problems that call for many heuristic ideas to solve the unsolvable. Most of the good testers I know are constantly honing their craft, learning the science and art of software.

There are many kinds of software in the "traditional" or information technology (IT) software world, and now the mobile and embedded worlds are actually seeing many of the problems of those software, but they will continue having all of the traditional embedded or mobile software space problems as well. This book offers an attack basis for testing mobile and embedded systems in a mix-and-match approach, but keep in mind that there is no simple "cookbook approach" to testing.

This book deals with testing attacks that can be used by individuals as well as teams. The attacks are aimed at bugs as well as providing information about the software product under test. Many testers focus only on demonstrating that a product works and therefore end up with "green light" testing, a method that can miss many errors. In attack testing, we try to show that a software product does *not* work (has bugs), and when the software "passes" all of our attacks, we have increased confidence that the embedded product will work, which should make customers happy. However, happy customers are not the only people who should be satisfied that a product delivers on promises. But happy customers can mean the difference in a company staying in business or folding. Attack-based testing can provide information to make testers, development personnel, management, and other stakeholders happy too. After all, life is too short to be unhappy.

I encourage testers to take their software, these attacks, and go forth and attempt to break mobile and embedded systems.

Acknowledgments

I WOULD LIKE TO THANK my many mentors, colleagues, and teachers who have taught me so much through the years. There are many ideas in this book from nearly all of them, which I cite throughout the book. We are all standing on each other's shoulders. Also, I certainly would like to thank the reviewers who helped to refine this book. However, my largest acknowledgment goes to my wife, Laura—a software-systems geek. Without her dedication, tenacity, and skills, this book would not have come to fruition. She has reviewed, input, edited, dealt with bad software, and commented on every part of this book. More importantly, she has tolerated my constant changes and, at times, my sick humor. She is my partner in life and in this book. With all of these people, the references in this book, and, hopefully, the book itself, I hope the worlds of mobile and embedded software become safer, more secure, and fun to use.

Copyright and Trademarks Declaration Page

Be it known that neither Jon Duncan Hagar nor Grand Software Testing, an independent company, are affiliated with, nor have they been authorized, sponsored, or otherwise approved by Research In Motion Limited, owner of the BlackBerry trademarks, nor have they been authorized, sponsored, or otherwise approved by Nintendo of America, Inc. owner of the Wii™ trademark, nor have they been authorized, sponsored, or otherwise approved by Google Inc., owner of the YouTube™ and Android™ trademarks, nor have they been authorized, sponsored, or otherwise approved by Apple, Inc. owner of the iPad®, iPhone® trademarks, nor have they been authorized, sponsored, or otherwise approved by Microsoft Corporation owner of the Windows™ or Excel^{SM} or XBox™ trademarks, nor have they been authorized, sponsored, or otherwise approved by Amazon, Inc., owner of Amazon, Kindle, Kindle Fire, the Amazon Kindle logo and the Kindle Fire logos and trademarks, nor have they been authorized, sponsored, or otherwise approved by Apache Micro Peripherals owner of the Apache™ trademark, nor have they been authorized, sponsored, or otherwise approved by Free Stream Media Corp owner of SAMBA™ trademark.

Introduction

THIS BOOK is not about pointing out specific people's or even groups of people's mistakes and berating them. However, we should all acknowledge that, as humans, we all make mistakes.

In building systems of any kind, mistakes can and will be made whether in requirements, design, code, hardware, and even documentation, and those mistakes equate to bugs (errors that can create failures). Perfection, after all, was never part of any stipulation or standard, though with some irony, it seems that many customers expect perfection.

The attacks in this book are meant to help ensure that whatever the system's function, the system will work as desired by the customer or user. I hope that you will simply take my advice to run the attacks and therefore make your products as "good" as they can be.

Further, the title of this book and the attack-based testing approach should not be construed as anyone, particularly me, wanting to create an adversarial relationship within teams between testers and developers. Testers should be providing information, only part of which is identification of bugs. Communication within teams or to groups about testing or bugs should be done with respect. As a tester within my test team, much of the time, my attitude was one of trying my hardest to break or show the software does *not* work by finding bugs, as opposed to just trying to show the software "works." Once I found a bug, other useful information, or in some cases no bugs, I reported this quickly and fairly to the interested stakeholders.

The most successful teams of testers and developers I have been fortunate to work on were those where everyone was treated with respect. Part of the respect I received was the fact I could find bugs quickly and effectively while providing other helpful data, which in the end created a good product and made everyone successful.

The tester's attitude of finding bugs by "attacking software" can become a negative if the tester is not careful. This can result in bugs escaping into the field just as testers who only do what some people call minimal "happy path" or "happy day" testing (tests created *not* to find bugs).

Testing attacks can start at the beginning of a product's life cycle based on developer efforts or low-level developer attacks. These attacks may be applied by testers or developers, but the taxonomy data shows that they should be performed by someone. From here, we continue into the control, logic, and fault tolerance areas, where many bugs seem to cluster. This clustering is likely due to complexity and misunderstanding by humans. As the story of attacks unfolds, readers will find some attacks are embedded, some are embedded mobile, and others mobile/mobile, and many are a combination of all three. The attacks will clarify these divisions. The story also includes special sections for the unique software–hardware and software–system aspects of these devices. The combination of hardware and software makes up embedded systems. The book continues with attacks on the interfaces: the user interface (UI), graphical user interface (GUI), and other interfaces with and to humans. Then, there are attacks that address fraud (theft of and fraudulent use of other people's information for illegal purposes), security threats, alarms, and the real world that these devices interface with and to. The book ends at the beginning for many testers with a consideration of general attacks applicable to the mobile or embedded space and test environment concepts. Finally, the appendices wrap up with some useful reference information, including how projects might avoid errors in the first place.

I hope this book will aide software testers working on mobile and embedded software systems in finding bugs. To make the best use of this book, testers will, of course, need to have some basic idea of what computers, systems, and software are and how they work.

Chapter 1 provides the details and definitions of mobile and embedded devices. This is followed by a series of chapters containing approaches to testing, called attacks. Attack-based testing is just one way of finding bugs and testing software. I have included a series of appendices on major concepts to support testing of mobile and embedded systems such as test labs.

Mobile and embedded devices contain embedded software, and these devices have exploded in use and position in the global marketplace. In fact, it is hard to find electronic or mechanical products that *do not* contain embedded software these days. With increasing dependence on such products, in addition to the complexities of the software in them, the issue of errors, or as I call them in this book, bugs, has grown. Bugs and errors can cause faults and even system failures, which may negatively affect people's daily lives (banking apps), damage things (rockets destroyed), and even kill people (medical devices).

This book includes a few stories about bugs, what caused them, and their effects as well as many references for your continued reading. Upon reflection by me and other authors, it seems that the number of bugs are increasing as the amount of code grows while the numbers of people trained to find them is not increasing fast enough.

GETTING STARTED

The attack concept has been popularized in the software testing field recently. Although approaches such as requirements verification–based testing can find some bugs, attack-based testing provides an alternative approach for testers. Attack-based testing targets known bug types with patterns of activities that have the intent of "breaking" the software (i.e., exposing a bug). Attack-based testing has been popularized by the work of Dr. James Whittaker, who is extensively referenced in this book.

Attacks are patterns for testing: Like different kinds of software design patterns, which must be matched to a design problem, attacks must be recognized as being applicable to a particular (local) context. They come with the pros and cons of heuristics. This book and other books on attack-based testing reveal the more common attack patterns—but not the only ones. You should treat the attacks in this book simply as a starting point to begin formulating your own attacks, customizing any attack in this book to fit your local context.

Attacks are patterns of behavior: There are no cookbooks to give a tester every step and input for every test. Like a pattern for making cloth, the tester must modify, adjust, use tools, and most importantly *think* to create and run a test that finds a bug. The attack patterns in the book do provide information on when an attack might be applied, what makes an attack work, where the attack can be done, and a basic set of activities, which are the starting points for testing mobile and embedded software. And just as a tailor must tweak a pattern for a particular customer or fabric, so must the tester adjust each attack for the nature or context of the software or product. A tester will not run the same attacks the same way for every piece of software on any mobile device. For example, a tester who is testing pacemaker software (to keep a person alive) would use white-box computational coverage (or structural) testing (see Chapter 2, Attack 2) in a different way from someone testing the latest game (see Chapter 8, Attack 26). While they may use some of the same attacks and ideas, much tailoring is needed, as well as determining what is "good enough" for a game is likely not "good enough" for life-critical software.

The software world has viewed patterns [1] for quite a while as a method to help solve problems and do reuse. A pattern in software can be defined as a known and working solution to a problem that repeats within a segment of the industry. Patterns have history, some proof of working, a common design, which involves an algorithm or set of logic, set of programming steps, and places of appropriate usage. Their use is another form of heuristic. Patterns can be mixed and matched to help in coding and avoid "reinventing the wheel."

The audience for this book includes software testers, developer testers, and other team-mates involved in testing the software for mobile and/or embedded devices. These people can be on the development team or separate from the development team. For example, they may be in a group doing independent testing or may be working for themselves in what is now called "crowd source" testing.

Skill levels for the attacks: There are many different kinds of testers and skill levels. For example, a tester may be a developer doing low-level program testing, while another tester may work at an integrated system level, where all of the hardware and software are in play. This book is structured to help each of these kinds of testers with different attacks. Then there is the tester's skill level. In this book, I define the following tester skill levels as a simple starting point. Each attack is correlated to these roles and skill levels. A graphic indicating the skill level of the attack accompanies each attack chapter.

- ○ **Novice**—a beginning tester, someone with perhaps 0–3 years of software and testing experience.

- ☐ **Intermediate**—someone who has been testing software for a few years and understands many different testing concepts.

- ◇ **Advanced**—someone who has been working for many years as a tester on different types of software and has a grasp of both basic and advanced testing concepts.

- ◇◇ **Expert**—someone with all of the characteristics of an advanced tester who can also mentor and teach testing. Moreover, experts are probably "researching" the next new ideas in testing in addition to being lifelong learners in testing.

Additionally, I hope you find the tables at the beginning of each chapter helpful with attack references inside the chapter to (what I have determined are) four different types of embedded software products (see Chapter 1).

Bug definitions: A bug can have many definitions, including

- A software feature that does not meet requirements

- An unexpected result or outcome

- A result that does not meet consumer expectations

- An output that does not meet legal or government regulation

It is possible to have different opinions on what a bug is or is not. Testers must be advocates for bugs to ensure that bugs get fixed, but not dogmatically so, since sometimes we deliver software with known bugs. Ultimately though, the tester's job is to identify bugs as a service of providing information to whoever is willing to pay for the information about the qualities of a mobile or embedded device. If the team builds software with enough positive qualities and few negative qualities, the product stands a good chance of being accepted (paid for) by the consumer, which is usually the goal (to make money). Then the software can be called "good enough."

Bug hunting: The reasons to find bugs should be obvious—we all want products that work, yet bugs can stop products from working. But it is surprising how many products and projects seem to have only small amounts of testing or no testing whatsoever, particularly in the mobile app domain. Testing provides information about the qualities of a product. Qualities can include functionality, usefulness, performance, safety, security, and other factors. In *Quality Software Management: Systems Thinking* [2], Gerald Weinberg defines quality as "value to some person." People are willing to value (pay and use) software despite it having bugs. A bug is a quality of software, albeit a negative quality in most cases. While people will tolerate many bugs, in the long run users will tend to value better quality (fewer bugs), and this situation will drive embedded and mobile software. They will vote for a device in the marketplace (through a purchase) or software that has "good" qualities, yet not buy a device they view as having too many bad qualities (i.e., having too many bugs). This does not say that good quality means zero bugs, but consumers can and do have some expectation of reasonably "good enough" software. "Good enough" can be defined as meeting the customer's expectations successfully for the price they are willing to pay (not necessarily bug free). Thus the reason for books like this one on software testing is to find bugs quickly and efficiently to get "good enough" software for a particular context.

Software and bugs are not easy topics, but they are not as magical as some people want to believe. Software is just communication to a machine in a language that the machine can understand. As in any communication, miscommunication can happen. In software, when this happens, you have a bug, which is one of the pieces of information a tester is looking to provide to stakeholders.

In this book, I talk about where to run attacks as well as how to run them. In fact, most humans are testers, all the time wherever they are. For example, you test your food to see if it is too hot to eat and evaluate your food to determine if it tastes good. You do this with tools—your senses. Software testing builds on this ability. To test software, you use different tools and skills such as the software attacks of this book. Some of these attacks can be run anywhere as long as you have the device and the software to be tested. Yes, the book talks about test labs, places where there are lots of special tools, but you can test a smartphone just by having the device, loading the software on the smartphone, and following one of the simpler attacks.

I have classified attacks based on the earlier mentioned skill levels. So, for example, a novice tester, new to mobile software testing, should probably start with Attack 33 toward the end of the book. This can be done at home in your spare time. If you are interested and get good at this kind of "play" testing, there are more attacks, classes, books, and places to go for you to become a more skilled tester. Additionally, there are companies now that will pay you to find bugs (see my website). There are other companies that will hire you for full-time work to do testing, if you have the right skill set. Is this for everyone? No, but many people can become software testers. However, being an advanced or expert tester may take some time and is very difficult.

Most of the attacks found within this book are self-contained. Some attacks can be done together in a test set or tour. Some attacks must be used in combination with other attacks or tests. I cover these relationships. How much to combine, when, what to change, and how to do these attacks varies with your skill level and the test problem at hand (local context), as well as your creative thinking skills and how much time and how many resources you have. Finally, some of the attacks are longer than other books covering attacks, because they cover topics that are "hard" or very complex, such as time and security testing.

If you are a novice, start reading this book by understanding this introduction and Chapter 1. Then pick an area such as testing a smartphone. Read Attack 33 (combinatorial testing, which can reduce the number of tests you may need to run) and then maybe parts of Chapter 8. Then go practice some test attacks. Learn. Repeat. Grow.

If you are an experienced tester, you may want to look for attacks in areas that you believe your software has risk. Review Table E.1 in Appendix E for the detailed mapping of attacks to domains. You do not need to read the whole book. You can add attacks. You can change attacks. Learn. Repeat. Grow.

The attacks most useful to development efforts are found in Chapter 2, since these efforts should happen first. The middle chapters contain important attacks for embedded devices and aspects of mobile software. The last part of the book focuses totally on mobile systems. The most general attacks are in Chapter 10. However, I suggest that most people review Chapter 1 to find the most interesting attacks for your local context.

Note: Many large and complex mobile or embedded system testers may want to review Chapter 11 on test labs, since most organizations need test labs to function with these attacks.

In closing, I offer Chapter 12 as some parting advice that I hope you find useful.

Classification of attacks: Classifications given here are not to be construed as black or white, since there is a continuum of systems across these categories. Using these and other contexts, however, a detailed table is provided in Appendix E of applicable attacks versus contexts that might be considered as a beginning test plan (see Table E.1). Appendix E is only an example to begin attack planning. In part, these make up the context of the embedded system, and there is no "one way" or "best practice" to plan and design testing and attacks. Testing takes much critical thinking. Review Table 1.1 to begin your strategy of test attacks.

Who Should Be Familiar with This Book

I believe that testers should read this book and many of the other referenced works to (1) become effective and talented testers and (2) to advance their careers. On many projects there are others who might find parts of this book useful, such as system engineers or business analysts who define the concepts, needs, and requirements for a system and the software. These people will benefit from many of the concepts early in this book as well as reading the attacks applicable to their domain. Managers and leads might do well to understand the concepts associated with attacks to create better teams. Also, there are sections in Chapter 2 aimed specifically at developers and programmers. Many teams now practice concepts associated with Agile or integrated product development in which both have teams spanning hardware, software, systems, and test. Particularly in these types of teams, attacks from the book will be of benefit as people do multiple jobs, such as testing and wanting to test quickly without detailed documentation. Finally, students and people studying testing who want a specialization in mobile and embedded software can read and practice the exercises provided.

What Is Not in This Book

No test effort is complete (exhaustive). No book on testing is complete. No author or presenter has a complete view on testing. Smart mobile and embedded software systems include areas that the attacks in this book may not fully address, including the following:

- Systems with complex user interfaces and integration issues

- All aspects of security (an area that is still unfolding)

- Test tools and automation

- Requirements verification and validation

- Change impact analysis and regression testing

- Configuration management

- Electronics and packaging

- Signal processing and networks

- Metrics

To understand all of these areas and their potential impacts on the type of systems they are working on, readers should be well-read on many topics.

Additionally, many mobile and embedded software–driven devices have regulatory issues to consider in the form of standards and compliance documents. Compliance with certification standards can be important and legally binding, whether it is Capability Maturity Model Integrated (CMMI [3]) for process, DO–178B [4] for avionics, EN–50128 [5] for railway transportation, MISRE/IEC 61508 [6] for automotive, or various military standards for Department of Defense (DOD) work, just to name a few. The standard itself and the publication must be used to create a complete test planning and production effort. The attacks in this book can be combined with the standards or compliance documents to create a more complete picture of test planning or test strategy.

Further, time in mobile and embedded systems is a huge topic, which is partially addressed with a few attacks. For a more complete treatment of time-related issues, I recommend timing analysis to look for issues, such as the following:

- Performance, scheduling, and processor utilization

- Priority inversion in task scheduling and orders

- Hard and soft real-time factors

- Input and output real-time jitter

- Interrupt or exception processing including event response times

- Watchdog timer issues

Teams dealing with time issues may want to consider specific tools and approaches. Some additional time concepts not fully covered in this book that will allow you to expand your research and to gain additional information include the following:

- Rate monotonic analysis (RMA) [7]

- Time modeling

- Simulation

Debugging: Attacks target bugs based on concepts like error taxonomies. Some attack results can be used to help debug or even isolate the problem in a system. However, many attacks will only indicate that a problem exists, but not exactly where or how to fix it. Fixing problems usually is the domain of developers, although testers should provide as much information as possible, up to and including suggestions for a possible fix. A good attack test is repeatable because the first question from a developer that a tester may get when reporting a potential bug is "What were you doing before you saw the bug?" The developer needs to be able to repeat the attack test to start the debug process, and the attack should provide as much information as it can to help in debugging. I acknowledge that debugging is important, but for me, debugging is a separate effort that builds on an attack.

Next, there are issues of levels of testing (unit, integration, system), functional versus nonfunctional tests, tester independence (who does the testing), black-box versus glass-box, oracles, and other heuristics that I leave for others to address. I acknowledge requirements and use them when they exist, but I can conduct attacks without them or, maybe, in spite of them.

The attacks within this book are based on errors and patterns of testing for errors. I assume that readers will be aware of the concepts of test approaches, objectives, strategies, planning and design using these, along with attacks to produce a viable test strategy for their product.

Finally, concepts in model-based testing are noteworthy and have books all their own. Many embedded domains, automotive, aerospace, and areas of medicine, now base much of their development on models using concepts within System Modeling Language (SysML) and Unified Modeling Language (UML). When these approaches and tools are in use, the context of embedded systems testing changes, as do aspects of the error taxonomy as well as the applicable attacks. The trend to use models in development and testing will continue. Automated check testing becomes an attractive option too, but future effort is needed to define attacks in these environments as they come into more common use.

Relationship to Other Test References and Works by Other Authors

This book relates to information contained in works by other authors and expands on those works to provide a broader perspective for testers. In preparing this book, I offer my own personal test experiences but also include a variety of reference sources. In general, this book is a part of the so-called attack-based concepts of software testing. With this concept, I am trying to find errors, first and foremost. In passing, I may be able to provide information about what is working, such as functional and nonfunctional requirements. However, the focus of this book is *not* demonstrating that requirements are met. The attack basis of testing is a mix between error (bug) patterns, test planning, and test design with different techniques.

I feel it is important to learn lessons from others personally and through their published works. Besides starting with this book, readers may want to consider the following books for additional learning. These and many other good references are cited throughout this book:

- *Agile Testing: A Practical Guide for Testers and Agile Teams* [8]
- *A Practical Guide to Testing Wireless Smartphone Applications* [9]
- *A Practitioner's Guide to Software Test Design* [10]
- *Black Box Testing: Techniques for Functional Testing of Software and Systems* [11]
- *Embedded Systems and Software Validation* [12]
- *Experiences of Test Automation* [13]
- *How to Break Software Series* [1–3]
- *Lessons Learned in Software Testing* [14]
- *The Art of Software Testing* [15]

- *Testing Embedded Software* [16]

- *Testing Computer Software* [17]

- *Testing Safety Related Software* [18]

- *Safeware: System Safety and Computers* [19]

- *Systematic Software Testing* [20]

- *Testing Complex and Embedded Systems* [21]

I believe that testers should be well-rounded by knowing more than just testing. I recommend having some knowledge about products from the domain in which you are testing (aerospace, medical, automobile manufacturing, airplanes, factory systems, robotics, and so on) in addition to a knowledge of hard sciences (math, physics, electronics, engineering, etc.), software sciences (psychology, philosophy, sociology, human factors [human–machine interface]), and the arts (music, literature, visual, etc.). Testers may only need to know one or two of these, but *good* testers will know several, and *great* testers will know and practice many of them during their career. Knowledge in many domains makes well-rounded individuals, and these skills are only gained with study and over long periods of time. Some of the best testers I have known had an understanding of one or more of the nontesting fields mentioned earlier.

I build on these books and concepts with citations where appropriate, so well-rounded testers would do well to have these other books in their reference library and to have read them. (I have over 300 books on testing and technical-related topics in my personal library.) Also, besides the books listed earlier, I wish to give credit to all of the great testers who have given me ideas, influenced my thinking, and helped me in my testing career. I would like to add a few other names to my gratitude list: Johanna Rothman, Elisabeth Hendrickson, Brian Marick, Boris Beizer, James Bach, Michael Bolton, Ross Collard, Danny Faught, Rex Black, Dale Perry, Jean Ann Harrison, many people I have worked with over the years, and many others I may have forgotten. I believe that a good project will employ a complementary set of ideas and approaches, including attack-based testing.

Testing is a broad and deep subject. This book can be a good starting point along with a couple of other references, but to become a really *good tester*, keep reading, keep expanding your reference materials, and keep practicing. To be a really *great tester*, you must understand many subjects as well as understand the characteristics of the bugs found in all kinds of software.

I find it interesting that so many publications seem to cite the same embedded failures (refer to Appendix A) and only a few mobile issues, which is probably due to a short history and limited public information. To prepare this book, I looked at these older references, but I did not stop there. I continue to look for more bugs as well as more references to provide data to drive attacks. It is likely that test concepts and attacks will change as we get more data and our industry matures. We can learn from well-known historic cases, but more importantly, testers should study local bugs and bug lists as they find them. Many senior testers in a domain have what I call mental models of mobile and embedded errors. These testers use skills from many knowledge domains, patterns of errors, and basic testing skills to create tests. I have met many testers who seem to just touch the software and it breaks. These people are using their intuition, critical thinking, and their mental models. Learning to trust and use mental models is good. However, one should also ask, "Where is my mental model weak?" One way to answer this question is to analyze the bugs you find against the ways they are found. Another way to expand mental models and knowledge bases is to have a large library, reading list, and attack list.

Attack-Based Testing and Different Approaches to Software Development

Attack-based testing can be used with any approach to developing software. Agile and iterative approaches are currently the most popular, and attacks can benefit testers working with developers and customers to produce Agile software. However, attacks can be used within traditional software development too. Attack-based testing can also be used on projects doing software as a service, service-oriented architectures, and even independent crowd source testing (independent testers). Finally, other testing approaches such as risk based or model based can also leverage attacks.

Why not get started today and have some bug-hunting fun?

A Final Introduction Warning

Testing consists of planning, types, approaches, and techniques—all of which are able to find bugs and provide information, but they do *not* guarantee that all bugs will be found and that the desired information will be right. This is a fundamental situation in most of engineering. The name is "heuristic." You can do an Internet search on this, but basically, a heuristic allows the solving of very hard problems. However, sometimes the heuristic fails. For example, attacks can find bugs, but sometimes bugs will slip by the attack. Also, not everyone will agree with every idea on testing or this book because these are heuristics and the testing industry is immature.

You will find managers and customers who expect testing to be perfect and the software to always work. This is not possible and these people have unrealistic expectations. Exhaustive testing is also not possible. Just because one test works does not mean that the next test will not fail. There will always be another test to run. The other reference works and standards talk briefly about this, but as you get started in attacks and testing, you should be aware of this situation. Do not promise that "it will work" or that "there are no bugs." You cannot meet this promise, and the other parties may become angry with you. You might wonder then, why bother doing any testing? Fair question. I know of cases where no testing has been done, really obvious high-risk bugs slip into the field, and customers react badly (they stop buying software or products and then companies go out of business). So some level of testing is a well-accepted practice, coming in many different forms. Testing is usually done within quality, cost, and schedule constraints. This book does not address these factors in detail—other books do. You cannot promise "bug-free" software. However, you can say, "We have done this testing, have found this information (including bugs) within a given schedule and budget, and some risks have been reduced."

REFERENCES

1. Software Design Pattern. http://en.wikipedia.org/wiki/Software_design_pattern (last accessed April 9, 2013).
2. Weinbergin, G. 1991. *Quality Software Management: Systems Thinking*, Dorset House, New York.
3. Capability Maturity Model Integration (CMMI). http://www.sei.cmu.edu/cmmi/ or http://cmmiinstitute.com/ (last accessed April 9, 2013).
4. DO-178C (or B). http://en.wikipedia.org/wiki/DO-178C (last accessed April 9, 2013).
5. EN-50128. http://de.wikipedia.org/wiki/EN_50128 (English translation last accessed on April 9, 2013).
6. MISRE/IEC 61508. http://en.wikipedia.org/wiki/IEC_61508 (last accessed on April 9, 2013).
7. Klein, M., Ralya, T., Pollak, B. et al. 1993. *A Practitioner's Handbook for Real-Time Analysis*, Carnegie-Mellon, Software Engineering Institute (SEI), Kluwer Academic Publishers, Norwell, MA.
8. Crispin, L. and Gregory, J. 2008. *Agile Testing: A Practical Guide for Testers and Agile Teams*, Addison-Wesley Professional, Boston, MA.
9. Harty, J. 2010. *A Practical Guide to Testing Wireless Smartphone Applications*, Morgan & Claypool Publishers, San Rafael, CA.
10. Copeland, L. 2003. *A Practitioner's Guide to Software Test Design*, Artech House Publishers Boston, MA.
11. Beizer, B. 1995. *Black Box Testing. Techniques for Functional Testing of Software and Systems*, John Wiley & Sons, New York.
12. Roychoudhury, A. 2009. *Embedded System and Software Validation*, Morgan Kaufmann, Burlington, MA.
13. Graham, D. and Fewster, M. 2012. *Experiences of Test Automation*, Pearson-Addison Wesley, Boston, MA.
14. Kaner, C., and Bach, J., and Pettichord, B. 2002. *Lessons Learned in Software Testing*, John Wiley & Sons, New York.
15. Myers, G. 1979. *The Art of Software Testing*, Wiley, New York.
16. Broekman, B. and Notenboom, E. 2003. *Testing Embedded Software*, Addison Wesley, Boston, MA.
17. Kaner, C., and Falk, J., and Nguyen, H. 1993. *Testing Computer Software*, 2nd edn., Van Nostrand Reinhold, New York.
18. Gardiner, S., ed. 1999. *Testing Safety–Related Software, a Practical Handbook*, Springer, New York.
19. Leveson, N. 1995. *Safeware: System Safety and Computers*, appendix A, Addison Wesley, Boston, MA.
20. Craig, R. and Jaskiel, S. 2002. *Systematic Software Testing*, Artech House Publishers, Boston, MA.
21. Pries, K. and Quigley, J. 2011. *Testing Complex and Embedded Systems*, CRC Press, Boca Raton, FL.

Author

Jon Duncan Hagar is a systems-software engineer and test consultant supporting software product integrity, verification, and validation. He has a specialization in embedded and mobile software systems. He has worked in software engineering, particularly testing, for over 30 years. He has supported many projects, including software control systems (Space Shuttle, avionics, robotic, and large rockets), spacecraft (including interplanetary vehicles), aircraft, smart mobile devices, and IT systems and has worked on attack testing on smartphones. He has also managed and built embedded test labs that use test automation. Jon publishes regularly, with over 50 presentations and papers, many of which can be found on the Internet. He received a best paper award at STARWEST in 2010 for his taxonomy (of which an updated version appears in Appendix A of this book). He is also an author on parts of three books with other authors, some of which appear as references in this book. Jon is also the lead editor and an author on the following standards: ISO 29119 software testing standard, the OMG UTP model–based test standard, and IEEE 1012 V&V plans.

Readers can follow Jon on testing and attacks on mobile and embedded devices on his website (http://breakingembeddedsoftware.com) or on his blog (http://breakingembedded software.wordpress.com).

Setting the Mobile and Embedded Framework

BEFORE WE DIVE INTO THE ATTACKS, let us look at some of the definitions and implications for mobile and embedded software and some general advice to testers.

OBJECTIVES OF TESTING MOBILE AND EMBEDDED SOFTWARE SYSTEMS

"The best defense is a good offense." Attack! How many times have you heard that phrase used? This book presents attack concepts to find common bugs found in mobile and embedded software, which testers of software systems are charged with finding. In James Whittaker's *How to Break Software* book series [1–3], he shared a "fault model" approach, presenting an attack concept from which to test software. This book emphasizes the attack approach to the handheld mobile and embedded software test space. I suggest that readers be familiar with the materials in the *How to Break Software* book series, since many of the attack concepts are applicable here.

A main objective in testing software systems is to demonstrate that the systems actually work. This approach is viewed by many testers as "checking," or making sure that a requirement in the software is meeting a requirement set forth in writing that has been agreed to with a customer (or to marketing). This can be illustrated, often using automation, for single benign cases, but then later, the system may fail in the real world because just checking to show something works, while necessary, is not sufficient information to produce a successful product. There are other types, methods, and techniques of testing to ensure a successful product. In the first book of the series *How to Break Software* [1], the many different reasons and purposes for testing were examined in Chapter 1. These reasons hold true for mobile and embedded software. Basically, testing provides information about the software, but what pieces of information do we need? Two pieces of that information are "When does it work?" (checking) and "When does it not work?," or in other words "What can or cannot break the software?"

Systems contain hardware, software, and operations driven by users (see Chapter 1 of the first *How to Break Software* book). There is no one way to test or attack. In this book, I outline

33 attacks on mobile and embedded software. These attacks are aimed at the software and system. The attacks form "hammers" that a test team can use to "beat on" their software to provide them with information. In part, these attacks attempt to show that the system does *not* work, and failing that, we have some confidence that the system and/or software *will* work in the hands of a user or the customer. Testers may not use all of the tools or attacks found in this book on every project. A bit of mixing and matching must take place to achieve desired results. This requires in-depth critical thinking about the context of the software systems as well as a diverse breadth of knowledge. Further, in the mobile and embedded world, attacks must consider the software, hardware, user, and interactions from a systems perspective.

One aspect of understanding how to mix and match is that you must know your system (what it should or must do) as well as the enemy (the bugs that can make it fail). Results from failures that arrive in the customer's hands show that by putting all of your test eggs in one basket, say only doing requirement–functional–acceptance–check, testing will miss errors. Testers must attack on many fronts to reveal the bugs to ensure that customers will be happy with a product. To this end, testers should know the patterns of errors—particularly their own (peruse to the taxonomy in Appendix A).

Press and industry reports show that it takes only one small bug to get system failures that make the news. No one wants to be the tester who failed to find the bug that is highlighted on the evening news. Mobile and embedded software involves risks, hazards, safety, communication, security, and/or controlling specialized hardware in the real world. To stay competitive, companies create new products with added features, implemented in ever-expanding software. This is a universal trend and means that the software tester's job becomes even more interesting and challenging. It also means that the likelihood of missing a bug increases, and once found in the real world, hindsight of what "we should have done" is usually 20/20, meaning that people will second-guess what we do or worse take legal action against the product's company.

Companies take risks as part of strategy since one cannot exhaustively test, but to stay competitive with better-quality products, testers will be called upon for better attacks to reduce the numbers of errors or bugs found in the field (in other words, found by the customer or end user of a product). Any set of defined attacks will need to be tailored to the local context; evolve as testers understand the software under test more; and be allocated based on concepts such as risk-based testing, as well as cost and schedule profiles. But before we get further along, you will need to understand what mobile and embedded software systems are and do.

WHAT IS EMBEDDED SOFTWARE?

The definition of an embedded software system is open to some consideration as it has many views.

From Wikipedia [4]:

> Embedded software is computer software which plays an integral role in the electronics it is supplied with. Embedded software's principal role is not Information Technology (IT), but rather the interaction with the physical world. It is written for machines that are not first and foremost, computers.

FIGURE 1.1   Examples of embedded devices.

Expanding on this, embedded software depends on unique hardware to solve a specialized problem interacting with and/or controlling the real world. It has been observed that if a human user of a system is only partially aware that they are interacting with a computer running software, then it is an embedded software system. Figure 1.1 shows a few examples of mobile–embedded devices, ranging from cell and smart phones to complex systems, such as airplanes and spacecraft.

Specialized hardware, as opposed to a generic computer system or IT, is intimate to a system being considered "embedded." Also, many embedded systems exist at different points within the dimensions outlined in Figure 1.2.

Many embedded systems run in hard or soft real time at very tight or short time performance intervals or frames (milli-, micro-, and nanoseconds). Often, they have less or limited user interfaces (UIs), compared to the extensive graphical user interfaces (GUIs) and large screen sizes of most personal computer (PC) applications. Even the new "smart" handheld or mobile devices in the embedded family still have limited UIs compared to IT systems. Finally, embedded software systems have limited or specialized features. Aspects of all of these are changing in the embedded world, meaning many of the bugs of IT systems are now being faced by mobile and embedded software and systems. This book's focus is on the "differences."

WHAT ARE "SMART" HANDHELD AND MOBILE SYSTEMS?

Handheld and mobile–embedded software systems share the characteristics of embedded systems but few of the characteristics from the IT world. They are a rapidly growing area of software, and with the commonality of both context domains, they share the patterns of bugs, and hence the attacks. Examples of these systems, shown in Figure 1.3, include personal digital assistants (PDAs) or IPAD, smartphones (IPHONE, ANDROID, BLACKBERRY),

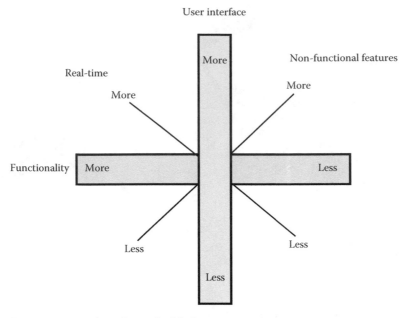

FIGURE 1.2   Dimensions within the embedded space.

FIGURE 1.3   Examples of cell phone and tablet devices.

portable medical devices, smart tools (laser checkers, scanners, etc.), and e-book readers (KINDLE, etc.).

These systems can be characterized by the size (usually small, handheld, and mobile), connectivity (Wi-Fi or cellular), as well as power issues (mostly battery based), limited memory, and simple UIs. They are also "smart," meaning there is software, firmware, or some kind of "programmed" logic functioning in them. If these characteristics sound like traditional embedded systems, they are. That is why many of the embedded attacks work, but at the same time, some IT attacks of earlier *How to Break* books such as on security, web, operating systems, and file or program issues can be applied to these handheld mobile devices. I offer mappings to the attacks found in the *How to Break Software* book series.

Many of these systems may not have classic software loaded the way IT software is loaded onto a PC. Software, like logic, is now contained in field-programmable gate arrays (FPGA), burned in read only memory (ROM), or put in application-specific integrated circuits (ASIC) to name a few, while having "classic" software elsewhere in a device. There are ongoing arguments as to if these "other" software constructs of logic follow software rules, including software testing, or something else, say hardware testing. However, the devices these constructs are contained in are "smart," meaning they have complex logic, and need to be tested. I believe that many of the attacks outlined in this book are applicable to these other logic constructs, and as with any test project, the attacks and techniques will need tailoring to the local context or problem at hand.

WHY MOBILE AND EMBEDDED ATTACKS?

Attacks provide an effective mechanism for finding bugs before the customers find them.

Today, these mobile and embedded systems are everywhere—in our homes, cars, offices, electronics, manufactured items, factories, trains, planes, and so forth. I see speculation that bugs are causing problems in all of these systems. The cost of testing these systems, the cost of fixing buggy systems, and the bad public relations that bugs cause are increasing concerns both to the public's and to the producer's demands. Further, some embedded systems are killing people while costing millions of dollars! So having good information, much of it from testing, is a big deal.

But why cannot we just use the testing common to other software and systems? In the next section, you will discover the differences. Testers must have additional attacks in their toolboxes with all of the differences and unique bug patterns. The tester may use attacks from the *How to Break Software* book series, but new attacks should also be considered. This book gives references to existing books and provides new attacks. That does not mean that a tester can just pick up these books, run the attacks, and become an effective mobile and embedded tester. It takes practice, work experience, and critical thinking in combination to become an effective mobile and embedded tester. I recommend that you practice with the exercises in this book. Find an embedded system to practice on. Critical thinking is the most challenging skill for anyone. Keep in mind that books are a good starting point to help you gain the knowledge to become a good tester, but nothing replaces practical experience.

Watch advertising in the media closely and listen to their data or talking points about safety, reliability, lower cost, newer technology, and so forth. Car manufacturers advertise about crash test dummies (producing thousands of data points versus millions of data points) and how safe their cars are; manufacturers hype their product's use and quality, but how do they actually *know* about the level of quality of their products? Cell phone advertising asks us, "Can you hear me now?" Because dropped calls or calls where neither party can hear the other makes people (customers) very unhappy with products. Providing information to stakeholders about such negative product qualities can directly or indirectly involve testing.

The extensive commercial competition in the world means that various factors of quality are being raised almost daily. Cost, functionality, schedule, and quality are all at play. Testing provides information about functionality and quality, but adding more testing impacts cost and schedule. Testing needs to be effective and efficient in the embedded space, but there are not many books focusing on the problems specific to mobile–embedded testing. I offer the attacks in this book to help in finding bugs before customers find them.

FRAMEWORK FOR ATTACKS

Mobile and embedded software testing is both different and the same as IT testing. What I mean is that many of the errors and, therefore, the attacks one might practice are the same because coders (programmers) make patterns of mistakes. Plus, management makes specific strategic decisions about software (or products) that directly affect the quality of the completed product. Many of the attacks in the *How to Break Software* book series [1–3] apply and can be adapted to embedded systems. Taxonomy and attacks in this book add clusters of common patterns for mobile and embedded system failures. I base the attacks in this book on these patterns and how they relate to other books for mobile and embedded software devices. Some attacks in this book will be longer that those found in the *How to Break Software* book series, because it is harder to test for the bugs. Also other attacks in this book are combinations of attack and test concepts, again to expose a bug, and, therefore, I found that more effort was needed.

Patterns of errors for mobile and embedded software systems indicate that many attacks should be based on the hardware, time, and "special" features of embedded software such as fault tolerance and "control" of the world. This book presents attacks that are outside of the "black box" of the first *How to Break Software* [1] book. My taxonomy indicates that many common errors can be detected through developer-level testing, static analysis, analysis, or other efforts that may be outside of the box for many traditional testers. Often, better testing happens by stepping outside of the box. However, whether an attack is done by analysts, developers, or testers is driven by project context. The key is that the attacks get done.

BEGINNING YOUR TEST STRATEGY

To get testers started thinking about which attacks apply to any context such as embedded or mobile, Table 1.1 provides a mapping of the chapters of this book to (my definitions of) four different types of devices for this context.

Mobile and smart wireless devices—These devices are everywhere today, and their use is growing daily by the thousands. They have large segments of software (millions of lines of code) as well as smaller pieces of software often called "apps" (applications). They

TABLE 1.1 Chapter Map to the Different Types of Mobile and Embedded Software

Context → Chapter ↓	Mobile and Smart Wireless Devices	Embedded (Simple) Devices	Critical Mobile Devices (Could Be Embedded, Small but Important)	Critical Large Embedded Devices (Could Be Mobile)
Chapter 2—Developer based	Yes	Yes	Yes	Yes
Chapter 3—Control system	Partial	Yes	Yes	Yes
Chapter 4—Hardware–Software	Partial	Yes	Yes	Yes
Chapter 5—Software	Partial	Yes	Yes	Yes
Chapter 6—Time	Partial	Yes	Yes	Yes
Chapter 7—User interface	Yes	Partial	Partial	Partial
Chapter 8—Mobile	Yes	No	Partial	Partial
Chapter 9—Security	Yes	Yes	Yes	Yes
Chapter 10—Generic attacks	Yes	Yes	Yes	Yes

Note: Yes means the attacks in a chapter can be applied, No means likely the attacks do not apply, and partial means that some of the attacks could be applied in some contexts.

include tablets, phones, smartphones, and a variety of mobile wireless devices for generic and specialized usages. The key is that they have network or communication access. They have aspects of the PC and/or the IT world, but on top of this, they add the classic limitations (limited size, power, etc.) of the more "pure" embedded world. Those testing the software in these devices face all of the test problems of the PC, IT, and embedded contexts. Specific usages include personal communications (cell devices), tablets (these are "almost" PCs), smaller noncritical medical (implanted and handheld) devices, transportation (ground, sea, and air), and support areas (factory, military, and many others). A factor here is that a software bug in this kind of system has less impact than that with the critical systems. A specific growth area now is the merging of mobile and smart with traditional embedded devices. I am seeing growth in areas like the so-called smart grid, smart office, smart home, and almost any other area you care to add "smart" to the name. For example, in a smart home with switches, sensors, appliances, cameras, door locks, thermostats, and many other smart devices where there is a blending of wireless and embedded software devices, their commonality is that all of these things "control" and "communicate."

Embedded software devices—These software systems are also everywhere these days, though often the users do not know that software is running the device. They are like a light switch, a brake system on cars (or other vehicles), and they run many electronic devices including refrigerators and washers. This software fits all of the definitions of a class of embedded systems. Many software testers have learned how to test these systems through the school of hard knocks. However, the attacks of this book work well, when applied correctly. The embedded world sometimes merges into the wireless, mobile, and IT worlds.

Critical mobile devices (including embedded)—These pieces of software are seen less in day-to-day life, unless you work in fields such as medical, factory, and service industries. These systems help save lives in medicine, make our assembly lines run, including aspects of robots, control the inputs to our lives (power, water, waste), control our phone systems, service or diagnose our cars, and other activities where usually there is a "technician" in the loop as "the user." Like the "mobile smart" devices, these systems take on some of the characteristics of the PC world but come with the embedded limitations. They are important but not as large (as much software and hardware) as the next category, so test budgets may be smaller. The usages here share the same areas as "mobile and smart wireless devices," but a bug will have larger financial, equipment, or human loss or impact, for example, people can be killed.

Critical "large" embedded devices (might be mobile)—These software systems are usually part of a larger system where all of the embedded parts work together. The embedded software may be in a series of processors. These devices and systems fly our airplanes, run our ships and factories, and control rockets and military systems and many other components where large risks, money concerns, and even life are at risk. These systems benefit from embedded attacks but often have few of the concerns of the PC world such as a UI, security, and integrated network communications. The usages can be the same, but the costs of these systems are usually in multimillions of dollars for many systems and subsystems, such as power grids or nuclear plants.

ATTACKS ON MOBILE AND EMBEDDED SOFTWARE

This book follows the attack approach brought forth by James Whittaker's *How to Break Software* book series, but here are some additional twists:

- Operating system attacks [1]
- Input/output attacks [1]
- File/data storage attacks [1]
- Hardware attacks
- UI attacks [1]
- Functional attacks (partially in basic book [1])
- Time-based attacks
- Fault-tolerance attacks
- Developer-based attacks
- Security attacks

And now some general observations that embedded and smart mobile software system testers need to keep in mind:

- Software systems usually do not fail to form a "single" bug, but instead cascade over a series of "triggers." One should look for these by finding the small bugs and think about "how big can it go." Further, it is interesting that many reports on system failures describe different aspects of this sequence, often citing slightly different root causes. When a tester is looking at bug lists and root causes, look down the "rabbit hole" for other connections.

- Test like the system is going to be used. The attacks should be realistic. This does not mean that the attacks are "nice," because the old truism of "if it *can* happen, it *will* happen" needs to be in the attack. Testers should look at the environment, reuse wherever possible, review user manuals and concepts of operations, use cases, or other information sources that define "use," and then think "abuse."

- Developer test efforts. The tester needs to know and influence what the analyst and coders are doing in testing. The taxonomy in Appendix A indicates that in some contexts, 20%–40% of errors could be found using developer testing. Attacking the implementation starts to address what the other team members should be doing, but coverage with these attacks and other methods cannot be complete, so errors or bugs will get missed. Additionally, if a test team realizes that developer attacks are weak, the test team likely will need to strengthen their attacks.

- Operations and logistics can impact the system. Software that works in one environment or with one set of embedded hardware may fail when these factors change. Test teams need to watch for changes and provide information to management that says "we need to test and attack these changing areas."

- Different attacks need to be employed over a complete development cycle to find errors early and avoid the "pesticide paradox," where tests get old and stale and lose error-finding ability unless testing changes.

- Testing is not only a filter but an information provider, giving the team information used throughout the life cycle.

IF YOU ARE NEW TO TESTING

I find many people who want to get into software testing as a career or to make some money. There are many ways to do this: the standard way of obtaining a degree, standard high school education plus often years of self-education and work experience. The most valuable is continuous experience and self-education in testing software. Also many people feel you must have a degree, often in hard science: math or engineering, and while this certainly is one path into testing, I find that many testers have degrees in liberal arts, software science, and even music. Further, many testers are self-taught, coming from the user's experience, and/or taken from nontraditional education approaches. If you are reading this book as a first or near first starting point into software testing, you will need to do a bit more work than what the typical techno geek (i.e., someone who already knows computing systems fairly well) might have to do. Don't give up! You bring a different perspective, and in testing, this can be an asset. However, you will need to do some homework to become an effective tester. Your homework can include the following:

- Review the reference books given in this book and pay attention to works (blogs, papers, etc.) by the authors cited in this book.

- Follow the test video training found on sites supplied through Internet searches, on YouTube, from BBST [5] and others.

- Learn the basics of computers and software such as the concepts found on Wikipedia and About.com. These are good places to start before moving to other references.

- Read and learn things about the software application domain you will be testing in such as factories, smartphones, sales, and so on.

- Get training from providers—again start with an Internet search. There are training providers around the world found easily through the Internet. Or get a recommendation of training from another tester.

- Engage in testing discussions, attend forums and conferences such as STAREAST and STARWEST and again attend some training sessions.

All of this takes lots of time and effort, maybe as much as that of a traditional degree. However, those of you that have taken the self-education path will continue to provide insights into testing and will earn a living in software testing by being a part of groups such as uTtest, Mob4Hire, or working for other companies that need testers.

Some people who are new to testing but do not have a college degree attempt to get credentials such as certifications. There are several certification programs in the test industry easily found on the Internet. While there are mixed viewpoints on the value and validity of such certifications, some companies do use certifications and/or degrees as filters when they are looking to hire, both as a positive (they like to hire people with a certification) and a negative (they dislike people with any certification). It is possible that the degree or certification is not the important factor. However, self-education and a keen interest in testing are both very important.

AN ENLIGHTENED TESTER MAKES A BETTER TESTER

To help guide the exploration and understanding of which attacks and tests to apply, it is good to understand what the developers are doing and other project context issues. There is *no single set* of approaches, methods, tools, techniques, or attacks that apply to all mobile and embedded software systems. Test teams should make sure that they plan appropriately for the time to do the "required" test tasks and then apply appropriate attack time because the testing space is infinite. The embedded context guides the nature of the plans and types of attacks. One must apply deep thought up front and throughout the project in order to select meaningful, information-providing tests. Life and testing are never static.

There is a dynamic interaction that takes place during attack testing that involves the mind, body, and universe, which includes the hardware, software, environment, and operations. When doing testing, the mind needs to be alert to look for what is expected as well as unexpected. The mind and senses need to be engaged. If the body of a tester is getting tired or bored, it may be time to take a break. Also if the tester's mind and body start becoming complacent, it may be time to bring in different testers to mix things up. The reason the mind and body need to be engaged is that bugs can happen anywhere (hardware, software, system) and anytime (start-up, during functions, after long durations of time, etc.). It has been described that *testers need to follow the "scent of the bug."* What this means is that often a bug is not a hard crash or something obvious. In embedded software, it can be a slightly late command, an analog signal that is trending in a slightly wrong direction, a missing or extra bit, and other things where the software works but maybe is "just not right." This means that during the attacks, all of the inputs, happenings, and outputs need to be critically watched by an observer—meaning the tester. Observers can be humans, recording and analysis equipment, or even another piece of software in the test lab. Following your instincts on something that just does not seem right will allow you to find a "bigger bug" that is more difficult to find.

> **Takeaway note:** Do not be a one-skilled tester. Good testers use many domain skills and lots of creative and critical thinking.

The final part of being "enlightened" is having skills and knowledge. This book talks about the knowledge a tester needs; however, if you want to be a really good tester, you will need knowledge of many domains (what some call "being well-rounded), including the following:

- Software programming and engineering

- System engineering and architecture

- Hardware and electrical engineering

- User or domain experience (how the app, software, or game is going to be used)

- Other domains whose knowledge or experience might useful include art, music, logic, philosophy, human factors, sociology, psychology, and many other activities that take creative thought.

The key to these many skills is critical thinking, which an enlightened tester uses every day. The context is now set. The next step is yours to take.

EXERCISES (ANSWERS ARE ON MY WEBSITE)

1. Define three reasons why you would test any piece of software.

2. Define three bases for testing.

3. Define three examples of embedded software devices.

4. List the changes you want to make in

    a. Your testing

    b. Your product testing

REFERENCES

1. Whittaker, J. 2003. *How to Break Software*, Pearson Addison-Wesley, Boston, MA.
2. Whittaker, J. and Thompson, H. 2004. *How to Break Software Security*, Pearson Addison-Wesley, Boston, MA.
3. Andrews, M. and Whittaker, J. 2006. *How to Break Web Software*, Pearson Addison-Wesley, Boston, MA.
4. Embedded Software. http://en.wikipedia.org/wiki/Embedded_software (last accessed April 9, 2013).
5. BBST Foundation Classes. http://www.testingeducation.org/BBST/foundations/.

CHAPTER 2

Developer Attacks

Taking the Code Head On

Thou shall covet thy developer's source code.

JAMES WHITTAKER
Ten Commandments of Software Testing [1]

I am not talking here about taking on your least favorite developer (or programmer). I am talking about getting developers to test the software or code that they write.

This chapter has attacks that many testers think should be "owned" by the developers. Some developers think that all testing is owned by a "test group." These views are at odds. Ownership implies responsibility. The *team* owns the qualities of any product. Therefore, if a developer runs the attacks of this chapter, finds their own bugs, and fixes them, then the quality of the product is improved. In many cases, the developers are busy or otherwise unable to run the attacks found in this chapter. In this case, an alternate option is to have someone independent from the developers conduct these attacks—as long as they have access to the code. It should not matter who runs the attacks of this chapter as long as the attacks are run and resulting bugs fixed, the result of which is to produce a better-quality end product.

These attacks are considered low level and/or structural testing, also known as white-box testing, because they are (or should be) run early in the development life cycle by someone with knowledge of the code. Table 2.1 defines attacks that are applicable to different mobile and embedded general contexts.

Traditional development teams will develop the software and then run these attacks. On Agile projects, the attacks might be run within the development team, for example, in a Scrum group or as part of test-driven development (TDD). The number of attacks and how to perform them will depend on the context of the project. For instance, a project developing a low-risk app system might just run attacks to reach simple code statement coverage (Attack 3). An embedded project for a large commercial jet aircraft will have many levels of test coverage (Attacks 1, 2, and 3).

TABLE 2.1 Map of Developer Attacks to Mobile and Embedded Contexts

Context→ Attack↓	Mobile and Smart Wireless Devices	Embedded (Simple) Devices	Critical Mobile Devices (Could Be Embedded, Small but Important)	Critical Large Embedded Devices (Could Be Mobile)
Attack 1	Yes	Yes	Yes	Yes
Attack 2	Yes	Yes	Yes	Yes
Attack 3	Yes	Yes	Yes	Yes

Notes: Seldom, seen only in specialized cases infrequently; sometimes, seen often in many different contexts; frequent, seen regularly in a variety of contexts; yes, should be considered for most contexts; no, generally not applicable.

However, not every test team can conduct developer types of testing. For example, when third-party commercial software is included in the system, testers are unlikely to have access to the code for this software. Sometimes the team needs to accept third-party software without the project team being able to perform these attacks. If so, different and more attacks may be needed in other areas to compensate. Alternatively, a project can ask for some "proof of quality" or test activities such as test reports, metrics, reliability numbers, or other quality information. The taxonomy in Appendix A reveals that operating systems (OSs) or third-party software was a component of failures. So remember the old Ronald Regan adage and Russian Proverb, "trust but verify."

Note: Basic code structural coverage concepts are considered throughout this chapter with a final general discussion in a subsection at the end of Attack 3.

Another approach for finding errors in software is to use code peer reviews or inspections, which are not classified as attacks, because they are considered methods of quality. Many books, articles, and standards present code peer review as an early line of defense in keeping errors out of products. I recommend *Peer Reviews in Software: A Practical Guide* [2] for further reading on these methods and to assist your team in doing peer reviews. I do believe that code *should* be subjected to peer-review. After all, code is just a language of instructions to a computer. Every book, including this one, undergoes extensive peer reviews, edits, and other changes before the end user reads it. This approach to publishing goes back hundreds of years. Why should a developer's code be approached any differently? Enough said.

ATTACK 1: STATIC CODE ANALYSIS

☐ Intermediate

> Static code analysis is the analysis of computer software that is performed without actually executing programs built from that software. In most cases the analysis is performed on some version of the source code and in the other cases some from of the object code. Wikipedia 2010 [3]

Static code analysis (SCA) searches for classes of errors and produces "messages" much like compiler output or warnings. SCA finds bugs that other testing often misses or at best, encounters by luck [4] (see examples of potential bugs in Table 2.2). SCA tools automate the work, but a human must still review the output to actually find the errors deemed relevant and worth addressing. SCA tool capabilities can vary from simple (e.g., Lint [5]) to

TABLE 2.2 Example Warnings Reported by SCA Tool Showing Potential Bugs

Message/Warning Type/Potential Bugs	First Run of Tool	Actual Errors Found	Note
Buffer overrun	11	2	Leads to memory leak
Uninitialized variable	149	36	Very common bug
Unreachable code	41	20	Dead code
Ignored return value	14		Why does code exist?
Division by zero	4	1	Machine fault
No return statement	13	1	Bad programming
Null pointer reference	72	28	Leads to memory leak
Redundant condition	121	4	Extra code
Unused value	23		Sloppy coding
Useless assignment	101	9	Such as a variable set, same variable set, variable now used in same logical sequence

fully complex (e.g., security analysis and heavy-duty commercial tools). Some of these tools provide measurements and other useful information such as graphical pictures of the code to aid in understanding or error identification. This attack implements an SCA against the project's source code.

Figure 2.1 shows an example listing that a tool might produce with flagged code constructs that might be in error. Each tool has a different way to make such notations and annotations; however, each flagged finding needs investigation during the attack.

As the SCA tools get more sophisticated, they move from simple checks (is a variable initialized) to complex rule checking and what is known as "formal" analysis methods, where the tool starts to consider meanings and semantics. Formal methods analyze system, hardware, and/or software for errors using rigorous mathematical methods such as predicate calculus, semantics, graph theory, and models. SCA tool abilities are progressing rapidly.

When to Apply This Attack

These tools are typically used during the coding efforts to target more complex syntax and semantic types of errors that developers can make. In recent years, a new generation of "compilers on steroids" has become available in the form of SCA tools, and it is surprising that they are not in more common use, since most developers rely on compilers or interpreters so readily (to catch their classic mistakes). There are a variety of vendors and even open source tools that can launch this attack.

There are several good times to launch this attack, but sooner is better: just after coding and before any peer reviews; at each integration or build cycle as the interaction between different coder's programs comes into play; and before releasing the whole code base to the world. Since the code is not executing, testers will not see failures in software execution. Also depending on the type of tool being used, the attacker will get a plethora of "messages" instead of a single failure about possible bugs. A limitation with this approach is the

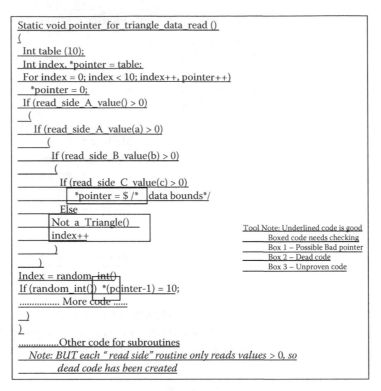

FIGURE 2.1   SCA flags the good, bad, and ugly example.

large number of what can be called "false positives" or messages, which indicate an error, when in fact there is no error. An SCA attacker will need to review these messages to weed out the false positives. Better tools may mean fewer messages, and as tools progress, the numbers of false positives will decrease but will likely never become zero. While weeding out the false positives, the attacker should review the message, locate the associated segment of code, and determine if the message is really an error in the software (see Figure 2.2, for examples).

Finally, the attacker should keep things reasonable. Not all checks or messages need to be run all the time. This may mean that the project has several message check configuration files to make the tools run. Also the messages checked will change over time and iterations. Some tools allow customized checks (messages) that can be added or changed over time, which can be a very useful (time-saving) feature to consider.

What Faults Make This Attack Successful?

Each tool has checks, which many of them report as "messages" or "warnings." For example, some vendors report as many as 50 or more different checks. The taxonomy prepared for this book indicates that the following common mobile or embedded "errors," which resulted in outside software failures, could be found by this attack. As SCA tool capability increases, more and more checks become possible. Current generic abilities include the following:

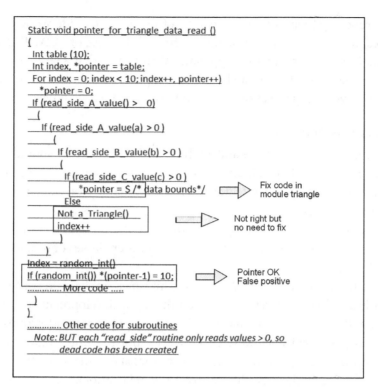

```
Static void pointer_for_triangle_data_read ()
{
 Int table (10);
 Int index, *pointer = table;
 For index = 0; index < 10; index++, pointer++)
   *pointer = 0;
 If (read_side_A_value() >  0)
 {
   If (read_side_A_value(a) > 0 )
   {
      If (read_side_B_value(b) > 0 )
      {
         If (read_side_C_value(c) > 0 )
            *pointer = $ /* data bounds*/          Fix code in
         Else                                      module triangle
            Not_a_Triangle()                       Not right but
            index++                                no need to fix
      }
   }
 Index = random_int()
 If (random_int()) *(pointer-1) = 10;             Pointer OK
 ............More code .....                        False positive
 }
}
 ............Other code for subroutines
 Note: BUT each "read_side" routine only reads values > 0, so
            dead code has been created
```

FIGURE 2.2   SCA flags the good, bad, and ugly resolution example.

- Memory and resource management—memory allocation

- Program data management—corruption, initialization

- Buffer overflow—pointers, dynamic

- Concurrency violations

- Vulnerable coding practices

- Platform support

- "Understandability" review, maintenance, test

- Clean or abnormal termination

- Code path and logic problems

- Data or variable initialization problems (or lack thereof)

Table 2.2 is a hypothetical sample set of messages found during early development builds from an embedded device before a full system build using an SCA tool.

One approach to these types of errors is to prevent them in code through the use of coding standards or restrictions on compilers (see Appendix B on coding standards). However, even with coding standards, bugs can exist in the code and SCA may not find them. For example, one report [6] states that up to 70% of errors in C code were due to pointer/memory issues, many of which can be found through the use of SCA. This report is not alone. The data in the taxonomy (see Appendix A) indicates that a large number of errors in mobile and embedded software can be found using SCA. I believe that this is an efficient attack, even when weighing the effort required to remove false positives.

Who Conducts This Attack?

Developers should be the first to use and apply SCA. The developers should be using these tools from day 1, even as they edit and peer-review their code, and continue using SCA until the final system software is integrated. However, many projects have found that testers need to proactively advocate the introduction and use of SCA tools. Further, some of these errors (memory leaks, for example) are difficult to find by any attack, so it is much more effective to find them in static analysis, leaving other classes of problems to be found by different attacks.

Testers on many teams run the SCA tool, do a first-pass analysis on the tool results, and then work with developers to get things fixed during a development (or programming) phase. Additionally, some vendors have static analysis services and groups that offer to perform this first-pass service. However, even with a first-pass filter, many developers may complain about this attack, and the false-positive messages they need to deal with, since it costs them coding time (here is where a cost-to-benefit rationale comes in). Testers need to be advocates for this attack, since it finds bugs, and, if needed, to help developers with the SCA runs. If someone has to wade through 100–200 messages to find just 40–50 "real" errors, the amount of time to do this is cheap compared to how long it takes to find the errors using other attacks, or worse, having a failure in the field.

Where Is This Attack Conducted?

SCA tools typically run on a host development system, where the source code is. No special hardware or processor associated with the embedded or mobile system is required, so once coding has started and there is a clean compile, this attack can begin. A few tools exist that can operate on the object code, but not many mobile or embedded projects run a static code analysis on object code.

The project does need an SCA tool to run this attack. Mobile and embedded tuned SCA tools are becoming readily available or improved, including ones with security, application or app, and interface check abilities. A quick Internet search for the phrase "static code analysis" will yield many candidates. Testers can select tool(s) from identified candidates based on the following types of context considerations:

- Programming language in use (the tool(s) must support the language(s) in use)
- Methodologies in use (some tools may be better suited for object or model based software)
- Criticality of the application/software (more critical software may mean more expensive tools and/or multiple tools)

- Ease of use and completeness of messages

- Security threats or other risk factors (some tools are tuned for certain aspects such as security) and

- Cost and schedule (do not buy a tool you cannot afford or do not have time to use)

How to Determine If the Attack Exposes Failures

As the software code comes into existence, the SCA attack can begin. The SCA tool is run against the code and messages are generated. Then, someone needs to do filtering analysis on the messages. This analysis needs to be tailored to the language, tool, and context and then determine which messages are real (need to be fixed), which ones might be left (lived with), and which ones are false. A classification system using a database or spreadsheet may be of benefit when dealing with large numbers of messages. For example, classification lists might include any of the following:

- Bugs—errors to be fixed. Consider bugs in terms of criticality—how important and how critical and how much damage they can cause.

- PIF—messages to go into a product improvement file (PIF), which is a bug that can be lived with and fixed when time or budgets allow such as dead code. Consider messages in terms of criticality (same as bugs).

- Open concern—unable to determine correctness and needs development to analyze for a final classification.

- Generic classes of errors—this class of errors can occur over and over and may indicate a process improvement area so that these types of messages do not reoccur.

- Disregard (false positive)—an invalid conclusion by the SCA tool about the code that can be ignored.

These messages must be analyzed within the context of the project (see Figure 2.2). The SCA tool does not have knowledge of them, so a human must make the determination. Likewise, open SCA issues must be analyzed by a human team within the context of the software and project. In the analysis of the significance of a message, consider its possible effects on the mobile/mobile or embedded software, hardware, and system, including trustworthiness, maintenance, testability, and failure impact. Finally, care in placing a message into a false-positive classification is needed so that the message is not misclassified, only to have it "bite you again later."

Takeaway note: SCA of embedded code finds important errors, which can be hard to find otherwise, with just a little added work. Plus, the SCA tools are getting better all the time.

For each tool warning in the list of generated messages, analyze each one into the categories (mentioned previously) with notes for later reference. Use caution when filtering out messages already seen while not missing new ones. Analysts may expand the databases or tracking systems to aid in this, though projects may not want to mix these databases or tracking systems into the official bug database to keep the project's statistics manageable. Tripping over the same message over and over is frustrating, but worse, risks losing real bugs in the forest of false positives, which may tend to get ignored.

Finally, as the software "matures," some error messages may no longer need to be checked during later passes of static analysis. As mentioned earlier, many SCA tools support turning checks off, but care needs to be taken to ensure that unchecked messages do not "sneak" back in.

How to Conduct This Attack

First, a project must obtain an SCA tool. There are static analysis tools available from the open source world. Additionally, more and more commercial software comes on the market every year or bundled for development environments. Project context factors to consider include cost of the tools, criticality of the project, language used, development processes in uses, applicable standards, and resources to support SCA. Projects must review and do decision analyses on which of the many static tools would work based on project context. Progressive organizations have established and recommended vendors or "libraries" of static analysis tools, which can be checked out and used as needed (within licensing agreements).

The SCA attack goes as follows:

1. Obtain a static analysis tool(s) for the programming language(s) you are using.

2. Configure the tool to check for specific messages (see your tool's user guide).

3. Obtain access to the code once it is ready.

4. Run the tool.

5. Review and analyze the messages for false positives.

6. Document "real" errors so that developers can fix them.
   (*Note:* If this tool is used as part of peer reviews on a module basis, include the information with the code review.)

7. Rerun for code updates.

8. Rerun across integrated multiple code segment boundaries.
   (*Note:* The tool may need message tailoring.)

9. Rerun the tool once the whole system comes together.

10. Document the results (at each iteration) to feed forward into later uses.

Whether you are running these tools for just one developer's segment of code, a group, or the whole project's code, consider which messages are being checked. Remember, more checks equal more time and effort, but may mean more errors are found. Some work has been done that shows using multiple different SCA tools on software finds different errors, and so can be complementary, but this does add cost.

Once this attack is accomplished, continue with other attacks or peer reviews.

ATTACK 2: FINDING WHITE-BOX DATA COMPUTATION BUGS

O Novice / □ Intermediate

Attack 2 targets the lowest white-box level of errors (sometimes called developer's errors). White-box testing is more commonly called structural testing (IEEE/ISO 29119 [7]). This kind of testing targets errors within software computations associated with data—what I refer to here as data computation bugs. This attack is closely related to Attack 3, which addresses the classic code coverage such as branch, decision, and path. Mistakes in these statements are usually made by the developers at the lowest level of the implementation.

In Attack 2, the tester selects data to drive developer testing. Data selection test techniques that can be used in setting up this attack include decision points, boundary analysis, Boolean logic structures, equivalence classes of data, range and resolution of numbers, and non-powers of two numbers, for example, 1/3. While some of these test techniques are thought of as "black-box" testing techniques, such ideas are, in part, arbitrary. Regardless of who is running this attack, interesting data values should be selected. Data values closely link with Attack 3 on logic path coverage. Up front test plans or project standards should define the details of data coverage and recommended techniques for this attack so that testing is consistent [11,14]. For example, if you select data inputs to execute the code structures at a superficial data level (nice powers of 2), these tests may be not very good at finding bugs. Specifically, take the following code statement:

```
X = y / z
```

Input $y = 4$ and $z = 2$. The output of x will be 2. These inputs and outputs are nice powers of two numbers and would achieve data–statement coverage. However, a more interesting case might be to input 5 and 3. Additionally, what about a value of $z = 0$; is that possible? A divide by zero is undefined and bad for computers, because strange things can happen. If a divide by zero is possible, this would be a bug. So, "simple" data coverage may not be so simple and may require critical thinking about the inputs, complex data values, and even the automation of data values (see Attack 32).

Another (C code snippet) code example to consider is the following:

```
if (jumpvalue < 8) jumpvalue++;
    if (jumpvalue > 11) value_y = value_y - 2;
    jumpvalue = jumpvalue/value_y;
```

If we are attack testing with an initial value of `jumpvalue = 9`, all three statements would be exercised, which would make Attack 3 coverage "satisfied" at a statement level, but the new assignments for `jumpvalue` (first line) and `value _ y` are never made. Thus, it is possible to achieve statement coverage and never exercise the data assignment logic shown in these statements. I recommend other values be used as jumpvalue to reach better data computation coverage.

This is an example of the kind of problem single types of structural coverage in Attack 3 can have and why running multiple attacks should be considered, such as running both Attacks 2 and 3.

When to Apply This Attack

The ability to detect low-level developer errors at the nearest time point to where they have been introduced saves time and money associated with finding and fixing them later. Many organizations may ask if this attack can be applied as part of test first, TDD, code then test (CTT), or hybrid approaches to code design. The answer is "Yes."

The time to run Attack 2 is when the code comes into existence, likely after the static analysis attack (see Attack 1), and sometime before a code peer review. While not without some controversy, studies have shown that waiting to run developer-level testing until the code has moved into integration or later in the life cycle means that testing can cost extra budget. (Running an Internet search using Google with the qualifiers cost+software+defects+graph and reviewing the sheer numbers of graphs will give you an idea of the associated costs of not finding bugs sooner in the life cycle of a product.)

Finally, if during the programming effort, this low-level testing is not done by someone, how can anyone say the code is "working" and "ready" to move to the next stage? A good series of developer-level tests are the starting backbone of testing attacks. This attack can be applied to mobile and traditional embedded devices.

Takeaway note: Developer test attacks are the first line of defense.

Who Conducts This Attack?

In the traditional view of developer testing, good developers run Attack 2 themselves as part of development efforts—like TDD [8], but it can be done by independent testers. The question of who should run the attack is decided on a case-by-case basis. Many developers will insist on doing it themselves. Yet others refuse to ever run the attack. When this situation happens, bugs "live on" until later in the product's life cycle or they can be missed altogether. If a developer does not run this test, someone other than the developer can run the attack—provided they have access to the code. But the key is this: ensure that attack testing gets done.

Developers can use Attack 2 to expand their view, understanding, and belief in the correctness of what they have done. This makes better code, but many developers are "too busy" to run the attack. Another option is to have an independent person run the attack in place of the developer, but this person must understand concepts from both the programming and structural testing worlds. This test person can save developer's time and has the advantage of an independent set of eyes. While not as effective as developer's testing their own work, the independent attack can find bugs and provide other helpful information that developer's might miss. What you call the people who are independent of development testers, who they report to, how closely they work with developers, the tools they use, the levels of automation, or the techniques they use, and other variations are all things unique to the team.

Key point: This attack is a second line of defense in finding bugs. The first is static analysis with peer checking.

Where Is This Attack Conducted?

Embedded system software typically gets cross compiled into assembly language for a target processor. Cross compilers make developer testing hard because running on a host development system may yield results quite different from those found on the target or end computer. Two basic approaches exist to deal with this: (1) Apply this attack on an interpretive computer simulator that emulates the intended environment on a host; and (2) run the attack on the target processor. Simulators need to exist or be created to do the first option. Some development environments include them. Target processors can be supported using a special configuration, often called a single-board computer, which has limited hardware. Or if the target environment has enough input/output ports, it may support developer testing by allowing specialized input/output. All factors need to be defined and set up prior to the beginning of this attack. Variations on the theme exist, and vendors offer some solutions (see Chapter 11 for more information on emulation and single-board computers). Risk occurs if developer testing is skipped in favor of doing "low-level" testing during higher levels of attack. Developer testing can take a "divide and conquer" approach to achieve testing in a timely fashion.

What Faults Make This Attack Successful?

Much has been written on white-box or structural testing, including Capers Jones [9], who report that 25%–50% of errors could be found using developer-level testing and another 10%–20% with a combination of peer review and developer-level testing. Developer testers should be familiar with this report, as well as standards such as ISO 29119 Part 4 [14]. Developers are human, make mistakes, and have blind spots [10] as well, so they miss their own mistakes. Testers can be seen as developer's assistants, having the good of the project at heart. But, we are all human and we all make mistakes. Correcting and learning from mistakes—whether our own or someone else's—is a part of life, in my view.

The following is a sampling of data bugs that developers can test from the taxonomy (see Appendix A):

- Off by one data and decision points
- Index values

- Equation data values

- Data (wrong value, type, initialization, conversion, used in calculations, etc.)

- Stack or location access (pointers)

- Divide by zero or other missing "guard" of data points

- Input/output parameters—for example, wrong data parameters passed

- Local data that may not be initialized has the wrong value, stored improperly, or sized incorrectly

How to Determine If This Attack Exposes Failures

White-box or structural testing (synonymous with structural testing) as defined by Copeland and Marrick [11,12] defines that the tester has access to the code. As part of setting up Attack 2, a tester determines the criteria that can be checked. Many developer-level test tools support some kind of function that can take expected results and compare to actual results of executing the code within a test. The number of checks depends on the following contexts: number of variables, criticality of software, design considerations, and cost. A problem at this point will be determining the value(s) to check against. Here, many testers will use requirements, design information, hand calculation, or "best guesses" (see Chapter 11 oracle discussion). Once a value to check against is determined, the tester uses the value in the check function of the tool. If the check of the expected value to code computed value fails (meaning, the value of the check function does not equal the value computed by the executing code), then a bug may have been found. At this point, it is possible that the expected test value is wrong; something in the test set-up (inputs or environment) is wrong; or there is a bug in the code.

Note 2.1: Many developer-level test tools have a function to compare actual to expected results. Each tool's function may use names such as "check," "compare," and/or "assert" with sub features such as "equals," "less than," and others. In Chapter 2 Attacks 2 and 3, I refer to the comparator feature provided by many developer tools as checking or a check. Throughout rest of this book, I may refer to "checking" or "checks" for bugs, where the tester "checks" expected results and other expectations. In general, outside of Attacks 2 and 3, the use of the word "check" is not meant to refer to a tool or function of a compactor tool, but a generic action.

For more information on comparator "checks," see http://istqbexamcertification.com/what-is-test-comparators-in-software-testing/

How to Conduct This Attack

It is expected that some kind of tooling will be used to conduct Attack 2 with automation of some kind. Questions as to the use of a debugger, simulator, emulator, or single-board computer must be addressed, as well as whether a commercial or open source development test tool will be used. Tooling and set-up for automation are issues for any developer-level test effort. Once tooling is set up, the attacker must determine the functions, logic, and data of the code to be tested. The basics of Attack 2 should follow this flow.

1. Obtain and set up the developer-level test tool.

2. Determine technique(s) including a level of coverage.

3. Construct drivers, and stubs, (mock objects) as needed.

4. Create input data to drive the test technique and functions.

5. Based on input, define resultant checks, which are needed. Some tools have functions to do this for you.

6. Obtain the code.

7. Compile, link, and run the developer tests.

8. Measure coverage and success criteria.

9. Fix any errors found.

10. Repeat until desired levels of data coverage are reached and checks passed.

11. Final review of coverage and checks to make sure they are complete.

12. Place the code into a release control system, to ensure that it is ready for software integration efforts.

13. Repeat for code changes if/when they happen.

Typically, Attacks 2 and 3 are performed in combination in the final development test product. Of course, this depends on what the code looks like. Software that is small and simple will have fewer attacks, while complex logic will require more attacks and maybe test automation [13].

ATTACK 3: WHITE-BOX STRUCTURAL LOGIC FLOW COVERAGE
O Novice / □ Intermediate

Building on Attack 2, this developer-level attack focuses on the logic paths within the code. Attacks 2 and 3 can be run separately or together. Historic logic coverage measures include statement, branch points, logic decision constructs, and paths, just to name a few. Many years back, research defined many coverage metrics to the point of getting us lost in the forest of logic coverage points. Additionally, some standards (DO–178 series [15]) define coverage criteria. I recommend that project context be considered when selecting the logic coverage types before low-level testing begins so that the testing is done consistently. There are many tools that will measure and report a variety of coverage levels from statement to definition use pair coverage, but just "meeting a number" may lead to dysfunction [16]. A measure may help in being consistent in developer testing, but the measure should be reasonable for the context. And attacking the developer-level data (Attack 2) may reach meaningful levels of logic structure coverage (Attack 3), meaning that we can integrate these two attacks in many cases.

Who Conducts This Attack and When to Apply This Attack?

Attack 3 is best done by the developers but can, in some cases, be done by independent testers.

What Faults Make This Attack Successful?

Attack 3 covers code statements. Table 2.3 cites examples from the embedded taxonomy of bugs that made it into the field (to the customer) and that could have been stopped during developer attacks.

This list is based on common modes of failure from the taxonomy (see Appendix A), but is only a partial list of basic mistakes made in the code. Plus, there can be logic structural issues from language, code constructs, and development methodologies such as in object-oriented (OO) versus structured programming languages.

How to Determine If This Attack Exposes Failures

Most tools in this area have the ability to define checks for specific success criteria (functions, computation results, coverage levels, and others). A tester defines the success criteria before executing the attack based on requirements or design information and enters this information into the tool. The success criteria value is entered into a check function, as provided by a tool. A check function will compare the success criteria to an actual result from running the piece of code. Check functions vary, but typically will allow the comparison, including factoring resolution values into the compare. The tester runs a series of

TABLE 2.3   Sample Code Constructs with Types of Bugs to Look for Indicated

Code Structure	What to Look For	Examples
Entry–exit points	Multiple points	Nonstructured code
Statements	Not meeting success check	Equation missing term
Initialization	Variable used but not set	Variable x used but never set
Return code	Wrong value set	Wrong bit data value set
Loops/iterations	Loops and termination of loop	To many or few iterations that works sometimes not others
Control point (if test)	Incorrect branch or to wrong place	"End" statement in wrong place
Boundary control point missed	$<, >, =, <=, >=$ and parameters	missing condition structures
Dead code	Code cannot be reached	Logic excludes a path
Optimized condition	Compiler optimization	Compiler removes "binary" code
Too complex	Code complexity	"I just do not understand this"
Math/equation error	Missing or wrong math symbols	A multiplication used instead of an addition
Exceptions	Test the exception handler under many call cases	Unhandled exception statement

attack cases to achieve coverage and meet all planned success criteria. When a check fails, the bug hunt begins. Many times the developer will need to put things into a debugger to actually isolate the bug. This is a back-and-forth process between development, coding, analysis, and testing. The key is to define the check(s) specific to the code, language, design, and system.

Note: Some tools may support checking and bug finding at object code level (0 and 1s). How this works and finds bugs is tool specific and is not specifically addressed in this book.

How to Conduct This Attack

First, the data is selected to drive Attack 3 into the logic structures. As already mentioned, the data and logic path coverage (Attacks 2 and 3) can be run in parallel, so data may be "reused" from Attack 2 to help get logic coverage started. I assume some use of tools and automation in this attack (Chapter 11 discusses tools and test environments).

If logic coverage is the goal, and the data attack (Attack 2) has not been run, the tester must identify the logic structures of the code and the data to drive this coverage. Many developer or support test tools can create logic flows using approaches such as directed flow graphs or production measurements on coverage. These tools help testers to visualize and understand the logic structures from which the needed test case data can be determined. Using tools, the tester can observe the logic structures in the code. For example, covering 100 percent of the statements at the superficial level may be relatively easy. Although it is surprising how few developers or testers actually achieve even simple coverage of 100% of the statements. Other coverage levels will require more and different data test cases.

If Attack 2 has been run, then Attack 3 should start by assessing how complete the structural coverage is using the data from Attack 2. If the coverage meets the intended targets, no additional test cases will be needed. If the target coverage numbers are not met, additional test cases will be needed.

SIDEBAR 2.1

Space Shuttle: Years back, I was testing branches and decision coverage on a module of code on a space shuttle subsystem. The piece of code looked like this:

```
if (x = 1 or (z = 10 and not t) or (t = 2 and m);
where x, z, and t are integers and m is a Boolean
```

You may realize that this does not make sense, particularly because t was not a Boolean but an integer. (Only Booleans or a Boolean expression can have a value of true or false.) It should not have compiled. But years back, compilers made a lot of mistakes (yes, they make a few now and are still not perfect). It did compile and made it out of developer testing and into system testing. I was an independent code level tester. When I created a truth table and tried to reach decision coverage measures, I could not make full coverage on this decision because the "not t" part of the statement could not be evaluated logically, which was the bug.

The saying "it is not that easy" applies with branch coverage. Just inputting values to get the true/false paths will get you coverage and find a few errors, but what are the interesting values to drive the true/false cases? And if we expand our cases to include decision coverage by constructing a truth or decision table [17], we can get much more interesting test cases, which will find more bugs. Each type of coverage should be constructed not only for coverage percentages, but also to find "interesting" testing values.

Finally, past some of the traditional coverage types, if the code has a looping structure (while, for, until, and so on), ask yourself how many iterations through the loop are needed: 0, 1, 19 on up to as many as the maximum number. From a path viewpoint, each iteration and exit number is a path, so a "while" statement with a max of 1000 iterations would need at least 1001 tests. Ouch! Several heuristics have been developed to deal with this. One heuristic is to test the loop 0, 1, to the max number of times, plus reach statement coverage internal to the loop. This is sometimes called loop approximation coverage. I recommend this heuristic.

There are many developer-based test heuristics, tools, and concepts to consider during Attack 3. In support of developer-based testing, there are many useful constructs such as mock objects, drivers, stubs, templates, patterns, and tools. These concepts and constructs can be found in good references such as *Test Driven Development for Embedded C* [18] as well as *A Practical Guide to Testing Wireless Smartphone Applications* [13], which also discusses automation approaches. Additionally, for this attack, there are tools to support these efforts, so automation usually makes sense (see the check and instrumentation section). When running Attack 3, testers should know the material in these references as much as other project-related standards.

Test cases are added to Attack 3 until the desired level(s) of coverage are reached. Also as part of the attack cases, checks (a feature of many tools) should be done on the inputs and testers should ensure that the expected outputs are met. The use of automation features and tools means that the attack cases can be built up over a series of efforts until the checks and coverage levels satisfy local context criteria.

TEST COVERAGE CONCEPTS FOR WHITE-BOX STRUCTURAL TESTING

A concept coming out of Attack 3 is that the results can be measured and coverage numbers calculated. Many organizations use the measurement as indicators of test completeness or goodness such that high coverage numbers are better. There are many kinds of white-box structural testing that one can measure. The ISO 29119 software testing standard (due out in late 2013) defines coverage for many of these statement types. However, basic coverage levels that I recommend teams consider includes

- Statement—hit each statement in the code at least once

- Branch—take each branch path at least once (not all paths in the code)

- Decision condition—exercise each logic element at the decision points (elements of truth tables)

- Loop testing—exercise each loop zero (if possible), once, and max number of times

- Path—measure coverage level of number of paths in a piece of code (usually not 100%)

ISO and many books [11] define these coverage types. Also, some standards such as DO–178B (that a project might need to comply with) also define coverage for compliance.

However, while these references define and advocate coverage measures, testers should keep in mind that a coverage number should not be taken as an absolute. For example, the combination of attacks, multiple attacks and reaching a test coverage measurement such as "100 percent of all statements tested," can fool managers and customers into thinking that the tests are "complete" and that no bugs remain." No attack, tool, coverage measure, or technique can equate that testing has been exhaustive (meaning that all errors have been found). Coverage measures are only data point indicators of what has been done. These measures cannot tell you how well or how complete testing is as a whole.

> **Takeaway note:** Warning, full coverage of data and logic (or any coverage measurement) does not mean that no bugs exist.

Projects should use coverage to define goals, then meeting these goals will improve the confidence level for testing on your project. As already mentioned, however, hitting 100% coverage does not mean that testing and attacks are complete.

Figure 2.3 is an example of what code that has been instrumented might look like. It is not from any specific tool but is provided to show the conceptual idea of developer testing with code checks and instrumentation.

Note: This is not all of the checks and instrumentation that are possible. Only a small example of instrumented code (see underlines in Figure 2.3) is provided. Most tools are much more sophisticated.

If we were testing statement and/or branch coverage, the added code "flags" could be evaluated in the tool's processing to determine which statements and branch paths have or have not been taken. The tools typically "add" code elements into the piece of code under test, but there are issues with this. For example, it increases the amount of memory and time the piece of code uses, and the "check" function must be determined ahead of time and provided to the tool. Also the piece of code would need to be called by something. This is known as a driver piece of code. And the logic being called, such as "raise_error" and "read," would need to be linked in or stub coded so that the call works. The use of tools and frameworks has made many of these factors easier to deal with, but still the amount of "extra" code (that may be created) can be three to five times as much as the code under test.

NOTE OF CONCERN IN MOBILE AND EMBEDDED ENVIRONMENTS

Of special note in the embedded domain is the issue of cross compilation to an embedded processor and object code. Cross compilation is where a host system generates code

```
Program Read_Sensor_Data (return output_data: integer)
c: declare variables
Input_data, Loop_data : integer;
Path1a, Path 1b, Path2a, Path2b, Path2c : integer;
c: initialize variables
input_data:= 0;
output_data:= 0;
loop_data:=10;
C: instrumentation variable (added)
Path 1a = 0;
Path 1b = 0;
Path 2a = 0;
Path 2b = 0;
Path 2c = 0;
c: This code reads a sensor and passes information to calling programs or the OS
Begin
Read (sensor1, input_data); c: reads data from sensor1 and stores in input_data
Path 1a =1;
c: If initial read does not get any data, this loop reads sensor data for up to 10 cycles, which is enough time for data to be read or until data is read
Loop if (Loop_data > 0 and input_data<>0)
Begin loop
    Read (sensor1, input_data); c: reads data from sensor1 and stores in input_data
    Loop_data = loop_data-1; c: decrement loop
    Path 1b =1;
End loop;
If input_data< 0 then
    Raise_error ("sensor_error_message");
    c: sensor should not be returning negative data, so an error is raised (a system utility feature) with "sensor_error_message"
    path2a = 1;
Elseif input_data= 0 then
    Raise_error ("Sensor_not_responding_message");
    c: sensor has not read any data in 11 cycles, so an error is raised (a system utility feature) with "sensor_not_responding_message"
    path2b = 1;
Else
    Output data = input_data;
    path2c = 1;
End if;
Check (Output=21) c: call to tool check function where 21 is predetermined success criteria
End routine;
```

FIGURE 2.3   Sample code with instrumentation and checks.

to run on a different target processor. Usually, this final code that runs the computer is called the object code. It will not run on the host, so the code must be executed in a special environment such as on emulators or single-board computers, and/or with hardware in the loop. Emulators run nonnative computer programs on the host system, which allows the object code to be executed as if it were running on the actual processor. There are issues and concerns with an emulator including timing, interrupts, hardware interfaces, fidelity of the emulators, and others. So most teams run an emulator and then include some kind of testing with the hardware. There are a variety of emulators available for many common mobile and embedded environments. Hardware in the loop has a variety of options that can be expensive and difficult to set up because of things like limited input/output, hardware interfaces, and memory concerns. The simplest hardware in the loop is to mount the processor and some limited amount of memory on to a card and plug the card into interfaces on a host computer. This configuration will not have all the hardware interfaces of the real system, but is more realistic in areas such as memory, processing, and timing. This set up is usually referred to a single-board computer. Another configuration is to have the full-up hardware and processor, but to include some instrumentation and electronics, which allow sufficient interfaces to support the developer testing. Many embedded projects spend time setting up hardware in the loop environments (for more information on emulators and single-board computers, see Chapter 11). The cross compilation test environment is one large difference in the mobile–embedded worlds. The test team needs to be aware that if this kind of cross compilation developer testing is not done, bugs will hide until later attacks.

EXERCISES (ANSWERS ON MY WEBSITE)

1. You are hired into a project, and they want to improve their testing, which is part of why you have been hired. You find out that the developers are not doing good developer testing. List the reasons why they should improve their testing.

2. The team has decided to use a static analysis tool and checks for a game app. Define which items you might check from Table 2.2 and why.

3. Define a case where you can exercise a line of code, so you would have covered that line of code at a statement coverage level, but the line of code could still have an error in it.

4. Define how your team will handle false positives in SCA.

5. Define who will use SCA on your team.

6. Can you think of a case where low-level developer Attacks 2 or 3 are not worth running?

7. Explain the items needed for developer testing in mobile or embedded environments.

8. Define why a tester would want SCA to be done.

9. Explain two objections developers may have to SCA and what a tester can do to overcome these objections.

10. Define who can run developer-level testing.

REFERENCES

1. Whittaker, J. 2003. *The Ten Commandments of Software Testing, Part 1: Random Testing and "App Compat"* (April 21, 2003), from Web column. http://www.stickyminds.com/s.asp?F=S6392_COL_2
2. Wiegers, K. 2001. *Peer Reviews in Software: A Practical Guide*, Addison-Wesley, Boston, MA.
3. Static code analysis. http://en.wikipedia.org/wiki/Static_code_analysis (last accessed April 11, 2013).
4. Beatty, S. June 2007. Where testing fails, *Embedded Systems Programming*. http://www.embedded.com/electrical-engineers/education-training/tech-papers/4125741/Where-Testing-Fails
5. Lint (software). http://en.wikipedia.org/wiki/Lint_(software) (last accessed April 11, 2013).

6. Walls et al. 2006. *Embedded Software The Works*, Newnes, Boston, MA.
7. ISO29119 to be published in late 2013. http://www.iso.org/iso/catalogue_detail.htm?csnumber=45142 (last accessed April 11, 2013).
8. Beck, K. 2003. *Test Driven Development*, Addison-Wesley, Boston, MA.
9. Jones, C. 2008. *Applied Software Measurement: Global Analysis of Productivity and Quality*, 3rd edn., McGraw-Hill, New York, p. 472.
10. Kaner, C. 2008. *Software Testing as a Social Science* (presentation). http://www.kaner.com/pdfs/KanerSocialScienceSTEP.pdf
11. Copeland, L. 2003. *A Practitioner's Guide to Software Test Design*, Artech House Publishers, Norwood, MA.
12. Marrick, B. 1995. *The Craft of Software Testing*, Prentice Hall, Upper Saddle River, NJ.
13. Harty, J. 2010, *A Practical Guide to Testing Wireless Smartphone Applications*, Morgan & Claypool Publishers, San Rafael, CA.
14. ISO29119, Part 4. http://www.iso.org/iso/catalogue_detail.htm?csnumber=45142 (last accessed April 11, 2013).
15. DO–178 series (DO–178B and DO–178C). Software considerations in airborne systems and equipment certification, 1992 and 2012. http://en.wikipedia.org/wiki/DO-178C (last accessed April 11, 2013).
16. Kaner, C. Lecture 6—Introduction to Measurement http://www.testingeducation.org/BBST/foundations/ (accessed September 15, 2012).
17. Truth table. http://en.wikipedia.org/wiki/Truth_table (last accessed April 11, 2013).
18. Grenning, J. 2011. *Test Driven Development for Embedded C*, The Pragmatic Bookshelf, Raleigh, NC.

CHAPTER 3

Control System Attacks

M ANY EMBEDDED DEVICES—even mobile ones are controlling something (see the robot in Figure 3.1). In this book, a large number of attacks are on the control features of the embedded software. However, these kinds of attacks usually need to be done at the system level, meaning the hardware, environments, and user operations will need to be included for realistic results.

Table 3.1 lists the control system attacks and examples of where they might map into mobile and embedded contexts.

ATTACK 4: FINDING HARDWARE–SYSTEM UNHANDLED USES IN SOFTWARE

◇ Advanced

Mobile and embedded software systems interact and deal with the real world. The real world is immensely complicated, so the number of "cases" that the software and system needs to address can be quite large. And with the additional element of time involved, it becomes easy to miss test cases since their numbers basically approach infinity. Attack 4 goes after situations that may be unhandled in software. However, an independent person (the tester) may view the situation as important and something the software–system should handle. Attack 4 is related to Attack 5 in dealing with the analog or hardware world and Attack 13 when fault tolerance becomes a part of the system.

The attack is conducted based on the operations of the hardware, software, and system. This is not hardware testing, but the hardware must be included or accounted for within the test environments or labs as a minimum—or to the extent possible. While this sounds easy, the result is a level of combination complexity where it is easy to miss many possibilities of the real world.

> **Takeaway note:** In the embedded domain, many serious bugs in control systems can be missed when they have been overlooked in test scenarios and use cases and, therefore, not tested.

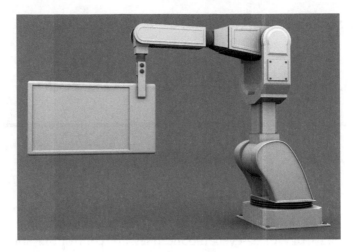

FIGURE 3.1   Example robot.

TABLE 3.1   Map of Control System Attacks to Types of Mobile and Embedded Software

Context → Attack ↓	Mobile and Smart Wireless Devices	Embedded (Simple) Devices	Critical Mobile Devices (Could Be Embedded, Small but Important)	Critical Large Embedded Devices (Could Be Mobile)
Attack 4	Seldom	Frequent	Yes	Yes
Attack 5	Seldom	Sometimes	Yes	Yes
Attack 6	Yes	Yes	Yes	Yes
Attack 7	Seldom	Frequent	Yes	Yes
Attack 8	Yes	Yes	Yes	Yes

Notes: Seldom, seen only in specialized cases and infrequently; sometimes, seen often in many different contexts; frequent, seen regularly in a variety of contexts; yes, should be considered for most contexts; no, generally not applicable.

When to Apply This Attack

Attack 4 should start as early as possible using analysis and design or code activities, before the hardware and software are finished. In fact, waiting to start thinking about this attack until the hardware and software exist can be too late, because bugs found later on may require extensive rework and can, therefore, have higher rework costs. Support items for Attack 4 can include

- Modeling concepts including use cases, scenarios, and other usage information;

- Simulations of hardware and software functionality;

- Prototypes and early versions of the software–system; and

- State charts and/or decision tables [1].

These factors can help with understanding what will drive specific test attacks. Attack areas to consider include the environment, power, communication lines, sensors, noises,

SIDEBAR 3.1

Therac-25: The Therac-25 was a radiation therapy machine produced by Atomic Energy of Canada Limited. This system was reported in six accidents between 1985 and 1987, in which patients were given massive overdoses of radiation, resulting in some deaths. The system has been heavily researched (Leveson [2]) and found to be an example of the risk in software-controlled embedded safety systems. The whole story of what happened (problem and root causes) is long and extensively reported by the researchers, but the basics include the following:

1. Poor software practice including no peer reviews, poor software testing (good testing was found to be almost "impossible"), questionable requirements and design, and bad implementation practices including the reuse of software from older models of the system without understanding or testing the system changes, for example, in hardware.
2. Removal of safety hardware interlocks and replaced by software interlocks (fault-detection/error recovery) without good testing.
3. Software interlock failures, which could occur due to timing problems (a race condition).
4. User interface–numeric overflow–timing combination bug, which was improbable but would happen under certain conditions, particularly as operators became more experienced. This combination caused the software interlock(s) to fail, resulting in overdosing.
5. Unclear system messages, such as "MALFUNCTION 32" with no documentation in operator help manuals as to the meaning or corrective actions for the operators to take.
6. No integrated hardware–software testing (no attacks in a test lab), until delivery at a customer site, and then only minimal delivery "checkout" tests were done.
7. No sensor (hardware)-to-software communication checks such as reporting a bad sensor.

Although this is an older case, Therac-25 is a classic and relates to many attacks in this book.

time, and any input that comes into the software. Variations of this attack should continue at each stage of the project's life cycle, where attacks are expanding and becoming more complex as a product evolves. For example, hardware changes, specifically the removal of interlocks, were associated in part with killing people in the Therac-25 [2] accident (see Sidebar 3.1), and changes to the Patriot missile [4] operational use profile (that was not tested) also killed people (see Sidebar 3.3).

What Faults Make This Attack Successful?

Errors in this hardware–software attack come when the analyst, programmer, or designer forgets and/or underestimates cases and contingencies. A recent example may be with various car companies having electronic stability control issues [3] and where a software update was used to "fix" the problem. In this situation, there was a slow skid possibility that the vehicle did not handle well and could have resulted in a vehicle rollover. The problem was likely a complex set of interactions between something in the environment, hardware, software, and user. The exact mix we may never know, but the fix was to update the software. A more optimal situation might have been to find the case in an attack and fix the software before the product made it to the field.

TABLE 3.2   Possible Missed Hardware Situations or Bugs

Buggy Situation	**Approach to Starting the Attack**
Systematic (predictable) noise on sensors, communication, inputs, electronic circuits, and device outputs	Identify, analyze, and introduce noise simulation into a system over a variety of levels looking for systematic source influences and unhandled cases
Random noise on sensor, circuits, and/or output, including EMI, sneak circuits, and spikes	Conduct risk and/or FMEA over a variety of levels looking for disturbance sources or influences and unhandled cases
Reasonable case where the system hardware misses a possible scenario that a user might introduce or encounter	Identify likely cases and conduct attacks to determine impacts
Unexpected interface patterns between hardware, software, user, or combinations thereof	Identify likely cases and conduct attacks to determine impacts

Table 3.2 presents some common hardware situations that appear frequently in the taxonomy provided in Appendix A.

The argument that software testers should not investigate these types of cases may be made, because these attacks are hard to set up and can be expensive, but fixing one "missed" case can be even more expensive. Testers with good product domain knowledge can attack the software in situations that developers overlook. This happens when the developer assumes things about the software or system or just overlooks a situation. Watch for such assumptions, find them, and attack.

Mobile and embedded devices assume many things about the state of the hardware and supporting network. Assumptions (which may be risks) to watch for include the following:

- Network bandwidth

- Signal characteristics

- Environment that the *system exists in* (e.g., hot, cold, wet, dusty, etc.)

- Power (e.g., batteries, usage levels, what is turned on or off, or nonbattery power options)

- Environment that the *system will see*

- Hardware characteristics

- User's actions

For example, I have seen mobile and embedded devices not respond consistently when the user's hands are cold; the device is in extremes of hot/cold conditions; when the signal undergoes a beam focus at a tower; an antenna gets covered by the user's hand position; and other cases where the hardware interplays with the software, producing results that users find less than desirable.

Who Conducts This Attack?

Testers, developers, and validation staff can start to apply this attack up front and throughout development in analysis and evaluation of prototypes. Once testing starts on the actual software or system, the staff, particularly crafty testers, should expand their understanding by identifying risks on which to build the attacks. Testers with domain history and/or user experience can be very effective at defining the right attacks since the problem space is large. The entire team, including management or other stakeholders, should consider the attack list and priorities, since the number of possible attack cases often will exceed budget and/or schedules. Thus, everyone needs to understand and agree to what is and is not possible.

Where Is This Attack Conducted?

Since aspects of Attack 4 can start early by using math models, simulators, and prototypes before the actual hardware and software exists, early "pre" attack efforts provide insight for development and testing. The level of risk, as well as factors of the product, determines the extent and nature of this early analysis, all of which can feed the attack.

As the products become available, this attack can be conducted. Variations of the attack can be done stepwise as the hardware and software are integrated and a "full" system emerges. This fits with integration attacks (Attack 12). Testers should define attack cases based on risk areas such as the following:

- Safety—when the well-being of humans is threatened.

- Security—data or information can be exposed to the bad guys.

- Hazard—damage to equipment or the environment is possible.

- Business concerns—bad computations generate wrong information and ends up impacting profits.

- Regulations—the product could result in lawsuits or be at odds with government regulations.

- Environment and input factors—hardware inputs for devices and electronics, which are susceptible to influence (noise) either systematic or random, including outside communication lines and characteristics; the "real world" (weather or conditions in the real world such as wet roads or rain); and even human operations.

- Output noise—outputs to devices and electronics that are susceptible to noise influences.

- Complexity—the size of the system or some aspect of the system makes missed cases likely.

The last three items, inputs, outputs, and complexity, will take the test team working with the rest of the development team to understand what the hardware and/or software are likely going to encounter. Ideally, complexity should be minimized, although it will never be eliminated, so attacks in areas of the hardware and software that are "complex" are important cases to define as soon as possible. Defining the risks and test cases in these areas takes imagination, creativity, and critical thinking (e.g., see the bug list in the taxonomy of Appendix A, as well as *Computer Related Risks* [4], and *Safeware: System Safety and Computers* [2]).

How to Determine If the Attack Exposes Failures

Testers should look for one or more of the following indicators:

- The system generates unexpected or unwanted responses.

- Small outputs or "bugs" that are not a major issue are wrong and, with a little work, expose a larger bug.

- System failures (halts, crashes, or simply stops responding altogether).

How to Conduct This Attack

Start with known cases and scenarios that detail how the system–software interacts with the world. The known cases may come from system use cases, operational use concept documents, industry history, and even the company's product line information. The known cases should be tested, but the problem with just testing the known cases is that there may be other important cases the software should handle. Attack 4 helps to identify these undocumented situations by using knowledge of the system and architecture.

Next, identify any test cases that might be missing using this information found in your research. Here, testers can use techniques such as risk analysis, failure mode and effects analysis (FMEA), failure mode, effects, and criticality analysis (FMECA), cause effect graphing, and brainstorming. (See *The Art of Software Testing* [5] and Wikipedia for more on brainstorming and cause–effect analysis. For more information on how to do FMEA or FMECA checks, see Wikipedia [6] and *Safeware: System Safety and Computers* [2], as well as Appendix A (error taxonomy) found in this book). The rest of this attack focuses on risk analysis with organized brainstorming, but substitutions using FMEA or cause–effect graphing are possible. Try conducting risk analysis following the materials found in Appendix G. Specific risk items should include missed error handling in hardware or system usages.

Now, design the attack to address the higher-priority risks. This can get into some interesting set-ups, environments, and test input and output cases. Forcing these cases can be hard. Attack support concepts to consider include the following:

- Special devices to seed hardware input ports with the required input in place of the hardware such as oscilloscopes, in-circuit emulators, generators, stimulators, and so on.

- Seeding the attack into the software–system as a special software function to make the software under attack react as if a special hardware or system condition existed. Examples include special builds with triggers, built-in software injection points, built-in software tests, and so on that may bypass the hardware input port(s) or allow an event to be triggered internally to the software.

- A wrapper tool in a test configured to act as part of the system, users, or hardware to force the attack into the software under test, such as

1. Holodeck-like tool [7], and

2. Models, simulations, and/or simulation environments [see Chapter 11 and Attack 17].

Many of these concepts are custom made or take special effort to set up. Each has advantages and limitations. Up front planning and allocations to the test lab are in order here.

Once the infrastructure is in place and the interesting risks for the attack have been defined, the data and environment to drive the attack have to be set up. This is the classic test design problem [1], and the same care, options, and thought are needed. A few seconds, different data points, sequences of events, or other options can make or break this attack's likelihood of finding a bug. Exploratory testing can be a good option here [1,8,9] with different tours [9] and sessions. The random nature of hardware factors complicates the testing, so tester judgment and critical thinking become necessary.

The test cases resulting from designing Attack 4 should be executed quickly. Learning from the results, then more attacks may be revealed. Iterations based on the risks, time, and budget continue until one or more of these factors are exhausted.

As this attack generates results, the tester should analyze the stored data in detail. Not all attacks like this will have a hard failure (i.e., a crash). Many a tester has not fully analyzed the data after an attack only to have a bug happen in the field, and upon looking over the historic attack results, find that the bug was hiding in old outputs. This is not a good situation and can be difficult to explain to those not directly involved in the testing. The test data should be preserved, and post test tools used to mine the data. Post test analyses can include the use of oscilloscopes, bus analyzers, PROBE or MATLAB® tools, other custom programs, graphics, or other methods. The level of bug hunting and checking is determined by the mobile and embedded system domain context. Keep in mind that finding a bug in historic attack data, after it has surfaced in the field, can also have legal ramifications.

ATTACK 5: HARDWARE-TO-SOFTWARE AND SOFTWARE-TO-HARDWARE SIGNAL INTERFACE BUGS

☐ Intermediate

One thing that makes many embedded devices different is the interface to and from the real world and the interaction between the real world, hardware, and software. Attack 5 focuses on finding bugs in these interfaces and interactions since study has shown that many bugs between the hardware and software can be found here.

Embedded devices get data from hardware about the world they operate in and provide outputs to hardware that often controls or otherwise influences the world. The real

TABLE 3.3   Examples of A2D and D2A Bug Possibilities

Type	Situation	Impact	Notes
A2D	A2D representation information is lost because measurement is not precise	Software computation is based on incorrect data	The number of bits used to store the analog converted data is not large enough or sampling rate to get bits is not correct
A2D	A2D information is contaminated with noise	Software computation use noise when it should not	The noise term may not be known, accounted for, or misrepresented
A2D	A2D information is calculated correctly	Computation has unknown error	Sources of error can come from calibrations used on variables, variables lacking initialization, or calculations are not done with enough accuracy (single versus double floating point)
D2A	D2A conversion losses "LSB" in conversions, but bits are, in fact, important because computer word sizes are too small	Output to analog device is wrong	The number of bits stored from the digital world to the analog world does not have enough precision, so analog data is incorrect
D2A	D2A information does not account for noise of the real world	Software computation does not include a factor for noise	The analog values are not correct given the noise of the real world (output data may be lost in the noise)
D2A	D2A information is calculated correctly because of internal factors	Computation has unknown error	Sources of error can come from calibrations used on variables, variables lacking initialization, or calculations are not done with enough accuracy (single versus double floating point)

world is not discrete (0's and 1's), but computer systems are. Digital divides the world into "chunks," which means that embedded devices must "convert" from analog-to-digital (A2D) for inputs or digital-to-analog (D2A) on outputs. In the measurement, conversion, storage, representation, division, computations, and reintegration back into the real world, odd or bad things can happen. Additionally, the real world includes unwanted input factors and disturbances such as noise, which can get used as meaningful information by a system when it should not. Information can be lost or added, resulting in a representation that is incomplete and computations that are in error. Table 3.3 defines some of these scenarios.

Further, the characteristics of a D2A input, output, or measurement data can and will be different over time. For example, on a brake system controller, shoe pads that are wearing out will change values within the braking variable over time as the pads progress from new to old. This means that the system (software) should adapt to changing values. Also the hardware sensors and actuators themselves can be different from vehicle to vehicle due to manufacturing differences. In many cases, the software system should be able to compensate and adapt to these situations. The programmers, systems people, and testers will need to take these translation and compensation scenarios into account during the engineering and test processes [10–12].

When to Apply This Attack

Apply Attack 5 when the software provides output to hardware or gets information from the real world via hardware. Testers should look for the following:

- Output devices: motors, actuators, engines, hardware controllers, and so on.

- Input measurement: sensors, scanners, receivers for signals, and so forth.

- Devices that convert from A2D or D2A.

These factors may be external to the processor and software such as the device connects via a wire or port to hardware, which may be fairly obvious. The interfaces between hardware and software may also be internal to the "box" such as cell phone devices with orientation sensors, radio receivers (cell), cameras, and satellite signals (global positioning system [GPS]), all in a tiny smart phone. These factors can have user interaction or may have D2A or A2D considerations, which testers should take into account. For example, early IPHONES had issues with antenna placement versus where the user's hand was placed, and this caused signal interfere, which degraded the phone's overall performance.

Once the input and output connections to the software are known, the job is not done. There can and will be differences from device to device. Variations mean that numbers such as calibration between devices may be different for some variables. This introduces the idea of device to device calibration concerns. Calibration and differences between device calibrations can be expected from causes such as the following:

1. Difference from hardware device to device such as variations in manufacturing, which make each piece of hardware "unique" due to differences such as wire length, machining, and other product process factors.

2. Environmental factors that are different or change over time such as physical noise, electronic interference, light, cold, water, dust, and so forth.

3. Timing and read characteristics such as wiring length or sensor response time in the connection between hardware and software.

4. Changes to any and all of these factors over time such as brake pads wearing out.

The software of many embedded devices needs to take one or more of these factors into account for the system to behave as the user wishes it to.

Finally, on embedded projects, problems come up in the system–hardware, when the hardware is "complete." A fix is needed, and someone will say "we'll fix that problem in the software because it is easier." When this happens, the software test team should have alarms going off in their collective heads, because while software can be changed faster than hardware, a small change—even one number in the software—can yield new bugs that ripple via coupling and cohesion throughout the software or system. Fast changes in the A2D and D2A software area because of last-minute hardware impacts likely mean that this attack should be conducted, even it has already been "completed."

What Faults Make This Attack Successful?

Faults in this area come when the software does not address analog and digital conditions, including the following:

- Control algorithms that are incorrect.

- A compensation noise term is wrong.

- A hardware device is not calibrated correctly.

- System changes over time have not been factored into the software.

- The noise or environmental conditions are worse or different than what was planned.

These types of faults are just a few of the embedded A2D and D2A situations to consider when conducting this attack. The question then becomes "What is reasonable for the system and software to handle and what is not reasonable?" *There is no one answer for every system.*

Who Conducts This Attack?

The test team, working with the systems and hardware engineers, conducts Attack 5. The team needs to know the A2D and D2A sides of the software. The team should consider normal usage as well as any possible variations. If devices, inputs, configurations, and calibrations are missed by the team during test set-up, the attack results in a false sense of security or, in other words, things work in the lab but not in the field. Often the test team needs specialized skills on the A2D and D2A conversions, nature of the hardware, and software algorithms.

Where Is This Attack Conducted?

Attack 5 should be conducted in the lab as soon as hardware and software are available. Analysis and test design can start sooner. However, many a project has waited only to find that there are A2D and D2A "problems" when system testing has begun or worse in field use before considering this attack area.

Start this testing in a lab where both the hardware and the software exist. These tests go by names such as hardware–software test labs, system integration lab (SIL), or system test lab (see Chapter 11). Whatever the name of the lab, it should be capable of testing the A2D and D2A interfaces. The fidelity of the lab and interfaces should also be considered. Here are a few questions that the test team should be asking of the lab.

- What kind of hardware is being used and how close is it to the "real thing"?

- What is different between the lab and the real world?

- What factors impact time and calibration elements in the lab?

- How long should these tests be run? (For these considerations, does the hardware have a chance to wear, do you need a longer duration test, or what factors in the system change over time?)

- Which things are "real," which are "simulated," and how do they impact the attack? (In many test labs, aspects of the real world need to be simulated such as with medical devices.)

Lab set-ups and fidelity are complex and have to reflect the interfaces and items under test. Therefore, the set-up time and analysis for any lab should be considered here.

A possible compensation for lab issues is to take the device into the "real world" and do some field testing. Many embedded devices are tested this way. Cars are tested in the mountains and on long-distance drives. Airplanes fly around the world on test flights. And teams put mobile devices into the "user–tester" hands in beta testing—before a device goes to the market. Gathering data this way may provide some useful information points, but field tests come with their own sets of problems including the following:

- Repeatability of any test

- Limited access to internal data to confirm or understand a failure

- Control of the field test environment

- Nature of the "user" in the field

- Configuration of the testing (hardware as well as software)

> **Takeaway note:** Since one definition of embedded software is a system where the user is only aware of the hardware and not some of the software, Attack 5 needs to be run where and when the hardware can interface with the software.

How to Determine If the Attack Exposes Failures

During this attack, the test team looks for the following types of clues to spot defects:

1. A signal, in or out, that does not match a specified or ideal value. Ideal values should be in a written engineering document with range and resolution. If values are not specified or available, limited analysis is possible, and this situation will require added research and work.

2. An input where noise may be included as part of the value measurement and the noise impacts the value of interest. Noise is a factor of measurement in many A2D devices, and there are many ways, both in hardware and software, to deal with noise. But sometimes the noise is not dealt with properly, which impacts the value and can lead to a bug.

3. Hardware that has variations from device to device caused by manufacturing, use, use over time, temperature, environments, and/or the software does not correctly compensate for the variations.

4. Incorrect numeric representations within the computer.

5. Very small errors that may over time be compounded into a failure. For example, an error in the least significant bit (LSB) of a floating point number that gets accumulated over a series of calculations can become a significant error.

6. For other ideas, see the notes in Table 3.3.

The net effect here can be things such as cars that do not stop, out-of-control devices, mobile device confusion such as bad map locations, as well as unhappy users. This is just a partial list. There are many kinds of A2D and D2A factors to look for and consider.

How to Conduct This Attack

Upfront data gathering and analysis are important beginnings. The test team needs to consider and understand that the issues noted previously may exist in their device. This early analysis can include the following steps; however, refer to Figure 3.2 for input or output considerations.

- Identify input devices with ranges and resolutions of values
- Identify output devices with ranges and resolutions of values
- Define the full range of input disturbances (unexpected system inputs)
- Define possible output disturbances (unexpected system outputs)

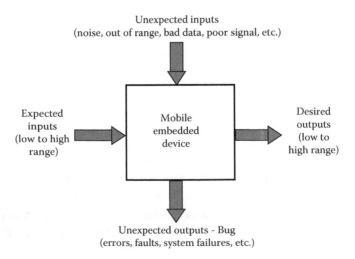

FIGURE 3.2   Inputs and outputs of a mobile and embedded device.

- Determine what is or is not possible in the test lab

- Conduct a risk analysis and/or software–system FMEA

Takeaway note: Testers should be aware that embedded systems have resource constraints in memory, central processing unit (CPU) usage, and time, which means many "nice to have" features will not happen, often complicating and impacting test design.

The test team uses information—particularly the risk or failure mode and effects analysis (FMEA) information, to first create a series of normal usage end-to-end tests from the input device, software functions, and output devices in the system. This testing should be done with "hardware in the loop." The aerospace industry has for years had a concept called "Test like you Fly and Fly like you Test." In conducting this attack, this testing concept is important to consider, but here are some other questions for your consideration.

- If the hardware is a prototype (not like what will be in the field), will that impact testing or test results?

- If a simulation is used, what bugs might be missed because actual hardware or software is not used?

- If the test inputs are not representative of the real world both in terms of expected and unexpected values, what risks will be acceptable?

- If the hardware is not understood, will testing be weak?

- If the major sources of "noise" are not defined, will the system be susceptible to impacts from unexpected input or outputs?

All of these questions will involve test tradeoffs, accepted risk, and compromise. As testers, our job is to provide as much information as we can to the rest of the team, to management, and to stakeholders. As testers, we should avoid assumptions and assuming test risk by ourselves. Normal tests are the starting points for these questions, but studies have shown that about 40% or less of tests end up being "normal" in a complete test plan. Most attack cases in this area should be "off-normal" or trying to break the software A2D or D2A. It is best to start with normal usage attacks and then proceed with off-normal usage attacks.

ATTACK 6: LONG-DURATION CONTROL ATTACK RUNS

☐ Intermediate

In the mobile and embedded control world, the time, state (current, past, and persistent), and accumulated results or data often come together to create a bug, but often not on the first execution of the code that contains them. Short-duration runs and single sets of input attacks are not enough to find these bugs. Many testers and test plans do not exercise

SIDEBAR 3.2

Mars Rover *Spirit*: *Spirit* [13] put itself into safe mode after about 2 weeks of surface operations. The errors involved a flash memory anomaly in a commercial operating system, onboard activities, and extended use. The rover recovered using safing software, which shut everything down and "called home" for help whereupon a patch was uplinked. The bug was missed during testing, but it might have been found by a long-duration attack and/or maybe some memory analysis (static analysis or dynamic testing).

SIDEBAR 3.3

Patriot Missile: Patriot missiles during the first Gulf War had a memory accumulation error [4], which resulted from long-term use (hundreds of hours). The loss of accuracy caused a system to malfunction resulting in 28 service members dying as stated in *Failure at Dhahran:* http://en.wikipedia.org/wiki/Patriot_missile#Failure_at_Dhahran [14]

long-duration testing, because it can be expensive and time consuming. Attack 6 defines long-duration tests that target these bugs including ways to work within limited project resources.

In Attack 6, there are questions of "How long is long? What inputs are needed? How do you vary these inputs realistically over time? How much test lab and equipment test time can be used?" Plus other questions can also come into play. With up-front planning and allocation of resources, this attack is not difficult to envision but can be a little hard to set up. Many errors that escaped into the field could have been found with this attack (see Sidebars 3.1 through 3.3). Further, some long-duration usage errors occurring in the field are difficult to diagnose because of limited data from the field. But these errors can be found with this type of attack. Many software test industry contexts appear to have a variation of the long-duration test, and most senior testers practice this attack in some form.

When to Apply This Attack

Conduct this test when the system can be left running and in use (not just turned on) for long time periods and has some precedent state or data or information that carries over and/or changes from time cycle to time cycle. Most testers think of this information as just in the software, but hardware sometimes has state, time, and input information that can change or last over long periods of time and, therefore, influences the software's actions. So hardware must also be factored into the testing when running this attack.

Consider using Attack 6 when long input command chains, sequences, or actions over time are possible. Software testers need to consider these other "sources" of time-dependent data.

Finally, a test team can run these types of attack tests toward the end of a test cycle after other attacks have been conducted and the system appears to be "stable."

What Faults Make This Attack Successful?

Faults do not immediately appear during the first execution of the code in which they live. They often involve subtle interactions and accumulating factors over time. These factors can include:

- Errors involving computer resources such as memory, disk space, variable sizes

- Data values accumulated and stored over time

- States of the system

- Time (days, dates, years, etc.)

The examples and errors cited in the sidebars of this chapter are not what many projects want to happen. These sidebars present example cases to allow you to justify some longer test runs, when the risks justify them.

Who Conducts This Attack?

This is a test team attack. The systems, software, and hardware staff should be consulted to discover what "long" means for the local context. Sometimes the new or junior staff can be used to care (initiate) and feed (provide inputs to) these attacks. They are often looking to make a name for themselves and so do not mind the long hours required to run these kinds of tests. Alternatively, remote computer operations and automation can be used instead of long tester hours (this can save budget), when and where this is possible. A final option is during field testing where special "long run" staff may take the system on a drive, flight, or some other "long-term" use.

Where to Conduct This Attack?

Attack 6 needs to be conducted under controlled circumstances in the lab or in a situation where there is continuous usage or time is allowed to pass. Consider running this over a weekend or holiday. There may be issues of staffing, so automation is handy here. Some industries will take test systems or vehicles for a long drive (usage) around the country or world, or into a production environment, recording lots of time and usage. The location needs to be realistic to the local context, though it may be controlled to some degree such as a special configuration or supported by "test equipment." For example, Detroit auto manufacturers use covered drapes to keep test vehicles from prying eyes and have been seen in Colorado's mountains and in the desert southwest—a very long way from engineering and manufacturing plants as well as cities. During these kinds of tests, the system (or auto) is tailed by vans carrying the "support" equipment. These trips are harsh, long usages, and make use of varied environments to achieve this attack. This may be a nice "vacation" for the testers or just a long test drive where the whole system is subject to the "wear" of time as well as many input situations. Lots of good data can come from these kinds of "long usage" tests.

How to Determine If the Attack Exposes Failures

During the long-duration run, system bugs may expose themselves in unexpected and unwanted behavior including the following:

- Data corruption

- Slow performance

- Dropped functions and tasks

- Running out of memory

- Crashing and/or failing of the system

Some of these results may be a bit hard to detect but not impossible with careful thought and some tooling (using memory leak detectors). Crashes and hard failures are much easier to notice.

How to Conduct This Attack

Start Attack 6 by combining or "stacking up" the tests, attacks, and efforts previously completed. These combinations can be collected in a longer-lasting (time-wise) procedure or automated script. In the test set-up, make sure that there is logging and that the correct information is being recorded since, if the test blows apart, the team is going to want to know what happened (for repeatability), and rerunning the test for a period of more days may not be desirable. Do not conduct this test by just leaving the system turned on, since this is usually not enough in many mobile or embedded devices, because idle loops may not change data or state. The attack usage should be close to how the system is going to see or run in the real world. A basic attack pattern may look like the following:

- Start by understanding the "long run" usages of the mobile or embedded system;

- Create a series of inputs and/or stack some existing attacks to create or emulate long time periods of system usage (normal usage, all aspects of the system in use, maybe some off-normal operations, and interesting outside influences such as noise, slow networks, communication drops, or stress cases);

- Define how the data and inputs to drive the long attack will be created, as well as when and where they come from;

- Consider what components—hardware and software—can be interacting and brought into the long-duration run;

- Identify if support automation, recording, and test resources are needed so that they can be created;

- Set up data recording, logging, instrumentation, and the automation to allow the long run attack to happen with minimal human care and feeding, or establish a tester who does not mind being the long-term guinea pig for the test;

- Start the attack (consider taking some well-deserved time off if over a weekend);

- Review and closely analyze the results looking not for the obvious crash but some subtle result that earlier tests or attacks did not expose; and

- Repeat the test with a different scenario or combinations of data, hardware, usage, and/or software, if time and budget allow.

In the case where a long attack is performed by using the system in the real world such as a test drive, a test flight, long hours with a beta user, and so on, care should be taken that realistic usage, information gathering, analysis, tooling, and expertise are in place. This often means that special product configurations (cars and planes with special instrumentation), support staff (escort vehicles), mobile labs (vans) with data recording and reduction abilities, and engineering staff that can "pay attention" throughout the test are all made available. In the mobile or smart device world, check for issues in task managers, response times, memory usages, and logs, as well as app conflicts.

> **Takeaway note:** It may be tempting because of cost or schedule to skip long-duration attacks, which are hard to set up, but the taxonomy showed many bugs that become public hide here. So ask yourself, "Does the risk of missing these bugs justify skipping this kind of test?"

ATTACK 7: BREAKING SOFTWARE LOGIC AND/OR CONTROL LAWS
◇ Advanced

In many software domains, the largest percentage of bugs lives in the logic flow of the software. Studies of embedded errors have shown that the major bugs cluster in the logic flow and control laws (see Appendix A). The control laws are often based in the physical reality of gravity, space, force, movement, size, weight, light, chemicals, biology, and so on. If these laws are not implemented completely and correctly in the logic, or get interfered with, then real physical errors can result. People are killed by radiation controlled by software [15], trains crash [16], and rockets blow up [17]. It has been observed that if a user of a smart device or system is either not aware or only partially aware that they are using software, then it is an embedded software device, but the user still expects the car, medical device, or other system to, in fact, be "controlled" and to work.

Attack 7 targets the software "brains" of the system (the control logic area) in an attempt to break the logic.

When to Apply This Attack

When a tester finds the device or system is controlling something, the attack on the control laws become possible. Embedded devices will usually have their control laws published and/or defined. If the laws are not defined, then there is a large problem for testers to work on, or in other words you will need to create the laws. As covered in other attacks, these laws can be attacked during analysis, design, implementation, integration, and system-level testing. Apply Attack 7 early, often, and consistently over the life cycle.

Certainly, after the hardware and software start to come together in a complete or near-complete embedded system within a test lab, Attack 7 needs to be high on the priority list. It is possible to apply the attack with prototypes or slices of the system, and then reuse these attacks when the completed system comes to fruition. The main items and features of the embedded system that are being controlled should be attacked. As any of these factors change or the product usage environment evolves, repeats of the attacks should be considered from a regression standpoint. However, watch wasting time with regression testing, since often the usefulness of just blindly repeating regression tests is overrated. A revised attack with new data points can be more effective than a pure regression test during updates and maintenance efforts.

What Faults Make This Attack Successful?

The complexity of the laws and algorithms means that bugs are likely in

- The complexity of the laws yields holes in algorithms, incomplete laws, and problems in implementation.

- Time factors and variations; the measurement of real-world factors, which must be converted and represented in the software world for the control laws to use.

- Conversion and output of the control signals to the real world (see Attack 5).

Who Performs This Attack?

Attack 7 is conducted by the test team, the control law staff, and input/output hardware experts. Hardware and systems analysts, who may be defining the control laws and requirements, should start the attack. At the lowest level of individual control points or statements, a developer may be able to conduct parts of this attack, but the end-to-end impact of the laws, inputs, and outputs needs to be tested across the hardware, computer, and software. The team conducts the end-to-end attack against the complexity of the hardware–software system to find the bugs.

Where to Run This Attack?

Depending on the size and context of the embedded software system, Attack 7 should be considered at a variety of levels and places, including the following:

- Analyses of requirements to design level—Use math simulations to analyze the laws, particularly for complex laws such as in avionics, medical, robotics, and factories;

- Implementation level—Use computer instruction simulators or emulators, target single-board computers or chips without all of the hardware, and/or debug environments;

- Integration level—Use test beds with slices of the hardware and/or software; and/or

- System level—Use software and hardware in a test environment and/or in the implemented system.

Math simulations can be used to run thousands or millions of data cases, where doing this at a system level would likely be impractical [see Chapter 11]. The simulation approach is used to establish the validity of the proposed control laws and can be done early in any life cycle, before the embedded hardware or software exists. Simulation can use models in the form of SysML or UML systems or a program. Specialized tools exist for this: systems modeling language (SysML) and unified modeling language (UML) have support in tools such as IBM's RHAPSODY, ICONIX, and others, and programming supported by tools such as MATHMATICA, MATLAB, MATRIXX, and just pure program code. This effort is usually the domain of the analyst and/or the systems engineers, and only occasionally testers. However, testers need to be aware of these activities and potentially leverage the results for interesting attack cases.

The results of this analysis usually are the "published" control laws in the form of requirements, models, or detailed code, which is given to the software implementation team. The implementation team completes the software code, hardware, and should conduct implementation-level attacks (Chapter 2). In some cases, these activities can verify the lowest level of requirements or control law implementation (Attacks 2 and 3). However, the whole system needs to be attacked before you know that the system laws are sound and the attacks would take place in the lab. Considerations for test lab implementations are defined in Chapter 11.

The specific control laws and logic flow are only a part of the story needed for success. The embedded devices are receiving inputs from the real world. This data will need to be measured by instruments—hardware and sensors. Instruments suffer accuracy and precision issues, as with any engineering equipment that may be subject to issues of calibration, misreading, inaccuracies, and so forth (see Attack 3). The data from the measurement instruments should then be stored in some data representation. Data representations in computers can vary from integer, to floating point, to high-precision floating point. Integer math is fast, but lacks precision. Floating point may be slower, but is more precise. Exact details of this kind of math are out of scope for this book [18], but a good embedded tester will have some grasp of these concepts. As an example, aspects of major system failures such as an Arianne rocket failure in part due to a numeric data value misrepresentation [19]. (Additionally, they reused software without testing it.)

Once the data is represented, the software can compute. Assuming analysis and implementation have happened correctly and are verified, the software control laws will then compute an answer for an output. Again, the system can have issues in numeric representation. This numeric representation makes its way from the D2A world (see Attack 5), where data is converted and sent to the output world, and controls the hardware and/or system in the real world. This complex dance of real-world inputs, digital world, and real-world outputs is not overseen or controlled by humans directly, it happens in real time at varying fast rates (seconds, parts of seconds, millisecond, nanoseconds, and even faster). The time aspect is the last complicating factor in this attack. The test environments and this attack need to capture all of the data. In fact, usually a single test environment is not enough. Testers may end up in software–hardware test labs, system test labs, and testing the system in the real world. Attacks of the control laws are possible at all levels and environments. The type of embedded system, risk, and any regulations, as well as cost and schedule likely govern where and how much of the control law attack can be accomplished.

How to Determine If the Attack Exposes Failures

Team members conducting this attack need to understand the level they are attacking and what the indicators mean. Examples of indicators to watch for include the following:

- System fault

- Unhandled (out-of-control) situations and cases

- Singularities, math problems, flow of data

- Performance issues (time)

- Buildup of error values because of numeric representation, accuracy, or precision problems

- Small input or output "fuzz" where the ideal is not met

Analysis of the results of this attack in the first two cases (fault and unhandled) is fairly easy since "expectations" (usually the control laws and requirements of the system) are not met. The success criteria are defined based on expectations. For many embedded systems, success criteria can be defined with tools and automated checks or filters (see Chapter 11). After the more obvious cases come the more subtle outputs, which, while not dramatic, may represent problems and even (later on) a system failure under the right (or wrong) situations. These results may be macroscopic in nature.

Control law, math, data representation, and flow computational checks represent microscopic logic. This attack targets bugs at the microscopic level. Such a low-level attack takes craft, understanding of math, nature, laws, science, and engineering—all large subjects unto themselves. Many embedded testers have a large reference library to leverage lessons learned into the attacks and analysis of the results.

The final complication of this attack for embedded control systems is time. The attacks need to be conducted with and over long periods of time. Long-duration runs are well known to find errors in this attack space (Attack 6). The problem may be large amounts of data to sift through and how to find that one wrong number on a trend line that is headed in the wrong direction (not wrong per se, but headed that way). Besides long times, embedded systems working in real time can have short time spans, which are wrong or produce an incorrect answer. Again, time and performance time analyses are subjects of whole books and classes. So a tester needs to consider time as it relates to the setup, in order to conduct and analyze this attack. To do this attack correctly, some extra effort and careful thought may be called for.

Takeaway note: If your software is controlling something, Attack 7 is for you, and you need to determine the risks and how to spot the bugs or the users will do it for you, which may not be a desirable outcome.

How to Conduct This Attack

Attack 7 is related to and can leverage several other attacks. The test team should be aware of other attacks and factor them into overall test planning:

- Analysis of time—data, simulations, time, modeling [20]

- Data inputs—see Attack 15

- Forced outputs—refer to [7]

- Long-duration runs to allow errors to build—see Attack 6 and Chapter 6

- A2D and D2A attack—see Attack 5

These attacks target bug in the control laws and flow inside of the software. The flow of this attack looks like this:

1. Obtain the requirement and/or laws and flow of the system.

2. Define the data ranges (command, values, inputs, etc.).

3. Design the attack to target the range end points using, for example, boundary value analysis [1] and random and/or combinatorial test data selection [1,21].

4. Run the attack (s).

5. Analyze the laws for stability and data problems.

6. Learn and repeat as necessary.

In Figure 3.3, I show a simple control law for a motor. To start, the system is in the "system off" state box. Turn it on using the "on command," and then the system goes into the "system on" state. You can issue a "move command," which first ramps power up to a motor in a "ramp up voltage to motor" state, so that you do not overstress the motor. Once ramped up, the system maintains an applied voltage in the "apply voltage to motor" state such that the motor keeps running. When the motor is running, a stop command can be issued, which ramps the voltage down and leaves the device in a "system on" state. You could create a control law test made up of three command sequences of "on command," "move command," and finally "stop command" to confirm that the motor enters each required state in sequence.

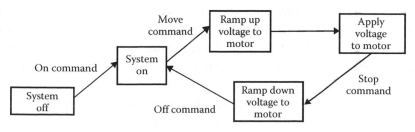

FIGURE 3.3   A simple block flow control state diagram.

There, you have a test that the motor works. But notice, if you create a test to turn the system off, there appears no way for this to happen as the "off command" has not transitioned. Is this a bug? Also should you have to create a test to issue the "stop command" in the "system on" state? What should happen here? During the "ramp up voltage to motor" state, what happens if you issue a "stop command," since this happens over some period of time? A lot of tests become possible; there are many unknowns to test; and a least one bug appears.

The keys to this attack are the laws, data values of the laws, and the flows that use these factors. Testers will need to understand these data points and then build the attack accordingly. Simple laws equal simple attacks. Complexity may make more attacks or more complex attacks. Testers should seek and covet all of the development and analyst's information that exists. When such information is lacking, testers will need to add to what is available, make assumptions, and use the tester's system knowledge to implement the attack.

ATTACK 8: FORCING THE UNUSUAL BUG CASES
□ Intermediate / ◇ Advanced

In the first of the *How to Break* book series, one section explored testers forcing output [8], instead of just focusing on inputs. Four attacks were proposed (see Attacks 7 through 10 in Whittaker's *How to Break Software* book). I believe that these attacks should be considered and then modified for mobile and embedded devices as a start. However, I introduce Attack 8 as an additional attack to Whittaker's work. Attack 8 addresses the uniqueness in the mobile-to-mobile world. This attack defines how to force mobile and embedded device outputs, which may only happen infrequently but still need to work.

When to Apply This Attack

Consider applying inputs that force output conditions and contingency actions within software and associated hardware when those outputs have not been realized during other testing. Of course, this means the tester will have some lists of all the actions and outputs that can be expected in both normal usage and off-normal usage situations. The coverage of the output is important. It implies that the tester should be using a spreadsheet, state map, or decision table to track the coverage, and when something has not been tested by earlier attacks, this attack should be revised to improve coverage.

What Faults Make This Attack Successful?

Many output situations do not happen very often, yet the software is created to address them. There are various heuristic rules with studies that show that some 80% of the software is used only 20% of the time. This means that "normal" testing inputs and associated outputs only cover about 20% of the code, leaving bugs hiding in the 80% area. This is a heuristic and not absolute, but should you plan to test the 80%? Much of the 80% of output situations can be difficult to create as they involve hardware, system states, sequences of inputs, strange inputs, user interactions, and other combinations of things that are difficult to set up. Most testers learn over time to conduct their bug hunts in the sparsely covered 80% areas. Outputs in the "80%" area to look for include the following:

- Alarms

- Fault recovery

- Command outputs

- Messages

- Hardware configurations, switch over, and error states, and

- Other hardware outputs

Who Conducts This Attack?

This is a test team attack, but consultation with developers, hardware, system, and even users may be needed to set up and conduct the attacks. Special team skills may be needed for the software, hardware, and operations necessary to implement this attack.

Where to Conduct This Attack?

This attack is best run in the lab with hardware and software fully in the loop. Simulated or software-only environments can be useful but can have limitations and may indicate that things work when there are actually still bugs to be found. System-level testing is preferred. Testing in the real world is maybe even better, but can take extra time and money, making labs more attractive.

Labs and test equipment may need special set-ups including hardware, environment, communication data inputs, and software changes to get a particular output to happen. Situations can include the following:

- Modify the software to have a "trigger" to make output get forced

- Special hardware

- Test data inputs

- Communication issues, sequence, and data values, and

- Operational environments (cold, wet, snowy, noisy, chemical, etc.)

These scenarios can be costly to set up and use, but not doing them can mean bugs escape. I have seen mobile devices that have problems with signals when cold (sensors reading incorrectly when frozen), during rainy or foul weather (wet equipment), and even with user characteristics (a user with cold fingers on a touch screen). How do you force such situations and outputs?

How to Determine If the Attack Exposes Failures

The desired output is known prior to conducting the attack. If the inputs that should "force" a particular output do not generate that output, you have a bug. A problem will be in knowing all the differing situations (inputs) that should result in a particular output. Getting the output once with the "right" input is good, but ask yourself, "How many other situations should be created to get that same output?" A missed set of conditions that does not get the appropriate output is a bug too. Also if the input does not generate the output as expected, you may need to ask yourself if the input was really "right" to generate that output.

How to Conduct This Attack

The tester starts by identifying the output states. This should be tracked in a list, decision table, matrix, or possibly a state map. Outputs and resulting information can come from the following types of sources: hardware references, requirements, design and systems information, user notes (help files), and other software documentation. There are "official" types of information, but also the unofficial information, such as history, tester knowledge, similar products, and even user experiences that may be entered into the lists. Then, every entry should be considered for coverage, though it is possible to not actually test an entry if risks associated with that entry are low. But remember that a bug may be hiding in those untested areas. For example, is it reasonable to expect an app to work offline, during a phone call, or a one-time alarm to work on New Year's Eve (e.g., the IPHONE alarm bug)? In the list, it is good practice to make annotations on any entry that is not covered, addressing why, why not, implications, and risks.

Next, define the conditions of the software, systems, and/or hardware to generate that output. This is a backward reasoning problem, which can get complex. It is not good enough just to find the single triggering point or set of points that generate an output. A series of inputs and triggers may be needed. In more complex cases, a state chart can be useful. In simpler cases, a spreadsheet or a decision table may be needed for set-up and tracking (refer to [1], Chapter 5 of this book, and [21]).

After this, the attack can be set up in a lab or test environment. Factors to consider are hardware status, software inputs, software state, and their interplay to get the desired output. Again, more complex systems may require state mapping. Also if the hardware or system conditions are difficult to set up, special attack preparation may be needed. For example, on one system, a special software build had to be created to set up a test that involved hardware failures where failing the hardware was not possible without great cost, as one would have to short or break very expensive electronics. In this case, a software build was created where, at a set time, the software issued a "trigger" (interrupt level), so that the fault process output logic was executed. This had risk, since it was "a special build" and not a pure demonstration of the system, but was better than breaking expensive hardware or not exercising this logic within a system-level test environment.

Once the lab and inputs have been set up, the tester executes the attack(s). The test team makes sure that all data from inputs and outputs, hardware and software, and test support information (lab equipment) gets recorded—often in great detail. In forcing output of 80% of code and associated outputs, strange things sometimes happen. This type of attack is often difficult to repeat or has results that raise many questions by developers, management, and even customers. Having good testing data recorded allows the team a chance to backtrack and figure out what happened, even if the attacks do not provide the same outputs. The results are reviewed for proper responses, not just expected output. Look for other outputs, late or early outputs, or other abnormal results. These results may be bugs, and further analysis may be warranted.

> **Takeaway note:** Forcing some outputs in embedded systems can be very hard such as you cannot break your hardware, which has software recovery logic, but you should not leave an output function untested just because it is "too hard," because sooner or later, a bug will be hiding in one of the untested areas.

EXERCISES (ANSWERS ARE ON MY WEBSITE)

1. List the environments that are different from a PC and how they could impact a smart phone with a touch screen display.

2. Define how long a long-duration test should be for a smart medical device used to record a person's heart and respiration rates.

3. Define where you have hardware-to-software or software-to-hardware interfaces that will need testing in a smart mobile phone with GPS, motion sensors, a touch screen, and Wi-Fi (or do this for your project's device).

4. You are testing an embedded control system that has two processors, and they reconfigure from one side to the other side when the hardware stops working. Make a list of the attacks you might apply and why.

5. You are testing a controller for a robotic car. How might a long-duration attack be applied?

6. You are testing a "fighting" robot for the games on your mobile device. Define which attacks you might apply and why.

7. Conduct risk analysis for a smart medical device used to record a person's heart and respiration rates.

REFERENCES

1. Copeland, L. 2003. *A Practitioner's Guide to Software Test Design*, Artech House Publishers, Boston, MA.
2. Leveson, N. 1995. *Safeware: System Safety and Computers*, Addison-Wesley, Boston, MA, Appendix A.
3. Electronic stability control. http://en.wikipedia.org/wiki/Electronic_stability_control (last accessed April 11, 2013).
4. Neumann, P. 1995. *Patriot Missile in Gulf War: Computer Related Risks*, ACM Press, New York.
5. Myers, G. 1979. *The Art of Software Testing*, Wiley, New York.
6. Failure modes, effects, and criticality analysis. http://en.wikipedia.org/wiki/Failure_mode,_effects,_and_criticality_analysis (last accessed April 11, 2013).
7. Whittaker, J. 2003. *How to Break Software*, Addison-Wesley, Boston, MA.
8. Kaner, C., Falk, L.J., and Nguyen, H.Q. 1993. *Testing Computer Software*, Van Nostrand Reinhold, New York.
9. Whittaker, J. 2009. *Exploratory Software Testing*, Addison-Wesley, Boston, MA.
10. Petschenik, N. 2005. *System Testing with an Attitude*, Dorset House Publishing, New York.

11. Labrosse, J., Ganssle, J. et al. 2008. *Embedded Software*, Newnes, Boston, MA.
12. Rorabaugh, B. 2010. *Notes on Digital Signal Processing: Practical Recipes for Design, Analysis and Implementation*, Prentice Hall, Upper Saddle River, NJ.
13. Mars Rover Spirit flash memory management anomaly, January 21, 2004. http://en.wikipedia.org/wiki/Spirit_rover_Spirit_Sol_18_flash_memory_management_error (last accessed April 11, 2013).
14. Failure at Dhahran: http://www.gao.gov/products/IMTEC-92-26 (last accessed April 11, 2013).
15. Pries, K. and Quigley, J. 2011. *Radiation Kills: Testing Complex and Embedded Systems*, CRC Press, Boca Raton, FL.
16. Roychoudhury, A. 2009. *Trains Crash: Embedded System and Software Validation*, Morgan Kaufmann, Boston, MA.
17. Gardiner, S. (ed.) 1999. Rockets blow up. In *Testing Safety–Related Software, A Practical Handbook*, Springer-Verlag, London, U.K.
18. Floating point. http://en.wikipedia.org/wiki/Floating_point (last accessed April 11, 2013).
19. ARIANE 5 Flight 501 Failure report by the inquiry board, Paris, 1996, http://esamultimedia.esa.int/docs/esa-x-1819eng.pdf
20. Lui, S., Klein, M. and Goodenough, J.B. 1991. *Rate Monotonic Analysis for Real Time Systems*, Pittsburgh, PA, CMU/SEI-91-TR-006, CMU tech reports.
21. ISO29119, Part 4. http://www.iso.org/iso/catalogue_detail.htm?csnumber=45142 (last accessed April 11, 2013).
22. Risk analysis. http://en.wikipedia.org/wiki/Risk_analysis_(engineering) (last accessed April 11, 2013).

CHAPTER 4

Hardware Software Attacks

T HIS SERIES OF ATTACKS are conducted on the software but aimed at bugs that have their origin in the hardware. So while they are not hardware testing per se because of the mobile and embedded hardware–software dependencies, you should "think" hardware first.

Here again, these attacks have differing applicability within the mobile and embedded contexts as shown in Table 4.1. These attacks are usually more applicable to "embedded" devices, but many mobile wireless devices have context, which demand these kinds of attacks such as mobile wireless medical devices.

ATTACK 9: BREAKING SOFTWARE WITH HARDWARE AND SYSTEM OPERATIONS

◇ Advanced

The unique hardware–software interaction associated with embedded systems must be tested within the context of the software and system operations. This is not hardware testing, but means the unique hardware–software interaction of the embedded system is attacked within the embedded software testing, to the extent possible, at a software system level. While this sounds easy, the result is a level of combinations and interactions between hardware–software–operations, which creates much complexity and where it is easy to miss the possibilities of the real world. The information technology (IT) world includes hardware during software attacks, since software must run on something (a generic computer). But often, there is more "trust" of the hardware, since it is in use in many places. The embedded world has a "generic" hardware side, plus it has the "unique" hardware and interdependency of hardware–software–real-world operations to contend with. This attack looks for bugs in the software by using the uniqueness of the hardware and its operations. This is really a pattern framework, which the testers must build upon.

When to Apply This Attack

Apply this attack when there is hardware and software interaction that is critical (e.g., medical, control, life-rated, etc.) or complex (e.g., hard to understand and important). You may

TABLE 4.1   Map of Hardware–Software Attacks to Mobile or Embedded Contexts

Context → Attack ↓	Mobile and Smart Wireless Devices	Embedded (Simple) Devices	Critical Mobile Devices (Could Be Embedded, Small but Important)	Critical Large Embedded Devices (Could Be Mobile)
Attack 9	Seldom	Frequent	Yes	Yes
Attack 10	Sometimes	Frequent	Yes	Yes
Attack 11	Sometimes	Yes	Yes	Yes
Attack 12	Yes	Yes	Yes	Critical
Attack 13	Sometimes	Sometimes	Yes	Critical

Notes: Seldom, seen only in specialized cases infrequently; sometimes, seen often in many different contexts; frequent, seen regularly in a variety of contexts; yes, should be considered for most contexts; no, generally not applicable; critical, important to do in most contexts.

know these things from requirements, design, risk assessment, and/or hardware–system operation concepts. The personal computer (PC) world has its "generic" hardware that can be attacked (see Whittaker's *How to Break Software* book [1]). When these hardware situations are found in the embedded world, with some modifications, conduct the attacks found in Whittaker's books. But embedded hardware will have other features and situations to attack. When you find your unique embedded hardware has some software dependencies, variations, and other specifics, you will need to conduct this attack.

What Faults Make This Attack Successful?

This attack is successful when a bug originates in the hardware–software interaction. The following situations may impact the software, creating bugs, and thus need to be tested.

- Fault tolerance is incomplete (see Attack 13)

- Power failure, shorts, or "brownouts" (low power supply or batteries) on all or part of the system

- Electronic hits and electromagnetic interference (EMI)

- Memory and register memory corruption

- CPU upset or factors that make the electronics jumble, from sources such as cosmic rays, power spikes, and so on

- Disk or long-term memory storage issues, such as slow disk/memory access, bad memory locations, full disk, or other memory storage issues (see [1])

- Problems from other complex software–hardware interfaces

Further, if you put the hardware and software into situations that were not anticipated by development or in the operational profile, bugs may appear (see Attack 4).

Who Conducts This Attack?

This is a test team attack with support from the hardware and system staff. A diverse team is good here, since no one person knows the hardware, operations, software, what can

SIDEBAR 4.1

Space Shuttle: Years ago, a space shuttle system that was designed to boost satellites into orbit experienced a hardware problem that the software was not designed to cope with. The system was to sequence hardware actions based on acceleration or the lack of acceleration (acceleration meant a rocket engine was firing or not). In this case, the hardware failed and put the system into a spin, which meant the accelerometer was sensing constant acceleration (centripetal). The software did not sense the end of final burn, which would have indicated to terminate the mission and release the spacecraft. Mission sequencing was "locked in place" and the mission could have been lost, if the system had not been commanded to terminate from ground controllers. This command uplink saved the mission. The bug or problem was a design flaw, which was a lack of redundancy in event sequencing. The solution was to add a backup timer feature to handle many different hardware problems.

SIDEBAR 4.2

Toyota: A few years back, a Toyota product was having problems, which some claimed was a software EMI problem. Toyota with assistance from NASA ruled out EMI as a problem on cars [2], but embedded systems are still being hit by EMI. For example, the airplane take-off and landing restrictions on devices such as cell phones. This is because there were (aircraft) crash cases (one or two) where embedded systems had issues and the only "trigger" seemed to be EMI from cell devices [3]. So embedded system designs benefit from being able to "accommodate" EMI, and testers may need to assess the EMI situations.

go wrong, when it can go wrong, and how the system should react. Here, the team needs the experience, thinking, and diversity to identify reasonable cases, create attacks with inputs that will create the cases, and then analyze the resulting test data, since the bugs are well hidden in this attack. Testers should never trust that the hardware and system will work as expected. The team will want to know that the software reacts appropriately to the hardware–system conditions both normally and abnormally.

Key point: Some have described this integrated team as needing to "run scared," which can be defined as being afraid of the things that can go wrong in the system. Consider that "Murphy is alive and well in the real world" ("Anything that can go wrong, will go wrong."—Murphy, a possibly fictitious character.)

Where to Conduct This Attack?

This attack can be performed early in a variety of environments, which can include the following:

- Simulation- and model-based environments

- Emulators or on

- Single-board computers

Note: It is optimal if error cases are run by developers in these first environments. These types of tests are usually driven with simulation tools, mocks, stubs, drivers, and so on, but it does not address the interactions between actual hardware and software. Attack testing will be needed in the following environments:

1. Integration lab with hardware–software test bed with special test tooling/simulation or stimulation/configurations [4]

2. Test beds and facilities with special simulation and stimulation

   a. Software in the loop

   b. Hardware in the loop

   c. System in the loop

3. Field testing with as-built, instrumented, or built-in test (BIT) configurations

Chapter 11 addresses how to set up these environments.

The team envisions the scary and/or missed hardware problems that the software should accommodate.

Finally, in these environments, many hardware situations are hard to set up in test labs. For example, if you are black box testing a complex pair of pairs computer processors (two computers running two CPUs as primary and backup software), where the hardware and software both have failure detection, isolation, and recovery abilities, how do you test the reconfiguration logic of the software when the hardware is "ultra-reliable" and you do not have the ability to fault the hardware? The "hardness" of testing hardware failures often means that special lab tooling and configurations are needed (refer to Chapter 11). In this testing, you may corrupt memory through a patch or force state changes and/or recovery logic. Triggering these kinds of bugs in testing may be hard. You may be able to use simulation or stimulation environments to do some of this, but sometimes you may even need to break hardware. Breaking hardware can be time consuming, difficult to do, and expensive. Good planning, critical thinking, and a good test design are needed here.

How to Determine If the Attack Exposes Failures

Here the test team is looking for bugs that show themselves in

- A failure such as the system crashes

- A user need, feature, or requirement that is not met

- Small differences between expected and actual results where a "risk" is realized

The first two may be obvious, but the third can be subtle and many a tester has missed a small difference in expected output results—not seeing the bug. If the attack had continued a little longer, had closer observation of the data output, or were tested with a slightly different situation, a real bug might have been revealed.

SIDEBAR 4.3

Long Duration Test: A group was testing a life-critical device that had complex hardware and intermix controls. Thousands of test cases were run. In this data, there were five tests that showed an intermix of a control law hardware–software bug, but the bug showed itself as an increasing trend line. The bug did not cause a failure unless the test case ran for a longer time or hit certain conditions, which were never run. If the team had noticed the bad trend line in the thousands of data points, they might have found the bug before it was seen in the field.

The Sidebar 4.3 example brings up a further complicating factor in finding errors in embedded attacks with hardware in the loop, in that embedded systems may generate large data files and bugs may be hard to find. I know of systems that generate megabytes even terabytes of data in a single hardware–software test. Here is the proverbial "finding a needle in a haystack" problem. During the attack design and environment set-up, consider how bugs will be detected to make finding them as easy as possible (see Chapter 11 for post test analysis information).

How to Conduct This Attack

For starters, the test team should have a copy of Whittaker's *How to Break Software* [1]. This book has many attacks that, with modification, can be applied to embedded systems in this area. Start with these types of attacks making sure that you push your system to its limits and beyond.

Next, consider the product's industry history such as bugs of the past, other products, and lessons learned. Which of these things are scary to the team? This knowledge should be captured as a risk-based testing exercise. A good risk exercise or even doing failure mode and effects analysis (FMEA) [5] and/or failure mode, effects, and criticality analysis (FMECA) [14] would also be good to do at this point (see Appendix G).

Further, conducting this attack may not be easy, and so many groups will run a couple of tests in a different environment (or two) and call it good. But you should pause here and consider the overall project risk. "Are a few tests acceptable?" Within many industries (medical, aero, and so on), finding negative consequences is not acceptable and so, more attacks of this type are needed.

To get details for this attack, use items such as feature lists, requirements, situation assessments, operational concept documents, risk lists, or other available information. Use these items to define the data and scenarios to drive this attack. Next, here are some factors to consider in designing this attack.

- Functionality (and nonfunctionality) of the system

- Special cases from the risk list

- Test design (how hard is it to set up, does the test environment support the attack, do inputs exist, etc.)

- Do you need special builds to simulate hardware cases and failures?

- Can you "break" the hardware (once or twice) within cost and schedule constraints and

- Do you have BIT features or test ports/points that you can leverage (see Chapter 11)?

As a tester, the job becomes testing what is likely and must work, and then testing what is not likely to happen, but also must work. Creation of "hardware" errors to drive into the software failure logic can be problematic in this attack. Tools such as Holodeck [1] and approaches such as using a fault injection tool can drive this attack.

Note 4.1: Built-in Test: In the embedded domain, the concept of building for test support or BIT can be important. In the electronics testing world, BIT is common and very useful, and many embedded systems extend BIT with points to access inputs and outputs of the software. This can be useful, but mostly thought of up front when hardware and software are being developed. It is usually "too late" to add BIT to the mix after the fact. This is where the embedded domain of testing turns the world upside down by defining requirements and design, which drives both hardware and software. Many resist because the added hardware and software can risk bugs through coupling and cohesion, and it increases costs. But if the cost and risk of doing BIT offsets the expense of not doing BIT, is that a valid tradeoff? The tester should be ready to make the case why it is better to do or not do BIT.

A variation of Note 4.1 is to have special hardware and system configurations built to support embedded testing. These configurations, while part of test labs, are not delivered and only exist for the role of testing. Examples in the embedded world include test airplanes, vehicles that will never go into production, jail broken or rooted cell devices, and others. There may be wiring, sensors, test points, and other changes that such special configurations can use to provide input, stimulation, coverage, and recording of the testing. These special configurations must be planned and implemented, have costs and schedules, which the test planning must consider. These test configurations also come with risk, because of "test like you fly" considerations. So these attacks may be complementary to the "pure" lab and system field attacks. And here again, the planning must address multiple aspects and risks of this approach. Additionally, the team must have the knowledge to define, build, and run these special hardware–system configurations.

A final option exists for testing when faced with hardware–software that is really not configured to be attacked. This is test in a "software only" mode during developer testing using software simulators and/or emulators. You can create developer tests with mock objects and stubs. This is of lower-level fidelity, but if used in combination with some of the previous test concepts (presented in this attack), perhaps the risks are acceptable. Although not very attractive, when compared to the earlier concepts, these kinds of tests may fit within cost or schedule.

In some ways, all of these ideas are "tricks" to make the attack work. Some testers may not be able to apply these attacks on embedded devices such as a separate independent tester who cannot do any of these things because they lack tools, hardware, and/or special configurations. For those groups who can do these things on their embedded devices, these tricks and examples can be invaluable in each lab and at each test level. They all have been seen in the mobile and embedded test environments.

Once you have the risks and test design activities done, it is time to run more tests. Run tests while watching for test repeatability caused by nondeterministic hardware as well as unverified failures (see Glossary). Also as the test happens, keep track if the scenarios are likely to happen (or not) as well as how critical a test might be. A scenario that does not happen very often, and only has a minor risk, should not be a high priority, whereas a scenario that kills someone or has other high risks would be worthy of an attack. These scenarios should be complete stories of inputs, system usage including the hardware, and outputs.

Reporting a bug and deciding to fix it can be tricky here. For example, we (the test team I was working on) once flew a missile into the ground in a lab simulation, and it went right to the center of the earth. When the system reached the center of the earth, the processor failed because of a divide by zero on the hardware, which was an unhandled situation. At first, we argued that the scenario and bug were not real, because you cannot fly to the center of the earth, so no fix was needed. But after some consideration, we examined a case where we could get the divide by zero bit pattern. The case involved memory corruption by any means to the offending value (e.g., bit flips caused by radiation), which then would have caused a larger "divide by zero" fault in the CPU. We did not have to fly into the ground to encounter the bug. The system was required to handle the memory corruption—and if we got corruption at this location, a divide by zero would happen and the system would fail. We decided that a fix (a guard on the divide statement) was a reasonable solution to increase the robustness and stability of the system. My advice is not to discount bugs in the hardware–software area too quickly.

SUB-ATTACK 9.1: BREAKING BATTERY POWER

☐ Intermediate

A specialized case of Attack 9 is dealing with hardware power, power management, and batteries. The basics of this attack remain applicable; you can follow the pattern, but you will need to add a focus on power.

More and more mobile and embedded devices use batteries. This sub-attack details the testing and risks of this power factor. In fact, most battery-powered devices should factor this attack into their test plans. It is a risk if your app or software draws the battery down to the point where users decide to uninstall the app, which will likely kill your sales. It may be worse if your mobile embedded medical device behave differently on low battery power than full, or somebody dies (e.g., an implanted defibrillator fails to "jump-start" someone's heart and that person dies).

If you do not believe power and batteries are a risk, try this. Turn off most everything (Wi-Fi, cell, global positioning system [GPS], etc.) on your cell phone and leave just the basic device running with no apps. Now, ascertain how long your batteries last. I ran this test on my ANDROID device, and my batteries lasted for days longer than normal (normal here is with all 70 of my apps loaded and the phone being turned on).

So in this sub-attack, your risk and focus will be on the software and how the battery and power use is impacted. The goal of this attack is to assess the impact of battery life and power drain in relation to running apps and using the device. The tester will design this attack to assess the impact of software in relation to the power use of the device. Keep in mind that one app may call another app or device feature (e.g., GPS), and another app, and another app, the combination of which could draw the battery down. The tester should use the device as a user would as defined in the test design, such as having many features on, several apps running, batteries only partial charged, and so forth.

In the test design, ensure that the following types of actions are tested. Turn features on. Turn them off. Run with different combinations of hardware and software configurations (see Attack 32). Run with high, medium, and low battery power. At each step, run the device long enough to measure battery use and the amount of time spent in the attack (wall clock time). You may need special hardware (a voltmeter and/or other kinds of timing devices in the lab), software (e.g., a task monitor), and lab test environment configurations to accurately set up, execute, and measure these tests. Record and report power readings (charge readings), use of apps and configurations, as well as time information, over a series of attacks.

For the mobile or embedded battery-driven world, this assessment and data can be very useful to the developers. The recorded data over the series of attacks may need to be plotted or organized into a report so that battery life usage trends can be seen by the stakeholders. Bugs here may be a configuration and/or apps that are found to excessively drain power or where the software or device behaves differently at various battery power levels.

Keep in mind that batteries are hardware. They can and do impact the software functions.

ATTACK 10: FINDING BUGS IN HARDWARE– SOFTWARE COMMUNICATIONS

☐ Intermediate

Just like IT systems, mobile and embedded devices typically have communication ports, both coming in and going out, which are hardware. But the software is involved too, containing logic for various hardware situations and commutation lines, including checking, protocols, retry logic, filters, switching, security factors, and others. Table 4.2 defines some basic situations with this hardware–software combination that testers must consider. This attack is conducted against these interfaces with a communication line focus to expose bugs associated with this aspect of the hardware–software system.

Attack 10 is related to Attacks 5 and 14. Attack 5 is very low level and relates more to A2D or D2A communications. Attack 14 is higher level and more purely digital related.

When to Apply This Attack

Attack 10 is applied when the software being tested has interfaces to or from hardware communication lines. Here are some situations to consider with communication input/ output when considering this attack.

Weak signals with slow bandwidth

Kind of signal or protocol (e.g., network, Wi-Fi, GPS, etc.) and/or information that is contained in the signal

TABLE 4.2   Communication Input/Output Situations

Situation	Approach to Testing	Expected Action (How to Detect Bug)
Communication line input/ output temporary failure with a single input/output line	Temporarily failed communication line and then restored	Assess. Retry logic, if present, and look for system response
Communication line input/ output failure with two lines, primary and backup redundant set	Three cases might be possible: (1) Fail primary line, and watch for switch over; (2) fail primary line, switch over, restore primary line, fail back up, watch for continued communication on primary, and (3) fail both lines and watch for retry/actions	Access switchover logic and retry in each case and look for degraded performance of system functionality issues
Communication input line slow	Create slow communication speeds	Look at system functional issues and slow performance, lockup, dropped data
Communication output line slow	Create slow communication output speeds	Look at system functional issues and slow performance, lockup, dropped data, and watch for data loss
Data communications that are corrupted going to or coming from the hardware	Create corrupt messages and/or signals	Corruption is identified and handled correctly

Interrupt processing associated with the communication line (such as normal, stressed, saturated)

Dead channel(s)

Slow channel

Location of device with respect to transmitters, e.g., line of site, near, far, fringe, or other

Channel at input/output capacity level

Dropout and dips (temporary loss)

Not all testers should care about this attack. For example, smart phone apps rely on the operating system (OS), basic phone features, and hardware to provide communications. But if you are testing aspects of the OS, you might care about this attack or if your app has some "special" action based on communication. Another key consideration when doing this attack is how to use the situations listed to create inputs for the test.

Takeaway note: For testers working in networked devices, Attack 10 can be very important to include when running tests called names like "Network Tests," "Location–based App Test," or "Communication Link Testing."

What Faults Make This Attack Successful?

From the perspective of the OS and/or input or output layer of the software, where usually the logic of communication issues are centralized, issues with input or output needs to be detected and acted on. The handshake between the application needing input/output, the OS, and the logic doing the input/output with the hardware can be complex and out of sync. What makes this attack successful is when software has missed a case or lacks detection of an input or output situation. Problems can include timing, performance delays, data corruption, and system failures. Apply this attack to digital, analog, hardwired, and nonwired (radio, infrared, laser, other) lines or ports.

> **Takeaway note:** Sometimes we think of communication as only computer–to–human or computer–to–computer, but there are many kinds of "users" each having or needing communication.

Who Conducts This Attack?

This is a test team attack with the assistance of the hardware and system staff. The team must define what is possible, what situations should be handled, and what is out of scope for the project. There are many types of input/output communication connections, so some specialized understanding is needed by the team.

Where Is This Attack Conducted?

The attack best starts in the lab at system level where the input/output hardware and software are present. To set these attacks up, special hardware and/or software may be necessary to simulate and stimulate the software under test with the various input/output scenarios. A variety of simulation or stimulation environments may be needed where different types of signals and connections can be applied and may be necessary. After the lab, full system test in the field should be considered. But here, control of the signals, input/output characteristics, and environment may be problematic, so the lab is the first line of controlled defense followed by use in the field to confirm lab findings. Many devices can be impacted by such things as weather, signal beam focusing, buildings, EMI, cross signals, noise, and other factors that the software must handle.

How to Determine If the Attack Exposes Failures

Testing for input/output communication errors requires the creation of a variety of cases and sequences. Table 4.2 defines a sampling of these situations. As the cases are run, look for error messages, lockups, slow performance, reconfigurations, missed cases, and functions not being met. These cases will indicate or expose a bug. Rorabaugh's *Notes on Digital Signal Processing: Practical Recipes for Design, Analysis and Implementation* is a good resource for this attack [6].

How to Conduct This Attack

Start the attack with the full test team by identifying all input/output points in the system. The team then conducts an analysis of the input/output to determine the cases for communication issues and risk areas. Remember to look to team members who know the hardware, signal, and environmental factors. The team defines the issues and risks, and then these will define the attack-specific case(s) and situations.

Try to force cases, configurations, weak signals, slow throughput, noise, and so forth. Forcing these hardware and/or environment situations can be hard (see Attack 17). You will likely need to design the tests to cover the cases. Once you have the attack design(s), you will need to set up a lab or field environment where the attack is executed. As Attack 17 and Chapter 11 indicate, special tooling may be needed to do this in the lab. Of course, you can go into the field, but how do you know where in the field these situations will happen?

Once you have the test design and a test environment set-up, conduct the test(s) following scenario or tour-based approaches (see Attack 33 or *Exploratory Software Testing* [7] concepts). Some scenarios may have performance and functional implications that must be checked during the attack. Keep good notes and/or have expert observers including test hardware, software that monitors the device, simulation or stimulation, test logs, electronic scopes, and tools to record the attack. Plan on doing test runs in several environments over some period of time and carefully record the results. Then, compare expectations looking for bugs in this area.

ATTACK 11: BREAKING SOFTWARE ERROR RECOVERY
◇ Advanced

Attack 11 attempts to break the software's ability to tolerate bugs, faults, and errors in the software. There exists something called the "80–20 rule," [8] which, in some versions of the rule, implies that about 80% of software exists to deal with off-normal situations, which, taken together, may happen only 20% of the time. This is really not a hard and fast rule but a notational concept and, therefore, is not exact. Not surprisingly, it does have grains of truth in that much of the code in mobile and embedded software deals with software-based faults and recovery from user interface (software side), internal data, OS, other software, and even the software under test itself. This attack builds on Attacks 4 and 9, and so is very similar, but has a focus on software fault recovery logic, which is a specialized area within the 80%, yet critical for many mobile and embedded systems. Since this attack is very similar, it will often be used after Attacks of 4 and 9 have been completed, using their information to feed this specialized attack. The bugs here are different than the bugs targeted by Attack 13, which will focus more at the system level of fault and error recovery.

In the embedded space, use cases and/or operational profiles must deal with the real physical world. Systems engineers or whoever defines these cases must refine them within the real-world context in which the system operates. These cases should be tested in the earlier attacks, but good testers are system/software thinkers in their own right. Think of software-related faults that are reasonable and then test these faults, in addition to the defined cases. Start with simple cases and expand to the ones that are off-normal and then move into Six Sigma (extreme and beyond cases). The problem is where to stop. You will miss cases, and hindsight is always 20/20. Analyses based in legalities, risk, standards, and industry experience can yield the cases you should focus on.

The experience, legal, and risk factors drive systems to incorporate fault tolerance where systems must be safe or prevent major damage. This specialized area has design concepts and standards to support it (see the list of standards in the reference section). When standards exist, focused testing using this attack should be considered.

Additionally, you should look to error taxonomies, industry history, "break it" ideas, and your history, to define the space of attacking fault-tolerant software. Your error taxonomy should come from your last project and your current project. For instance, take your bug database and the team's experience with bugs and then create a local bug taxonomy similar to the one I have created in Appendix A. Industry cases come from research and reviewing public data: What has the company's competitor done (or not done). And finally, in all of this, build your personal history and mental models. All testers have these mental models. Why not hone and expand yours in the fault tolerance area? Become aware of issues and mental models. Listen to the little voice in your head, then attack software faults inside and, if possible, outside of your own comfort zone, knowledge base, and world.

Takeaway note: When software has error-handling or recovery code, a focused attack should be practiced since this logic may not get executed often, but when it does, it needs to work.

When to Apply This Attack

Attack 11 is applicable whenever the software has fault tolerance logic. Many mobile and embedded systems have this logic. Testers should look for the "nonhappy day" variations, boundaries, extreme and really extreme software fault situations. Fault-tolerant logic can be implemented in many ways including recovery from the fault, continue working through the fault, accommodate the fault, or ignore the fault. The attack must consider variations on what is defined for inputs, time, and outputs, since all of these factors may be part of the tolerance. The attack can be conducted on a variety of levels, although it may be best if this is done when as much of the system as possible is present. Auto manufacturers test their cars on test tracks, simulators, and "test drives" to places like the mountains of Colorado, the Mojave Desert, and the traffic of downtown Los Angeles to test fault detection, isolation, and tolerances.

Small or unusual software bugs found during this attack should be considered by stakeholders and how they can be expanded to larger bugs [9]. It is possible to take a bug everyone is willing to discount, and with some critical thinking, make it into a bug everyone wants to fix. When testing software faults, consider stacking and varying attack fault cases, situations, and times. The problem, of course, is that in complex systems in the real world, variations and combinations are practically infinite. So when running these attacks, smart testers and teams take the cases and small implications and play detective—following clues can lead to bigger bugs.

SIDEBAR 4.4

Airplane Flies into Water: There was a war story about an aircraft that was actually flown into the water during fault testing using a simulation in a test lab. This is not really possible in the real world—at least it is not good for the airplane. Needless to say, in this test, the software failed. The cause—negative altitude data was not handled as a software fault case. (Altitude data is a compilation of sampling altitude taken over time.) There are many places and several airports on the planet that are actually below sea level, so an aircraft *could* get to negative altitude.

What Faults Make This Attack Successful?

The real world is many times more complex than the developer's test cases, models, requirements, and imaginations. Embedded software systems interact with hardware, operations, human, time, other software, and the real world. When development misses aspects of a software fault case, the system will behave in unexpected ways when the software-related fault actually happens. Also, since the logic associated with such faults is only executed infrequently (and maybe, therefore, fewer test cases), the bugs will tend to "hide" in these lesser tested areas. You will want to try to minimize the surprises and learn from where you miss something or run less testing.

Testers will attack the identified software fault cases in this testing. This attack should focus on the logic of the software fault tolerance, making sure it is well tested. This is key. However, as with Attacks 4, 9, and 13, searching for cases that are not handled can also be useful information for the stakeholders.

Note 4.2: Some parts of the mobile and embedded industry implement a system state called "safing" or "safe mode." This is common in spacecraft, embedded medical, aircraft, and control systems. The basic idea is that the system is put into a "safe" state as quickly as possible after something unexpected or unplanned happens. Examples might be stopping the movement of a controlled system and sounding alarms, or halting critical operations and notifying operators. What is "safe" varies among embedded systems, so engineering thought and extensive testing will be needed. Safe mode is often implemented in software in part or exclusively. Many of the cases in this attack may be against the so-called safe mode of operations (getting there as well as the safe processing itself).

Types of faults and examples that make this attack work are outlined in Table 4.3.

It should be noted that these bugs are different from the bugs that Attack 13 will be looking for, so I separate these two related attacks into different bug hunts. The attacks are, in fact, related, and many groups may choose to do one first, then the other—or both together depending on the bugs they see as a risk.

Who Conducts This Attack?

Attack 11 is a test team attack, but what are the specific skills of the test team? The expertise involved could be from a systems perspective, or maybe understanding project history,

TABLE 4.3   Examples of Bugs in Software Fault Tolerance

Generic Name	Example
Lock up or lock out in software	Hardware used beyond limits resulting in a nontrapped error state and software stops responding
Switch over stalemate	Two (or more) processors, the one in control declares itself bad at the same time the other processor declares itself bad (both tell the other to take control and neither is in control) and the system fails (this can happen in software-driven failures)
Complex interaction between two pieces of fault-tolerant software	Fault between software–to–software communications happens
Unclear fault detection and isolation	The system faults and this is detected, but the cause cannot be determined, so recovery is not possible (the answer here is often to "reboot," but that is hard to do in some cases such as in an aircraft in midflight, or on takeoff or landing)
Error conditions within the system are not trapped or processed	Totally missing cases, so no handling for errors (no case or exception handling). May run Attack 4

hardware, and how the system gets used. Some teams hire ex-users, hardware people, system people, and other nontesters who can support the efforts and then become testers while bringing in the knowledge of fault failures that the testing should target. Testers need to consider all users not just the human users of the system. Small test teams can get this knowledge by communicating with and leveraging critical thinking by other development staff members. A good approach for this is to develop your test plans and tests, and then subject them to a review by the engineering staff. Listen carefully to what they have to say about any of the tests.

Where Is This Attack Conducted?

Since Attack 11 involves interactions, testing at the developer or integration levels of just the software will likely not expose all errors of this type. The test team should conduct this attack at the integrated levels, including software-to-software, software-to-hardware, and full system. The attacks continue from the partial levels until the full system exists and can be placed in the "real world." Of course, the real world is huge, so cases must be selected judicially. While I recommend these levels, the use of simulations and modeling can identify software fault cases early on that software should handle, thus reducing the attack cases that must be "invented" during the end system-level testing. The known fault cases will be tested in the lab first. Even with all of this, expect to miss important cases or triggers for software faults, since the software and real world are complex. Ethically (or legally for some projects), you will want to demonstrate "due diligence" in defining these software fault cases and conducting the associated attacks.

How to Determine If the Attack Exposes Failures

The following types of behavior or use may indicate or expose a bug:

- Software fault case is not responded to (results look like another situation)

- Software fault case is responded to but not correctly (expectations are not met)

- System fails (crashes and burns, so the fault is totally "missed") or

- A small bug is exposed, which at first looks innocent, but with a little thinking, a scarier fault case can be defined

There can and will be arguments as to which of these fault-handling bugs are worth fixing in the software. Developers have been known to say "that fault case can never happen" or "that case happens so rarely that we don't need to fix it." Testers should learn to "never say never" and be alert when they hear these kinds of statements. Testers are the advocates for finding the bugs. This is not to say that every bug needs fixing, but good testers learn to follow their instincts. In the end, testers provide information to developers, management, and customers so that the team can assess the software fault tolerance (or lack of), the risk of a bug, and decide to live with it or fix the product. Keep this phrase in mind, "Do you want to hear something negative about your project on the nightly news?" This must be balanced with time to market, costs, and quality.

How to Conduct This Attack

Start Attack 11 by understanding what has been done in the other similar or related attacks, such as in Attacks 4, 6, 8, 9, and 10. The information from these attacks should flow into this attack. Next, define (or refine) the users of the system and the software faults they can trigger. Users could be human, hardware, other software, and/or operations. Next, the team should identify the software faulting concepts and cases of operation that the embedded software system will encounter with associated recovery actions. These faults come in normal, off-normal, and extreme modes. Much has been written about fault tolerance, detection, and isolation. A good tester should be familiar with all of the reference materials for the local embedded system [10]. It is good to consult the system, hardware, and user community (if they are not already on the test attack team). From this, you should be able to

1. Conduct a technical risk analysis for risk-based testing (see ISO29119 [11]), focusing on what can go wrong, cases, and faults.

2. Identify variations in the software of input, time, data, activities, users, and other factors on the software fault cases.

3. Build on the variations using history, industry data, experience, boundaries, and your imagination, applying such concepts as

   a. Boundaries/edge cases to look at

   b. Equivalent classes of data and/or

   c. Stress tests.
   Note: All of these factors are defined in *A Practitioner's Guide to Software Test Design* [12].

4. Define a list faulting triggers and recovery cases to attack.

These factors form the starting point for defining the cases to attack. Often the team will find large numbers of combinations and data points. Useful test techniques to have in your bag of tricks are combinatorial testing (see Attack 32), which can reduce the test case problem down to more manageable numbers. Once a set of cases have been created, consider team peer review plans with developers and if possible customers and/or users. This will likely allow you to modify, add to, or remove cases. The software faulting attack cases should be executed at this point. Take note of your lessons learned as the attack progresses. Each attack will have the fault triggers and expected outcomes or recovery information (see step 4 provided under "How to Conduct This Attack" for this attack). After this point, the steps in Attacks 4, 9, and 13 can be followed. During this attack series, new cases are likely to spring up and planned ones may fade. Further, exploratory testing and tours [7] should be used here, but should be modified as the attacks continue. Finally, project resource constraints (people, machine, labs, money, and time) must be balanced. Proactive test teams provide information to development and management from these attacks to manage the resource constraints. No one ever gets enough resources, so resource creativity is needed here.

ATTACK 12: INTERFACE AND INTEGRATION TESTING
☐ Intermediate

A large number of problems in embedded systems and systems in general have been seen at the interface and/or at the time of integrating the embedded device. The interface points can be at the hardware, software, hardware–software, data exchange, and combinations thereof. Additionally, negative, extreme, and boundary situations crossing these interfaces may influence the app or software under test. Many organizations call this attack "integration testing." Initial integration-related attacks were addressed as part of Chapter 3. Attack 12 extends the attacks of Chapter 3 to full integration problems, although most projects would use a series of these attacks in combination. Attack 12 is useful during the stages of the project where the hardware, software, and system elements are being brought together for the first time since it targets bugs at the interfaces.

Some view integration testing as a phase. Others see it as an approach. Many just ignore it until they have problems. What exactly to call it or where to classify it is not that important. What is important is that when dealing with a system or application where there are numerous or complex interfaces, some variations of this attack are advised to be planned and implemented. During planning, the test team will need to consider the factors below to make this attack successful. More factors may mean more interface attacking. Here is another case where the plan (a document or artifact) is not so important, but the planning (the continuous repeated activity) of integration planning is important.

A good practice is to incrementally and iteratively build a system or app up. The more numerous and complex the elements and interfaces are, the more the need for increments and integrations. But this takes effort and time, up front. So many projects ignore this iterative attack, or try to do "big bang" integration testing—trying to do everything at once at the end, when all of the hardware and software exists. My research has found most large projects failing with the big bang of integration. I have uncovered many incompatibility stories ranging from the obvious, where the app just does not work at the interface, to memory leak bugs caused by the interfaces. This attack focuses on the software device and app side (software-to-software or software-to-hardware interfaces), but many software test teams may need to consider the outside world of hardware–system testing. A partial or complete system view in integration testing with other integration interfaces is out of scope for this book. This chapter offers only analysis and guidance on configuration and compatibility testing from a software viewpoint.

Some may view the increment, iterative, and even continuous integration as associated with Agile concepts. However, this approach predates Agile and can be practiced whenever interfaces may be a risk factor. In fact, most software is done incrementally and iteratively, through whatever life cycle is adopted by the project. So the idea of ongoing integration and interface attacks should be considered as the software and hardware get built.

> **Takeaway note:** Most books and articles do not talk about integration testing. There is lots of information on developer (or unit testing) and software testing. My experience is that many mobile and embedded devices miss integration problems and/or do a poor job in integration testing, meaning that many bugs are left to be found by testers or, worse, users.

When to Apply This Attack

Attack 12 should be considered with complex interfaces and integrations when they exist, even in the mobile/smart world. The basic attack outlined here is really a start. Many larger embedded systems such as airplanes, ships, spacecraft, and factories will have many activities and even multiple teams conducting integration testing. I know of cases (large airplanes, for example) where there is a vice president in charge and hundreds of people working integration tasks. Obviously, in these cases, there can be a lot more to this attack. I recommend that integration guides and planning be detailed enough to fit the scope of the situation. Finally, this attack may not be applicable to small simple apps in mobile and embedded when only a few interfaces exist.

Testing the integration, configuration, and compatibility of the software should be considered in initial test planning. This is a good time to think about what resources will be available and what the integration risks will be for more complete planning. More risks and interfaces equal more resources and effort in integration.

The actual testing should begin after the developers complete an integration cycle of their efforts and maybe after the first round of attacks, which hopefully have proven certain "easy to kill bugs" are not there. Testers do not want to be chasing the "ghosts of bad development" and missing functionality. Of course, if the team is trying to answer questions of integration associated with functionality, say an interface communication protocol or something that does not work with the new hardware, this attack may be used in combination with other attacks, if it is not done first. Keep in mind that *tests provide information*.

The integration attacks focus on the interface first, but may not forget the application's functionality or qualities. Once you have some history on the interfaces with early attacks, there is another level of integration attack that becomes a likely candidate. I call these configuration–integration sub-attacks, where you are assessing the software's ability to run on other configurations of hardware or with other software. Industry often calls these configuration tests and/or interoperability testing.

In configuration–integration sub-attacks, we know for certain that users will have many kinds of software on their systems and you want your software to run on different hardware and be comparable with other software and/or hardware configurations. Worse, there are also differing versions of software and hardware to worry about. Basically, there are infinite numbers of hardware and software configurations that you might need to test. Here you may want to combine this attack with a combinatorial attack (see Attack 32) to check different configurations that the team considers likely risky and/or important enough to run.

For example, networked mobile and embedded apps and games sometimes check the OS or other configuration items. They will report that the users need an update. This will avoid some interface bugs, but someone had to figure out what else was needed. Understanding compatibility is useful information for the development team, which this attack can provide. Additionally, there are several kinds of users: those who always want the latest updates all the time and those who never update because they think it is a pain. Another aim of this integration attack is to determine how the software will deal with each of these situations.

Finally, for those teams doing integration testing over an operations and maintenance effort, where new software and/or hardware are introduced, regression testing considerations may apply to the integration attacks. Just because an interface worked with the last revisions of hardware and/or software does not mean that they will work with new hardware or software, particularly if they have undergone a configuration update.

Who Conducts This Attack?

Attack 12 may be conducted by a specialized integration team. What makes the team specialized are the resources the team has access to and what they are testing against. In this attack, the team focuses the attack on the interfaces, so the team must know and understand them completely. The integration team will care less about functionality and other attacks and focus only on the interfaces, first. Sometimes integration attacks may be performed by the development team. Finally, in other cases, the attackers are just regular test team members focusing on the interfaces.

Where Is This Attack Conducted?

Many organizations have special facilities, differing phases, people, and other resources to conduct this attack. Smaller groups and teams may not have access to any special resources, other than their own critical thinking, the device under test, and time. Let us consider both.

If you are lucky enough to have a test lab with different hardware and software configurations, as well as tooling to help understand the interfaces, you will need to leverage them in your attacks. But even the best labs may have only a limited set of hardware or software. So the environment must be populated and tests run based on risk and likely known interface issues. The rest of us will be left with what we have, one or two pieces of hardware and whatever software configurations we can lay our hands on. In both of these cases, the attack process will be similar, and with more resources, more attacks may be possible. Labs with the ability to do interface configuration testing can get large and expensive, so a balance of cost versus risk must be struck.

It should be noted that in many mobile and embedded product lines, there are now test labs and vendors who provide the ability to attack on many different configuration platforms. Two such labs are Device Anywhere and PerfectoMobile. A quick search of the Internet will likely yield other candidate providers as well. These labs have hardware, software, and virtual features, which can allow many integration attacks to be done, though it is likely better (more cost effective) if you have automation to drive these attacks. Also some of these vendors actually provide crowd source testing options using resource such as the cloud or networked systems.

Next, it is convenient if, in the integration test lab, you have tools to support the attack. These tool types can include

- Interface probes to record the interface (usually hardware, but can be software)

- Analysis tools that will "read" the software and define the interfaces (see Attack 1, static analysis tools, some of which can define interfaces), if this has not been defined by development

- Automation (run attacks on lots of different configurations in parallel)

- Scopes and bus loggers to record these interfaces and

- Probes to stimulate inputs to the interface

Some of these tools are commercially available while some others are custom made. The design and configurations of tools and labs are driven by risk, cost, and schedule.

What Faults Make This Attack Successful?

Attack 12 works because software, hardware, and the system exchange data, and these interchange points (interfaces) can be very complex. The interchange and interface lead to something called "software coupling" (to learn more about the effects of this, search the Internet for the term). In coupling, something that happens in one place influences something somewhere else across the interface. If there were no interfaces, the coupling could not happen. But when the something and somewhere are spread across different pieces of hardware and/or software, a changed hardware and/or software version, or an added app (that uses the data) and presto, you have a bug. Some developers call this a "side effect" and tend not worry about it. You will hear statements like "well it works on my machine," "we do not have that interface here," or "we cannot test everywhere." These statements might be true, but keep these questions in mind: "Which interfaces should I test, in which configurations, and why?"

Consider these sources of interface and configuration issues where bugs may be found:

- Data variables inside of the device
- OS and support resources (libraries)
- Stack data
- Library extensions and add-ons
- Browser variations
- Device security features
- Blocking software for cookies or childproofing
- Sharing data between apps (e.g., voice recognition running in a task list)
- Certain kinds of scripting
- The actual interfaces
- Internet/Intranet
- Web server problems (virtual, director, data, SQL)
- Communication lines (see Attacks 5, 10, and 14)
- Firewalls
- The hardware

The integration/test team should get a list of these sources of interface and configuration issues from development staff or create and maintain it. Additionally, many industrial organizations maintain information on configuration and interfaces. For example, in the mobile world, testers can check on the following kinds of information, by doing a little research:

- Carrier compatibility (Verizon, Sprint, Orange, O2, AirTel, to name a few)
- Backward compatibility of software by vendor release information
- Hardware compatibility listed by vendor

How to Conduct This Attack

If you are supporting the development and integration team, start Attack 12 during planning by identifying the integration elements, or by what some people call "integration threads." An integration thread is a logical grouping of hardware and/or software. A thread is smaller than a system or subsystem group, but larger than the smallest element such as a unit or class. Threads are combinations of hardware, hardware–software, and software. For example, in a smart phone, it might be a grouping of off-the-shelf hardware, OS, screen display utility, and a part of an app. Another integration attack might add a set of data to the screen for display, but not the communication package of our software that takes the app to a cloud database. You could integrate one thread and then the other, while identifying the inputs, outputs, and display interfaces within the attack. You might have to simulate the data to be displayed, since the package does not exist, but you have attacked interfaces incrementally. This incremental and iterative thread approach works for all development teams.

After defining the threads, understanding and identifying the interfaces become important. Now if you are working the development side of test, you can use tools such as static analysis tools, which help define interfaces. Or if you are lucky to have interfaces documented or modeled, these items can be mined from the documentation or model diagrams. Teams can put these lists into a spreadsheet or database to track all of the interfaces. Once you have the integration threads, you may need to map between the threads and interfaces to make sure that you have coverage of the interfaces. Untested interfaces will need to be addressed using additional attacks.

On large programs, usually the integration team must plan and schedule which interfaces and threads can be tested and when. If a piece of hardware or software is not going to be available until late in the development cycle, you cannot easily test it. Therefore, these items need to be identified and factored into all plans and schedules. If you are external to the development team, you can still produce an interface table, although it will be shorter and not have insight into the software internals.

An interface table should list the following:

- User interface input options (menu items, input fields, clicks and wiggles, etc.)

- External interface input (communication inputs)

- Hardware options

- Software to include and/or that is running concurrently

Next, design an attack that exercises one or more of these interfaces. Start simple and build up. Mark off the interfaces as they are exercised in the integration table. Recognize that some interfaces may have multiple options, boundaries, and classes. In this case, these factors are really separate interface entries in the table, and each one needs to be exercised. Given this assumption, it is likely that the interface table will grow and must be incrementally populated and expanded but will be bound by cost and schedule constraints.

Further, the interface integration table may want to address topics such as

- Tested on a prototype, early version, final version

- Blocking factors such as needing hardware, software, or test equipment

- Notes and assumptions about the interfaces

The interface table and testing of it should be as complete (coverage of its elements) as time and budget allow. As the team builds the interface integration table, a concern is will the users have the same visual and functional experience irrespective of the interfaces? If the experience is different, you may have a bug.

Keep in mind that an interface that is not tested may be a "risk item" for later testing.

SUB-ATTACK 12.1: CONFIGURATION INTEGRATION EVALUATION
☐ Intermediate

The configuration test is a well-known test concept practiced by many teams and projects. The IT world has used with it for years. Have you ever asked yourself how many variations of hardware and software are there in the PC world? Many projects have seen a piece of software work well on one configuration of hardware and software, while failing on another.

This sub-attack considers the integration of a piece of software on different configurations of hardware or software. The basic pattern of Attack 12 can be repeated, but in this variation sub-attack, the tester will list configurations instead of the interfaces in a table. These patterns can be configurations of hardware and/or software. Just as in the PC world, the number of configurations can be basically infinite, but Attack 32 can help set up the configuration to run tests on. If using Attack 32 with this sub–attack, test automation to cover different configurations quickly can be a good thing to consider.

ATTACK 13: FINDING PROBLEMS IN SOFTWARE–SYSTEM FAULT TOLERANCE
◇ Advanced

Takeaway note: Testers should be focusing on attacking faults rather than focusing on "happy day" or "happy path" testing (tests that are designed *not* to find bugs).

Attack 13 builds on earlier attacks (4 and 11), but specifically targets embedded hardware–system errors contained in the software fault-handling logic, which are called "fault-tolerant systems." Such software logic is designed to deal with problems in and coming from the hardware, although there can be fault tolerance for problems in software (Attack 11) or coming from external users. Fault-tolerant systems are designed to keep working under one or more failure situations [10]. For some embedded devices, it is not acceptable when the software crashes for the system to stop working or to be rebooted. Fault tolerance is needed in critical, safety, hazardous, or other systems, where it is desirable to have the system attempt to keep providing some level of functions even when faced with faults. Examples of such systems include aircraft, spacecraft, automotive, critical control systems (power, factories, water treatment plants, etc.), medical, and now some mobile smart devices.

Attack 13 will seek to force faults, invalid cases, inputs, and/or outputs, which will trigger the use of the fault-handling logic. \This is closely related to Attack 4, missed hardware situations and Attack 11 testing, but has a very specific focus on hardware–system fault tolerance when such features are implemented in the software. The unique faults of hardware–software–system must be attacked within the context of the software and operational testing. This is not hardware testing, but software testing with hardware in the loop.

When to Apply This Attack

Apply Attack 13 when your system has fault tolerance features. Features may be in customized code or using features provided by the OS. Attacking the code under test is pretty obvious (Attack 11), but generic OS features where data is used to drive fault-handling logic provided by many OSs should also be exercised with this attack. Additionally, the inclusion of commercial off–the–shelf hardware "or smart hardware," which contain fault-handling abilities, may need to be evaluated within the overall software testing. This is needed even if the hardware has industry use history or where a stakeholder wants to "trust" hardware because of its history. My experience is that it is better to test the unique aspects of the hardware within the context of your embedded system. The time for testing with this attack may extend to considering regression testing (or re–testing) when some element of the hardware or software changes external to your development efforts. New or changed electronics, added or removed hardware, software updates, and even small changes such as an updated transistor or third level hardware interlock change have been known to negatively impact a system's ability to handle faults. Regression testing the software and fault tolerance when hardware changes may sound strange, but then, this is the mobile and embedded world.

Finally, besides the hardware and software fault handling, you should consider Attack 13 when humans can interact with the fault tolerance of the system to create unsafe or hazardous conditions. In Leveson's *Safeware: System Safety and Computers* [13], software safety engineering has shown cases (Sidebar 3.1 Therac–25) where "human error" triggered faults in combination with poor hardware and software, and while the systems probably should have tolerated faults, they did not. As systems become more complex, "smart," and feature rich, the more likely they will have human errors. If the system is really smart and/ or fault tolerant, it should handle faults, and developers and testers should test these faults to find the bugs.

Therefore, this attack must include testing within operational human, hardware, and software environments. This includes the edge, boundary, extreme, and complete failure cases.

What Faults Make This Attack Successful?

Attack 13 is separated out as a special attack because research from the taxonomy (see Appendix A) indicates that many mobile and embedded bugs exist in the system's fault tolerance logic. This is likely because in these systems, often much software is devoted to handling the faults, while at the same time this logic only gets called when something is "faulty," which may be infrequent, so it does not get exercised or tested sufficiently. Therefore, bugs can live in the code for long periods of time, going undetected, until certain trigger conditions happen. This attack targets those areas.

Specific situations, faults, and conditions, which make this fault likely, are listed in Table 4.4.

Much work and research in software engineering over the years have indicated that many systems are not designed from the beginning to be safe or to tolerate faults. Adding such features in after the fact is hard (sometimes impossible) so some aspects of this attack

TABLE 4.4   Common Faulting Situations

Situation	Starting Approach	Expected Action	How to Detect a Bug	Notes
Electrical system failures	State chart coverage to drive failure cases and show system action, driving into a risk analysis matrix	Test coverage and high risk such as safing code, unsafe cases, watch for cascading failures	Design tests to trigger faults and watch for correct handling	Conditions to trigger include power, EMI, spurious power signals, etc.
Mechanical system failures	State chart coverage to drive failure cases and show system actions	Test coverage and high-risk areas such as brake failure, noise, computer crash	Design tests to trigger faults and watch for correct handling	Conditions to trigger include slick roads, noise on hardware, crash on computer, etc.
No hardware or input signal response	Identify all interfaces and inputs from hardware into software	Stop signals and inputs, corrupt input, test high/low, out of range inputs	Watch for fault tolerance processing and failures	Requires ability to input expectations in test lab (may be hard in field test)
Safing logic—exception processing where the software recognizes a "fault" and goes into a system-level "safe mode." Here the system may shut down or continue to run but in a state where further damage is minimized (see Attack 11 note)	Identify if safing logic is present. If it is, define safing triggers and risks for it. Define which safing modes have been tested already, as these attacks may not need to be repeated or can serve as a start for new attacks	Define untested safing areas and risk. Implement attacks to target these areas. Safing logic is usually built to be very reliable	If any of the attacks cause safing logic to be triggered unexpectedly, you may have the start of a bug	Most safing logic is reused from project to project. The logic is small and "critical." Testers need to consider these factors in planning the attacks
Fault tolerance for software bugs—some systems can detect and "recover" from bugs in the software	Understand the nature of the detection, how the recovery works, and from this, the kinds of attacks that might be needed	Design an attack to test this "bug safing" logic. This can be hard	Consider: triggers, mutation testing [15], special builds (see Chapter 11), and creative tests	See Attacks 11 and 18 (interrupts)
Human inputs to hardware/system that causes a fault, which the system can detect and react to saving a system failure	Identify human interfaces and inputs into software	Define inputs, corrupt inputs, test highs/lows, out of range inputs, etc.	Watch for fault tolerance processing and failures	See Chapters 7 and 8

may need to start in early prototypes of the hardware, software, and/or system. If testers can provide data to assist in the development, much better systems may be the result. Besides prototypes, tests can look for and/or help define

- Fault and uses cases that are not complete

- Missing complex requirements where faults can slip by and

- Real world faults such as noise, EMI, signal problems, and so forth

It is not usually possible to consider every case and situation that would trigger a fault or where a bug could hide. However, the team should look for and attack the high-risk areas.

Who Conducts This Attack?

The test team will conduct these tests working with the hardware developers and system engineers to define and understand each fault case, as well as the priorities of what to attack. Testers need to engage with the team working safety and/or fault analysis. Or if this has not been done (and should have been done), the test team may need to do their own analysis. The less analysis done up front, the more testing may be needed to address faults and "safe" critical elements later. The "who" of testers will include people in the lab as well as field testers, depending on the criticality of the faults and environments where testing will be done.

Where Is This Attack Conducted?

This attack on the software must be conducted when it is connected to the hardware and can consider the environments and/or users. Some of these elements may be part of a test lab (see Chapter 11). More realistic testing is possible in the field where "faults" are more apt to be real world like, but the field may not be able to introduce all fault cases easily, which would move your testing back into the lab.

Attack 13 involves having the hardware "input" data, as well as conditions that include faults or failures. Many pieces of hardware are designed not to fail, but systems still have a chance of failure over time. And, therefore, fault tolerance is designed into software. The tester's challenge becomes how to make the software receive the failure data for processing. Some failures may be easy to create or set up. Others can be hard. Here, concepts such as generators, simulators, stimulators, special software builds, and even "failed' hardware can be used. However, damaging a piece of hardware, which could be expensive or one of a kind, introduces a concern that testers should think clearly about. Other items may lack certain realism, but the option of not doing a test because it is "too hard" potentially leaves parts of the software and system untested. In a legal situation, this might not meet the "due diligence" criteria, leaving projects open to legal action.

Chapter 11 defines tools, simulation, or stimulation systems and other approaches to get test data into the system under test to trigger faults or events leading to faults. Test facilities should be developed to provide these capabilities when needed. Also several versions of the system can be created with "test" instrumentation and/or input channels to allow faults to be simulated in the software. Where this is done, there are several concerns that testers need to be aware of:

- The test instrumentation can influence the test, both positively and negatively (through hiding errors and indicating false positives).

- Changed hardware or software with these "special" inputs channels might not be realistic.

- Long duration test runs on hardware and/or in the lab may not be possible due to schedule.

- The cost of setting up such test environments and systems can be very high (meaning it might not get done).

- The complexity of such test environments can make the facilities development as hard as developing and testing the embedded system.

Setting up labs can be very time consuming and expensive for fault tolerance attacking. This may tempt some to try to do everything in the "real world." As already covered, some faults are not very likely to happen in the real world, so you need both. Real-world field testing can be used to confirm results from the lab work (a cross check on the lab). Field testing may also be used to address additional fault tolerance cases not covered in the lab. Lab attacks cover cases and faults the real world cannot.

From these environments and attacks, the test team can indicate where unsafe and unhandled faulty cases exist. The areas to attack must be prioritized as the combinations, cases, and risks usually exceed budget and schedule in either the lab or field. Additionally, as outlined previously, setting up these tests may require help from specialized engineers with knowledge of the hardware.

How to Determine If the Attack Exposes Failures

The following types of behavior may indicate or expose a bug:

- Hardware response is not correct

- Software response is not correct

- Operations of the system are compromised

- System fails (crash and burn) or

- A small "bug" is exposed, which at first looks innocent, but with a little thinking (and maybe more testing), a more severe case can be defined

As in other attacks, there may be debate about whether a bug exists or not. Here again, the team must weigh what is worth fixing, what is not, and when. Not all faults should or can be tolerated. It is not a tester's job to make a call about when or if to fix a bug, but it is our job to provide information about the fault situation such as what is tolerated and what is not. The team may need to consider the software, hardware, and user operations to determine if there is a real bug or one worth fixing.

How to Conduct This Attack

Attack 13 is similar to Attack 11 but has a hardware flavor and some twists from environment and humans. You should be familiar with Attack 11 and Table 4.4 to proceed. Here is the pattern outline to begin Attack 13.

1. Identify hardware and software elements.

2. Identify the cases of operation that the embedded software with hardware will encounter, including the fault tolerance situations, and place in a list.

3. Conduct (or participate in) an FMECA [14] or FMEA to identify the risk situations and cases, which may not get handled and create a matrix.

4. Transform this matrix into a risk matrix.

5. Identify valid, boundary, and invalid cases for fault tolerance such as input, output, time, corruptions, and other factors and add these items to the matrix.

6. Determine where human input can be and what the boundary and invalid cases can be and add these items to the matrix.

7. Build on the matrix using history, industry data, experience, boundaries, and your imagination.

8. Using the matrix, define conditions of normal, off-normal, stress, risk, and extreme test cases to attack (likely a series of tests).

9. Consider using Combinations (see Attack 32) to select which conditions to attack (the more the better).

10. Define how the hardware or system condition will be input to make the failure (this can be tricky, see Chapter 11).

11. Peer review the specific test plans with development and, if possible, the customers and or users to get their agreement on the matrix, priority, conditions, and attacks.

12. Finalize and run attacks.

13. Learn from tests and consider new runs within cost and schedule.

14. Continue with variations of this attack as project constraints permit, involving the stakeholders as needed.

EXERCISES (ANSWERS ARE ON MY WEBSITE)

1. A manufacturer of diesel tractors with GPS positioning controllers has implemented a series of software-driven controls (engine, GPS, automatic positioning, farmer display). Why would they go to test places outside of the lab and their field in the mid-West? Give specific examples of where they might go and why.

2. What communication channels can a smart cell phone have these days and can you list any protocols these channels might use?

3. Define the fault cases for a smart cell phone.

4. Define safing logic for a car.

   a. How would you test this car?

5. Define the different "users" of your embedded software system (or if you do not have this, define it for a car).

6. Define the different interface points of your embedded software system (or if you do not have this, define it for your car's antilock braking system (you can research the basics of such a system on the Internet).

7. Define risk factors (hardware, software, faulting, etc.) for one or more of the following:

   a. Smart phone being used to monitor a person's heart rate

   b. Commercial passenger airplane flight control system

   c. An embedded device controlling the fuel rods in a nuclear plant

   d. Factory assembly line robotic welder embedded system

      i. Put risk factors in priority order

      ii. Define one or more attacks for each of the top five risks

REFERENCES

1. Whittaker, J. 2003. *How to Break Software*, Addison-Wesley, Boston, MA.
2. National Aeronautics and Space Administration (NASA). http://www-odi.nhtsa.dot.gov/cars/problems/defect/defectresults.cfm; *National Highway Traffic Safety Administration (NHTSA) Toyota Unintended Acceleration Investigation.*, NHTSA Action Number: RQ10003 NESC Assessment No. TI-10-00618 Chapter 11 (January 18, 2011), NHTSA's report is entitled: *Technical Assessment of Toyota Electronic Throttle Control (ETC) Systems*" (February 2011), NASA's report is entitled: *Technical Support to the National Highway Traffic Safety Administration (NHTSA) on the Reported Toyota Motor Corporation (TMC) Unintended Acceleration (UA) Investigation*, Both reports should be read in conjunction with each other.
3. Mobile phone. http://en.wikipedia.org/wiki/Mobile_phones_on_aircraft (last accessed April 11, 2013).
4. Broekman, B. and Notenboom, E. 2003. *Testing Embedded Software*, Addison-Wesley, Boston, MA.

5. Failure_mode,_effects, and criticality analysis. http://en.wikipedia.org/wiki/Failure_mode_ and_effects_analysis (last accessed April 11, 2013).
6. Rorabaugh, B. 2010. *Notes on Digital Signal Processing: Practical Recipes for Design, Analysis and Implementation*, Prentice Hall, Upper Saddle River, NJ.
7. Whittaker, J. 2009. *Exploratory Software Testing*, Addison-Wesley, Boston, MA.
8. http://en.wikipedia.org/wiki/Pareto_principle (aka 80–20 rule).
9. Kaner, C., and Falk, J., and Nguyen, H. 1993. *Testing Computer Software*, 2nd edn., Van Nostrand Reinhold, New York.
10. Marcus, E. and Stern, H. 2000. *Blueprints for High Availability: Designing Resilient Distributed Systems*, Wiley, New York.
11. ISO29119 to be published in late 2013. http://www.iso.org/iso/catalogue_detail. htm?csnumber=45142 (last accessed April 11, 2013).
12. Copeland, L. 2003. *A Practitioner's Guide to Software Test Design*, Artech House Publishers, Norwood, MA.
13. Leveson, N. 1995. *Safeware: System Safety and Computers*, Appendix A. Addison-Wesley, Boston, MA.
14. Failure mode, effects, and criticality analysis. http://en.wikipedia.org/wiki/Failure_mode,_ effects,_and_criticality_analysis (last accessed April 11, 2013).
15. Mutation testing. http://en.wikipedia.org/wiki/Mutation_testing (last accessed April 11, 2013).

CHAPTER 5

Mobile and Embedded Software Attacks

S O FAR, I HAVE PRESENTED ATTACKS that can be used on embedded systems with a focus on developer testing, controls, and system-hardware areas, as well as a few attacks that are "cross-over," in that they can be applied to either mobile or embedded smart devices. Now, let us start looking directly into the mobile and embedded domains. As I stated previously, the embedded world blends into the mobile smart domain, which in turn blends into the domains of software such as web, information technology (IT), and personal computer (PC). I expand this blending here.

Table 5.1 lists the attacks of this chapter and how they relate to different mobile and embedded contexts. If you are expecting perfect or clean boundaries, there are none. Each attack will need some customization and a lot of critical thinking.

ATTACK 14: BREAKING DIGITAL SOFTWARE COMMUNICATIONS
☐ Intermediate

So far in embedded software, we have looked at how the software interfaces to the real world via sensors and/or outputs to hardware-based signal paths (Attacks 5 and 10). Additionally, embedded software can send and receive digital commands or messages within a system to other pieces of software, or to the outside world. These commands, message structures, and formats range from the basic input/output data values for sensors and devices, to commands with more complexity and information content between pieces of software in the system. Communication has become increasing complex and information laden, especially for embedded devices. Attack 14 targets these digital communication streams. Commands and communications for this attack are digital in nature, so you will not see some of the problems seen in analog-to-digital (A2D) or digital-to-analog (D2A) attacks (Attack 5), yet they are more complex than the commanding seen in Attack 10.

TABLE 5.1 Attacks Mapped to Mobile and Embedded Contexts

Context → Attack ↓	Mobile and Smart Wireless Devices	Embedded (Simple) Devices	Critical Mobile Devices (Could Be Embedded, Small but Important)	Critical Large Embedded Devices (Could Be Mobile)
Attack 14	Yes	Frequent	Yes	Yes
Attack 15	Sometimes	Frequent	Yes	Yes
Attack 16	Sometimes	Frequent	Yes	Yes
Attack 17	Seldom	Sometimes	Yes	Critical

Notes: Seldom, seen only in specialized cases infrequently; sometimes, seen often in many different contexts; frequent, seen regularly in a variety of contexts; yes, should be considered for most contexts; critical, important to be done.

Takeaway note: There are many communication standards. As a tester, you should know the details of the communication standards that your system must be compliant to. You may find it helpful to be familiar with several communication standards.

Standards can be specified by industry, customer, company, or even project. They keep evolving and changing, so attackers should find applicable standards to their product, use them within this attack, and keep up with them as they evolve. But standards tend to be "generic," so the project will have to define the specifics for the development effort. Testers can use those specifics and the information found in the standards for the attack.

When to Apply This Attack

Apply Attack 14 when there are complex digital communications. The test team starts planning and designing this attack as soon as the standards and/or specific project interfaces to the commands start taking shape within the design of the system. Development elements to consider for this attack can include the production of interface specifications, interface control documents (ICDs), industry communication standards, identification of industry standards, and later the actual protocols.

If a complex command structure and interface exists, the team may want to consider using a modeled or simulated command interface to exercise the commands and associated software as they evolve. Modeling or simulations of the interface allow the following:

1. Early command inputs to the software under test when the associated hardware, network, or system are not available

2. Quick creation of commands and command sequences

3. Exact tester control over the commands and/or

4. Creation of unusual or off-normal commands

What Faults Make This Attack Successful?

As embedded systems have grown in complexity, the commands, messaging, and communications have likewise grown. More data, power, and information are transferred, which increases the complexity. This has led to standards and communication protocols. All of these items must factor into the testing. Just as in IT and web messaging systems, embedded communications can result in incorrect formatting, data content, and command selection. This attack is run on the software interface under test, unless you are doing system-level testing. This interface is usually a digital communication channel, which does not have the A2D and D2A issues of electromagnetic interference (EMI), noise, numeric representation, and other factors. However, you do get new interface areas to attack. Faults can be found in the communication itself, the timing of the communication, communication authentication, and/or local risk items the team may have defined in association with the communication.

Further, when a standard communication protocol has been defined in such things as an ICD, then items such as the commands, messages, structure, and data, defined in such documentation, should be tested. The documented communications must be sent and/or received. Variations of the communications may also be defined for testing. Where communication sequencing is possible, combinations of tests may be needed (see Attack 32). Items to consider include retry logic, fault-handling cases, overrides, and so on. Finally, as in Attacks 4 and 13, we are also looking for missed cases where a communication or communication sequence that does not exist, should exist, in other words, a case is missing.

Who Conducts This Attack?

Attack 14 will be conducted by the system test team who must understand the communications, standards, documentation, and protocols, as well as what can go wrong. The team should have skills in the hardware, system communication, and software. Communication within the test team and to other groups can become quite important for avoiding miscommunication, which might be translated into the software as bug. Many projects have a working group, whose focus is on the communications and interface, helping communications to be fully understood and hence tested. These groups may be called hardware–software working groups, communication teams, standards and protocol team, or other names, but the key is to have good communication between humans to drive communication data into the attacks.

Where Is This Attack Conducted?

Attack 14 is ultimately conducted at and on the hardware–software communication interface. This is often best done in the test lab environments, but field testing of communications is still possible.

In the lab, Attack 14 may start using models and/or simulations on one or both sides of the interface. Figure 5.1 shows a sample configuration of hardware, software, and

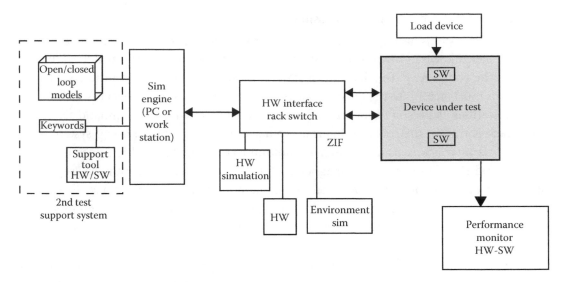

FIGURE 5.1 Generic lab picture with models and closed-loop simulation (see Chapter 11 for more on labs).

simulations. The hardware and simulation allow the software to be attacked with many communications.

In the lab with the whole system, Attack 14 should exercise

1. The communication channel (physical), whether it is a data bus, wire, or even a wireless connection

2. The requirements from the standard(s) and

3. As many of the logical communication interfaces (messages or commands) as is practical within cost and/or schedule

After the ability to provide inputs to or from the interface, a key item in the lab is the ability to record the communications and responses. These are lab systems, which are often complex and costly. Data recording systems and data bus analyzers may be needed to capture and analyze the communication data. Since many mobile and embedded systems have high communication and data rates, these recording facilities need to be capable of handling the rates of the system, time tagging information, as well as be able to properly store the information.

Note: Data rates from some embedded systems are high, resulting in data files no human can analyze in any reasonable amount of time. Test file sizes have been seen as high as terabytes so testers should have tools (see Chapter 11) that can analyze and/or graphically display data communications and results.

Finally, field testing using this attack should be considered, but problems mentioned (earlier) of how to input some communications and/or record results become problematic. I have seen test teams create special field prototype systems with special instrumentation and input abilities. Also, in the field, I have seen teams test only subsets of the communications only to confirm the results seen in the lab. There is no one right answer here on how to construct field testing for this attack. Tester and team creativity, as well as consideration of context, will drive lab and field testing of communications.

How to Determine If the Attack Exposes a Failure?

Attack 14 is based on communications, which should be understood and documented in such things as communication standards, project protocols, interface control information, requirements, or other documentation sources. If there is a difference between these things and what the software does, you have a bug. Of course, the communication documentation can be wrong or missing, so a tester needs to keep a wary eye even on the protocols and documents, viewing them as starting point for this attack—not an end. Essentially, you are looking for an incorrect software behavior based on the sent or received communication.

How to Conduct This Attack

Start Attack 14 by collecting existing communication documentation. Testers must understand each input and/or output location where one of these communications exists. The test team can capture this information in a table or database listing. Include in these tables or databases the following information: message structure, meanings and expected results, normal messages, off-normal messages, risk factors, and critical messages. Additionally, either in this table or in supplemental documentation, testers should ensure that communication sequencing and time factors are recorded.

If there is an interface working group, this team needs to consider the communication information in the documentation, modifying it as needed. Likely the interface and communications will change over time, so such table information will need to be kept up to date. If a working group does not exist, the tester will need to talk with the interface, hardware, systems, and software people to confirm and update the table. This communication can be tricky because many of these people will be busy and testers are often told, "don't interfere with development." In such a case, the tester should do the digging on their own, recognizing that they will miss things at first, so repeated updates become more frequent.

Next the test environment is defined. Consider the following test environment elements:

- What are the modeled, simulation, stimulation, and/or actual hardware interfaces?
- Are there full, partial, and/or evolved interfaces?
- What types of communication automation are possible?
- What are the needed interface recording abilities?
- What are the sequence and time considerations for the test environment?
- How realistic is the test environment?

The set-up and configuration of the communication test environment may take time and specialized engineering skills. In mobile and embedded labs, it is normal to find that the initial environment is not ready or able to support the communication attacks completely. For example, if a test environment did not have the right level of command recording ability, it may not be possible to trap communications as they come or go, meaning that testing is not well documented or repeatable. Thus, testers must be flexible and creative to devise workarounds in the lab or in the schedule to accommodate limitations as they arise. The communication test environment will mature as the project progresses.

Once a test environment is able to support some testing, a starting point for the attacks will be the normal and critical high-risk communications. Once normal and critical communications are shown to be working, then consider other risky and off-normal communication cases. This attack continues until the elements of the table or database are covered, ideally up to 100%. This coverage is usually single communications with limited worry about time, communication interactions, and sequencing, but in complex real-time embedded systems, many failures (see Appendix A) have their roots in time and the sequence of communications. As the communication attacks progress, the tester should consider including time, sequence, and combinatorial attacks (see Chapter 6, Attacks 20 and 32). Since basically these attacks can become infinite, a good handle on risk-based testing is needed at this point. Risk prioritization of attacks can be captured in a table or database. Additionally, as attacks progress, problems can expand to include how to force certain inputs, sequences, and output communications. The test team should test for a while, review the results, and consider other communication attack variations to explore as their understanding increases. Then, the attackers and support team should take time to understand the data using tools or other analysis programs.

In this attack and its iterations, a big question arises of when to stop attacking this area. The coverage of the table mentioned previously offers some measurable guidance, but also as mentioned, the factors of time and sequencing turn this attack into one where exhaustive testing is not possible. Here the test team, development, and other stakeholders must be consulted. Ask yourself such questions as "what are the risks; what are the regulations; how much time do we have; how much budget do we have; what does the history tell us," and so on. The answers to these questions will offer guidance on when to stop.

ATTACK 15: FINDING BUGS IN THE DATA

☐ Intermediate

Dr. Cem Kaner once wrote:

> I came into software development with a doctorate in experimental psychology. I did a lot of programming (and writing about code) as a student and was deeply interested in what made products learnable and usable. A specific interest was what made a person more or less likely to make a user error. I wrote several data entry programs for large sets of scientific data. It was remarkable how much my design choices influenced the types of errors people made. What people call "user errors" are at least as much a feature of the program's design as they are of the people who make the "errors." (Printed with permission from Dr. Cem Kaner.)

Attack 15 targets bugs existing in the "data." Computers work with instructions and data. Many tests and attack efforts are focused on finding logic bugs in instructions, but the embedded taxonomy (refer to Appendix A) shows that a large percentage of errors within mobile and embedded software are actually data related. Data errors in software have caused rockets to be lost (Titan B-32), [1,2]), space craft mishaps (Mars Rover *Spirit* [3]), medical system malfunctions [4], and many others due to the assumption "it's just the data, don't worry about it." As Dr. Cem Kaner has pointed out in his lectures, the other parts of the system influence data; therefore, data is worthy of an attack too.

Takeaway note: Don't forget to test your data!

When to Apply This Attack

As soon as a tester finds numbers (data) in the software or being input into the system, then data values should be attacked. Data appears as inputs via the user, or via exchanges between hardware and/or other systems or software. Data is also stored in a computer's memory either dynamically as variables, or statically as constants, during development and coding. Finally, data comes as outputs from the software to the world. This may also need attacking.

What Faults Make This Attack Successful?

Computers really are just fancy adding machines. They are data-driven logic machines. Humans and other "users" enter the data into the software and then use output data from the software. Values can be entered incorrectly into many places through various means. For example, the value of Pi had been entered incorrectly in the memory of a control system. The Pentium bug had the wrong result preloaded for division of two numbers [5]. A Titan rocket had the wrong exponent factor entered (value entered was −0.1992476; correct value was −1.992476), which resulted in no measurement from one side of an attitude control channel and subsequent loss of a satellite—as well as billions of dollars [1]. These examples demonstrate how erroneous data can negatively impact the software system, even when the "code logic" is correct. In the first examples, the data was wrong before the software went into use or execution. Data values can get entered by humans during the software's execution too. Examples of user data entry errors include the Therac-25 [6], location–direction reference data on a GPSsmart device app, or a user entering unexpected inputs on the speed control system of a car. Finally, data can be output by the software and used by the system incorrectly. For example, a slightly wrong count can make a robot turn too far (which could damage people, equipment or production lines); an output to a D2A converter is off in a calibration constant causing a wheel to turn too fast or too slow on a digital milling machine (wasting valuable resources); or a conversion from one set of units to another is missed between two systems [7], causing the loss of a spacecraft and wasting billions of dollars.

Further, research indicates that many failures are caused by the interaction of a few data parameters acting together [4]. Testing large sets of input data in combinations is time and labor intensive, and often projects do not do enough data-driven attacks, with the result being missed bugs. Many teams only test single data points in isolation, typically normal and "off-normal" or invalid case. Not testing various valid and invalid cases and/or the interactions leaves bugs unfound (see Attack 32 for more information on how to solve this problem).

Who Conducts This Attack?

Data attacks can begin with developers during their development level attacks (see Attack 2). However, this attack is owned by the testers. Following developer attacks, independence is in order with a test team who must look for the unusual data—what can be input, and then the unusual possibility—what can go wrong with any input. The independent tester defines the valid inputs and invalid entries. Often the invalid inputs and combinations are greater than the valid. The test team's independent eyes are a critical resource toward finding the bad data (wrong number), missed data, or assumption in the logic.

Where Is This Attack Conducted?

Attack 15 can be run on the hardware, in the lab, or in the field. Simulators or emulators can start the attack, but they may not represent the numbers exactly as the real hardware might. Additionally, Attack 15 should be applied at or during

- Developer level (see Attacks 2 and 3)
- Static code analysis (see Attack 1)
- Integration—at interfaces between software and hardware (see Attacks 5 and 12)
- System constant values (this attack) and
- System values—input, outputs human and nonhuman

Testers should attack the following data areas:

- Hard code constants or internal data not set in the code correctly
- Data structure errors in the database fields, records, or indices
- Data variable ranges and resolution (design information and information contained in some programming languages)
- Conversion of data between systems
- Accuracy and precision of input and output data and
- Data pointers or indices

How to Determine If the Attack Exposes Failures

When conducting the attacks, watch for the obvious, a failure or significant departure from expectations. Some data bugs appear as small differences such as a few bits being lost; a value off in least significant areas, which might accumulate over time; or a number that is "close" but not exact. Further, pay particular attention to data values that are hard to test because of embedded limitations. An embedded limitation example would be an internal data variable that is not externally available from some information stream such as telemetry. Finally, it may be necessary to look at post test data or inputs to historic ranges to spot "negative" trends such as in calibrations of input values or changes of compensation terms due to system usage.

How to Conduct This Attack

Start Attack 15 by reviewing all of the data within the mobile or embedded system. Look at where the data variables come from and where they are stored within the software. A data dictionary can be particularly useful for this. Data that is constant or has a hard coded value should be double checked for validity within the code.

Once data is known, test design can be started. Again, exhaustive testing in design is known to be impractical, but a variety of data selection approaches are widely reported in the test technique literature. Use design techniques such as the following examples:

- Boundary value analysis

- Equivalence classes

- Numerical analysis (Math)

- Combinatorial testing (see Attack 32)

- Data flow and

- Output forcing

These techniques can be used in manual testing or automated test execution situations. Testers should be familiar with the various definitions and variations of these concepts, being ready to use each one as indicated by the data's context.

Further, consider applying a combination of these techniques with the combinatorial attack (see Attack 32). A combinatorial attack can reduce the number of data sets while using more test techniques. This reduction allows meaningful testing using a fraction of the data sets needed for exhaustive testing. A combination of test techniques (see Attack 32) can be productive in finding bugs, but it takes careful thought and planning.

ATTACK 16: BUGS IN SYSTEM-SOFTWARE COMPUTATION
◇ Advanced

This is classic bug land. In mobile and embedded software, this is where the second most common errors come from in certain domains (see Appendix A). This is the logical heart of software that you must attack to expose these common bugs.

> **Takeaway note:** Since many bugs cluster in system-software computations, testers should exercise this attack frequently.

When to Apply This Attack

Basically, you will want to consider Attack 16 at all times and at all levels. You may be able to combine this attack with a variety of design considerations and information sources. Target the software computation attack based on the requirements, design, history, and risk. How much of this attack to apply is the main question. The answer lies within the project context. Many mobile and embedded systems must be verified or checked against the "shall" requirements (or statements) for legal and/or governance reasons. Most testers are familiar with testing requirements and functions (Attack 33), but too many testers just test requirements for the "go" or normal case to show what works with "nice" data. They do not test the "no go," hard, or invalid cases. The "go" cases are usually obvious, or at least are the easier ones to define. The "no go" cases may be restricted, which often means considering history and risk. These cases can be harder to set up, which is what this attack is about. When deciding how much of this testing to do, ask yourself "what would my mother or my lawyer say about this" if the bug gets out to the public. Yes, there are restrictions in budget and schedule, meaning that often these attacks are performed early in the schedule. Therefore, requirements can be verified, but only running Attack 16 with the nice "go" requirement cases can introduce risk into the product (see Attack 4, also [8–10]).

What Faults Make This Attack Successful?

Attack 16 is based on faults of

- Mistakes made by coders/programmers and compilers, in the code

- Designs that are incorrect but make their way into the code

- Software data input or output factors, for example, calibrations, filters, buffers, and others and

- Sequences and interplay with the hardware or system that are not understood within the computation because of complexity

Who Conducts This Attack?

Attacking the software computation should start with developers using such things as developer level testing, peer reviews, and other developer verification activities. After developer activities, to assess the "interplay" between pieces of code and/or the hardware, testers should conduct integration-level attacks. One approach is to exercise the developer-level attacks on the integrated items. Each of these attack areas will remove some percentage of the computation bugs, but not all bugs can be removed by the development and integration test teams. Thus, independent software testers should take the requirements, design, interface information, and risks into account to implement their own software computation attacks.

The "official" testers (independent from development and integration) will pick up with this attack, where other testing leaves off. If the other testing has been done with some rigor, a tester's job can be reasonably small focusing on verifying the requirements in both the normal and off-normal situations. This may be done very quickly. Once these are done, the testers should move quickly to other attacks in this book. If the development organization has been less than rigorous (not a recommended practice), the test team may want to focus more time on this attack and Attack 33. Unfortunately, some test teams never move much past this type of attack. Resist such a strategy.

Where to Conduct This Attack?

Attack 16 should be conducted at the following levels:

- Developer testing by coders or programmers

- Integration testing at all levels (software, software-system, hardware–software, etc.) by development staff, possibly reusing tests from one level to another and

- System level with the software, hardware, and expected operations, done by the test team

Developer-level tools and frameworks can be used for the first two. However, the test lab with hardware and/or the full system should be used for the last attack area. It is also an option to do aspects of this attack in the field on the live system. This should be done by building on the other levels of testing, but not as the only testing.

How to Determine If the Attack Exposes Failures

The clues to finding errors during this attack come from

- Requirements not being met

- Standards or other external information items not being satisfied

- Expectations not being met, where beyond the requirements, there are things that just "do not seem right" and

- System failures, in whole or part

Most testers focus on showing that the requirements are met and not enough time on looking for what is "just not right." It is possible to do both at the same time, but this takes critical thinking. The attacks should try to show that a requirement is not met and, by failing that, verifying that the requirement is met. A tester's "break it" viewpoint is important here. The expectation of what a piece of software should do come from the customer, industry, regulations, history, and the mental models that testers have. No one needs to be told that an airplane crashing is not an expected result. A definition exists in most courts of law as the "reasonable person" test [11], where, given all factors, if a reasonable person would say "this is wrong," then (construed) you can conclude that you have an instance of a bug. Testers should exercise this "jurist" view when using their history and mental models. Keep in mind that mobile and embedded system failures do not have to be spectacular to be wrong. They can be subtle trend lines of data that head in the wrong direction on an output that, once noticed, reveals a bug. Look for these.

Mobile and embedded systems can generate huge amounts of test output. Finding the smallest trends, even though they do not jump up and announce themselves as hard failures or crashes, requires a sifting of the data usually using tools to filter and cull the data. It is a very bad feeling to have the mobile or embedded software demonstrate a failure in the field that, on retrospect, exposes a bug that was overlooked on post analysis due to the sheer volume of data points. Worse, this sort of oversight can have major legal and/ or financial implications. The level of rigor on data analysis (when doing bug hunting) is determined by the context of the project.

How to Conduct This Attack

Attack 16 should be run after developer-level attacks have been accomplished (see Chapter 2). If developer attacks have not been complete, more variations of this attack may be needed.

A tester conducts mobile and embedded system software-level attacks with the following steps:

1. Obtain the requirements, operational concepts, user, and/or other reference information in existence for the software, hardware, and system. (Remember, *Knowledge is Power.*)

2. Review the available information and allocate key concepts or requirements to start defining Attack 16 in detail (see Attack 32 and [12]).

3. Determine what and how much the developers have attacked in these areas (less work by the developers means more work for the test team).

4. Expand the attack area with risks, but balance the attack with budget, schedule, and risk since it is better to have a concentrated good attack done quickly than an unfocused, bloated general swipe.

5. Design your attack data using one or more test techniques [13].

6. Run your attack.

7. Analyze the data and look for subtle errors and insight.

8. Learn.

9. Adjust or define new attacks in this area based on your improved understanding.

10. Repeat with added or new attacks in this area, until your plans are fulfilled in this attack area and/or budget or schedule allocations are exhausted.

These attacks can be expanded, organized, and in some cases even automated. Expansion can include adding more test data (sub-attacks) to reach more coverage of requirements, usage, and so on, as the attack repeats and your learning continues. A good approach to organization is to create stories, tours, and written scripts, although testers must be aware of the positives and negatives of these ideas (see Attack 20).

Design Attack 16 to go after the following areas: normal use (first or early), edge cases (valid and invalid), extreme points (valid and invalid), and combinations of some or all of these [14]. These are not done all at once but in iterations and versions of the attack. Some people have observed that a good test designer has the patience of a saint, the heart of an athlete for long sprints, the brains of an expert, and the craft of a fox to sneak up on the bugs. These traits are developed in mobile and embedded domains over long periods of practice and critical thinking. Testers need to be ready to go into unexpected directions to find bugs and identify new attack variations.

Once attacks are planned, designed, and executed, the job is not over. Output data should be reviewed by testers during the attack and then analyzed by experts after completion, both efforts looking for any bugs revealed by the data.

ATTACK 17: USING SIMULATION AND STIMULATION TO DRIVE SOFTWARE ATTACKS

◇◇ Expert

Understanding where the inputs come from in some mobile and many embedded environments is a common problem. This issue is commonly encountered in the labs and, to a limited extent, in field testing. For instance, if you are testing a smart braking system, an autonomous spacecraft, or a robotic factory controller in the lab, many of the real-world factors are not present. How do you create slippery surfaces for the brakes? How do you create zero-G settings on the ground for the spacecraft software? Can you create the noise and variability of a factory in your test lab? All of these "factors" can influence the system, software, and testing of the embedded or mobile device. The problem itself is hard enough, yet some test teams ignore these factors until the product encounters them, usually in the field with the customer. Other test groups realize that creating these factors as inputs into an attack is desirable, and they do this by creating simulations and/or stimulations, what some call "sim/stim." Simulation or stimulation drives inputs into the system or software under test.

Attack 17 outlines how to use simulation or stimulation to drive attacks. These attack concepts are usually used in combination with other testing efforts and attacks conducted within a lab test environment or on equipment in a lab environment (see Chapter 11 on how to set these up). The level of simulation or stimulation depends on factors such as risk, cost, and schedule. For some critical mobile or embedded devices and applications, more simulation or stimulation efforts are justified, while other efforts may not be justified. For example, in the aero and space industry, extensive real-time closed-loop simulations exist that allow testers to "fly" the vehicles in virtual worlds. This allows hundreds, thousands, and even tens of thousands of test attacks, including ones those that are very hard to do in the real world. The simulation or stimulation attacks combined with real-world field tests provide a more complete picture of the software system, without all of the costs of thousands of tests in the field. Other embedded or mobile test environments have simple simulation or stimulation set-ups. For example, in the mobile cell world, there are simple emulation environments and tools, which can quickly allow a tester to create inputs for virtual devices running apps. This is very convenient but has risk such as how "real" is the emulation–simulation.

When to Apply This Attack

You will want to consider applying the simulation and stimulation attack when your software has and uses sensor or hardware inputs. These inputs can involve complex input interactions with the outside world. Additionally, the inputs and interactions can be difficult to set up in testing. For example, in a test rocket system, many different input conditions might be encountered for an engine thrust output, such as invalid low, low, low-to-high, and invalid high in a range of 1000.0–9000.0 units over some period of time usage (real-time). But we cannot easily have a real engine blowing fire over these ranges in our test lab. Therefore, the solution is to create a simulation of the engine, the outputs, and the thrust units into the software where the tester can vary these in a test over ranges and time.

You may also want to apply Attack 17 when the costs of doing many tests in the real world become very high. Cost factors like expendables (fuel, tires, and so on), labor (test pilots, test drivers, test users), and equipment (numbers of smart phones, airplanes, cars) can all become expensive very quickly. Simulation or stimulation modeling can be cost effective when compared with the real-world costs.

> **Takeaway note:** Apply simulation or stimulation when the possibility of creating the real-world inputs into the test environment is too hard, too expensive, or too unrealistic and in combination with other attacks to achieve more robust testing.

Who Conducts This Attack?

The creation of simulation or stimulation environments, like creating a test lab, is a development effort (see Chapter 11). The person or persons creating the simulation or stimulation need to understand planning and development, often of both hardware and software, as well as the nature of the environment you are trying to simulate or stimulate. This often requires the person to be knowledgeable in such things as electronics, the hardware, the system, the real world, the interface to the device under test, real-time closed processing, and other engineering areas outside of basic tester knowledge. Further, the team needs to understand how the simulation or stimulation will be used to create this and associated attacks. Questions this person should consider include "Will it be data driven, keyword based, software, hardware, vary in time, need a feedback loop, or combinations of these things?" As the software-systems become more complex, the more complex the simulation or stimulation modeling can become.

Where Is This Attack Conducted?

The simulation and stimulation activity typically starts and exists in the test lab and facilities, but can, in some cases, move into the field (see Chapter 11 for more details). In the lab, simulation and stimulation systems exist in simple (just a data stream) and complex (closed-loop feedback driven computations) configurations. In a simple form, you could create a simulated data stream before the test attacks start. Then, feed this into the software under attack. The data stream may be time based and make assumptions about system-software action. The disadvantage here is that if your software-system is reactive and/or intelligent, it can be hard to judge ahead of time what a long pre-canned stream of data should look like. This leads many simulation and stimulation attack systems into becoming reactive themselves. Some simulation or stimulation systems in real time detect the output of the device under test and compute an appropriate response. This is called a "closed-loop real-time simulation." The real-time cycle can be very fast (in microseconds).

There are extensive embedded test labs where the computing power and lines of code are tens or hundreds of times more powerful than the device under test. As mentioned previously, this can get expensive, so some teams opt for simple lab simulation or stimulation environments and more field testing. Some groups use a hybrid approach where in the field, simulation and stimulation attacks can be introduced at test points in either the hardware or the software. This can be done using concepts such as "built-in test," field test configurations, and special system add-ons like instrumentation to allow hybrid testing to be done.

In all cases, the impact of the simulation or stimulation to the device or software under test needs to be considered. Is the simulation or stimulation realistic? Is the simulation or stimulation influencing the testing? Is the cost and schedule of the simulation or stimulation justified? What kind of the simulation or stimulation can or should be used?

What Faults Make This Attack Successful?

The faults that simulation or stimulation is looking to find are the boundary, hidden, and seldom traveled cases recommended in many other attacks of this book. Certainly, the system's "normal" usage will be touched early on. If only simple and normal data points are input, error cases are likely to be missed. In simulation and stimulation attacks, you provide mechanisms, hardware, and/or software that allow these other inputs to happen. The faults that might tend to hide in complex time-based data input streams can be detected with realistic simulation and stimulation attacks.

How to Determine If the Attack Exposes Failures

Since Attack 17 is normally used in concert with other attacks, these factors determine how the errors are found. In simulation or stimulation, you are inputting the whole spectrum of real-world inputs (valid and invalid) to trigger the errors in the system-software. You are looking for failures, system crashes, and incorrect responses.

How to Conduct This Attack

First in Attack 17 is the creation of the simulation or stimulation environment. This is associated with the lab. (See Chapter 11 for a detailed discussion of simulation, stimulation, and modeling to support attacks.) The lab is configured with the simulation and stimulation features. Often there will be multiple simulations corresponding to different parts of the system.

Modeling various simulation and stimulation features can be time consuming and expensive. Simple simulations and stimulations can be modeled in a data stream. The more complex simulations and stimulations modeled in a closed-loop real-time simulation will take time to create and even set the condition right. For example, how do you set up inputs for over 100 different stimulation variables, all running in 10 ms computation cycles for over 14 h of real-time rocket flight and events? This could be tens of thousands of inputs and take a long time to set up!

Once you have the simulation and stimulation environments set up, Attack 17 should proceed with some quick exploratory tests and then more detailed test planning or design (see [12,13]). I would run a normal use case (or two) and then some stressing cases [14,15]. I would follow these up with more specific attacks from Chapters 4 and 6, as well as specialized attacks from other parts of this book. Then, repeat and explore more simulation and stimulation cases. This attack provides the structures (simulations and stimulations) to support a variety of attacks. For many mobile and embedded systems, these structures are important to have in place and can be reused in many other tests.

Be careful, because some time problems in the simulation or the stimulation set-up and development can hide bugs or provide false positives. In complex lab environments with larger simulation or stimulation tools, I have seen as many bugs in the lab and tool environment as in the software under test. This may lead some to conclude that the simulation and stimulation attack is not worth the cost and effort. Certainly the benefits of simulation or stimulation versus the costs must be considered, but in many mobile and embedded environments, the benefits justify the costs. The tradeoff of how far to take simulation and stimulation attacks requires much critical thinking, analysis, and likely some agility.

EXERCISES (ANSWERS ARE ON MY WEBSITE)

1. Define attack tests for the following communication commands:

   - System "ok"

   - System "ready"

   - System "memory location corrupted"

2. You are testing the embedded software for a car, which can have the following different features: engine type/size, number of car doors, entertainment systems, communication systems, and seat configurations. Define data values that you can think of to check that the software might have and how you arrived at testing those values.

3. You are charged with testing a car's antilock braking system (which has software). Define some simulation or stimulation input factors for your tests.

4. You are testing a cell phone app that presents weather to its user based on communications from a server. Define some simulation or stimulation input factors for your tests.

REFERENCES

1. *Titan IV B-32 Mission Failure Review Board Report*, U.S. Air Force, 2000.
2. Pannikottu, A. February 22, 2010. American engineering group engineering failure analysis. http://www.engineering-group.com/papers/aeg_engineeringfailureanalysis_review.pdf (last accessed April 11, 2013).
3. Mars Rover Spirit flash memory management anomaly. January 21, 2004. http://en.wikipedia.org/wiki/Spirit_rover Spirit Sol 18 flash memory management error (last accessed April 11, 2013).
4. Wallace, D.R. and Kuhn, D. R. 2001. Failure modes in medical device software: An analysis of 15 years of recall data. *International Journal of Reliability, Quality and Safety Engineering* 8(4):351–371.
5. Halfhill, T.R. March 1995. The truth behind the Pentium bug, *Byte*, p. 163.
6. Leveson, N. 1995. *Safeware: System Safety and Computers*, Addison Wesley, Boston, MA, Appendix A.
7. Isbell, D., Hardin, M., and Underwood, J. September 30, 1999. MARS climate orbiter team finds likely cause of loss, Release 99-113. http://mars.jpl.nasa.gov/msp98/news/mco990930.html (last accessed April 11, 2013).
8. Risk-based testing. http://en.wikipedia.org/wiki/Risk_based_testing (last accessed April 11, 2013).
9. ISO29119 to be published in late 2013. http://www.iso.org/iso/catalogue_detail.htm?csnumber=45142 (last accessed April 11, 2013).
10. Risk analysis (engineering). http://en.wikipedia.org/wiki/Risk_analysis_(engineering) (last accessed April 11, 2013).
11. Reasonable person. http://en.wikipedia.org/wiki/Reasonable_person (last accessed April 11, 2013).
12. Craig, R. and Jaskiel, S. 2002. *Systematic Software Testing*, Artech House Publishers, Boston, MA.
13. Copeland, L. 2003. *A Practitioner's Guide to Software Test Design*, Artech House Publishers, Boston, MA, Chapter 3, 4, 9.
14. Buwalda, H. 2004. Soap opera testing, *Better Software Magazine*, 30–37.
15. Pries, K. and Quigley, J. 2011. *Testing Complex and Embedded Systems*, CRC Press, Boca Raton, FL.

CHAPTER 6

Time Attacks

"It's about Time"

THIS CHAPTER CONSIDERS TIME with respect to the system or other major elements in the system and software. Time is an interesting topic and is often called the fourth dimension, so these attacks only poke at that topic.

Embedded software systems often have potential "real-time" problems related to the "clock." Time can be wall clock time, a date, time between actions, internally represented time, and so on. Time can also be long (years, days, hours, minutes), short (micro- and nanoseconds), and/or combinations of these. When we say "real-time" and "time critical," we should immediately start asking more questions about what is meant by those references and then begin to examine implications of incorrect time. A few timing attacks are defined in this chapter, but the topic is very complex. On my website, I offer advice on time attacks as well as some further reading.

Table 6.1 maps the attacks of this chapter to different mobile and embedded contexts. Of particular note to the reader is that you should be concerned about nearly every context when using time-related attacks.

The error taxonomy (see Appendix A) included many time-related bugs. These bugs can be transient, hard to find, and difficult to reproduce with testing. A project's first lines of defense should be good architectures and design with support analysis. By using some analysis in a spreadsheet, modeling, or simulation development staff can avoid many errors, but some will remain for tester to find using attacks, tool checks, time-stamped log files, or other timing information. However, even with this information, learning to be extraordinarily observant and patient in recognizing time patterns is critical to finding bugs. The attacks of the chapter are longer than many other attacks in this book and other references covering attacks, largely because the subject is complex and some people would say "infinite."

TABLE 6.1   Time Attacks Mapped to Mobile and Embedded Contexts

Context → Attack ↓	Mobile and Smart Wireless Devices	Embedded (Simple) Devices	Critical Mobile Devices (Could Be Embedded, Small but Important)	Critical Large Embedded Devices (Could Be Mobile)
Attack 18	Sometimes	Frequent	Yes	Yes
Attack 19	Sometimes	Frequent	Yes	Yes
Attack 20	Frequently	Yes	Yes	Yes
Attack 21	Yes	Yes	Yes	Critical

Notes: Seldom, seen only in specialized cases infrequently; sometimes, seen often in many different contexts; frequent, seen regularly in a variety of contexts; yes, should be considered for most contexts; no, generally not applicable; critical, usually a vital aspect of the system and can contain hard to find bugs.

ATTACK 18: BUGS IN TIMING INTERRUPTS AND PRIORITY INVERSIONS
◇◇ Expert

The use of interrupts and timing are wild cards in embedded systems. The taxonomy (see Appendix A) shows that major problems have been associated with interrupts, while at the same time the bugs associated with them are hard to find. Interrupts are useful as triggers into and out of the real world for computer system processing. Interrupts can vary in time and are basically nondeterministic (we do not know exactly when they will happen). Their use is the root of the problem. It would be nice to avoid them, and some embedded systems do. However, many embedded systems rely on them because it solves problems associated with the nondeterministic nature of the real world. We just do not know when many things will happen in the future. Attack 18 looks for bugs and issues in interrupts and their timing. While being an advanced concept of attacking the software, the attack itself is a beginning and not an end, or in other words, just because you run this attack does not mean that you will not have bugs in this area.

When to Apply This Attack

If your embedded system is driven by interrupts or uses interrupts to drive time-related actions, Attack 18 should be considered. Attack 18 is run in two parts: analysis and testing. Ideally, the time performance analysis starts before the architecture and design of the system or software are finalized. The information from up-front analysis can become the performance requirements, which help define the implementation. These requirements will be used later by the test staff for confirmation checking of what was achieved versus the required performance. Unfortunately, many groups do not think that they will have performance issues until it bites them. This often results in large reengineering efforts of analysis, design changes, and even more testing, simply trying to fix the timing performance issues. In these cases, the systems people, developers, and testers are likely all working in parallel. The analysis and testing continue throughout the life cycle of the product [1].

Key point: Timing and performance problems can cost major rework efforts if found late in development. Teams can prevent this by practicing early and frequent timing attacks.

What Faults Make This Attack Successful?

In real-time embedded devices, faults come from the interaction between the interrupt (operations), hardware, software, and time. A single interrupt can cause a problem, but when multiple interrupts happen, issues of collision and contention come into play. Once the tester knows that interrupts are in use, they should look for the following kinds of situations:

- The interrupt is not contained within some deterministic architecture design (meaning you have nondeterministic aspects in play).

- The interrupt is not detected and processed correctly in time.

- The interrupt is incorrectly suppressed (masked) when needed.

- The interrupt is superseded by another interrupt before processing is complete (conflicting levels of interrupt processing) also known as priority inversion.

- The interrupt is detected, but the wrong time-based process(es) may be used.

- There are excessive interrupts allowed, resulting in saturation of processing responses and starvation.

- The interrupt is not detected and processed in time quickly, so subsequent processing is late or missed.

- The logic in parallel processing is not correct to handle sequential interrupts (a potential problem in multi-core processors).

- The computer is in the wrong state when an interrupt happens and so a wrong interrupt or no processing occurs.

- Time (time out) allocations are exceeded.

Many embedded systems use interrupts extensively. Interrupts refer to stimulus outside of the software to which hardware and software must respond. Most systems (in hardware) have a variety of interrupt levels [2]. The outside world, through the hardware, triggers the interrupt, and then, the software processing the interrupts must respond with correct actions and within a certain time frame. Software prefers actions or events to be deterministic, but interrupt-driven systems can have large amounts of time-based nondeterministic actions. The more interrupts used, the more work on interrupt timing and performance is needed, because there are more "unknowns" in the mix.

Takeaway note: What is a nondeterministic set of conditions? As one simple example, consider driving your car. Nondeterministic relates to not knowing ahead of time the following conditions or factors: what day and time you will step on your brakes; when you will travel on an icy road; when the electrical system in your car will not work correctly; and the combination of these things. When taken together, these combinations might produce a bug. If a series of interrupts were tied to these conditions, there is no way to know, when they will happen, or in other words, the conditions to produce a bug are nondeterministic. It is very hard to test and even plan for testing of nondeterministic behaviors, because testers can never know all of the time or system conditions for all possibilities. This basic problem in nondeterministic systems is knowing ahead of time the condition and timing to produce bug behavior. It is an important issue for many mobile–embedded systems. For those who want to know more about this topic, feel free to "wade into the theory of automata."

Who Conducts This Attack?

Attack 18 is run by a tester working with systems analysts. The tester supplies the data to drive the analysis, once real hardware and software become available for testing. The tester may even help in the analysis work. The tester's main role is to figure out how to create the needed interrupts and/or system–hardware faults, which trigger interrupts, interesting situations, or otherwise generate interrupt data to help the analysts. The people running this attack will most likely be senior level, since this is a very difficult and complex area and requires extensive experience and lots of creative thinking.

Where Is This Attack Conducted?

In the beginning, the performance analysis team starts Attack 18 with modeling or math-based methods to define and allocate time for processing. Time performance requirements will be general at the start and be refined by analysis during development. The analysts will use tools, models, simulations, spreadsheets, and methods like Rate Monotonic Analysis (RMA) [3]. The results of the analysis will be provided to the rest of the team including testers, the requirements team, and/or the developers. This is where the test team moves into the lab with performance testing tools that can be called on to provide the "real" data points for interrupt timing analysis.

Takeaway note: Complete timing analysis activities such as all aspects of RMA are important but out of scope for this book.

As testing happens, test runs in the lab will first address normal scenarios, then more stressing points, and finally breaking data points. The test results are fed back into the analysis and development. The test and analysis results may be used to modify the software, system, and sometimes, even the hardware.

There are features, capabilities, and tools in the test lab that are needed to support this attack. As you set up for an interrupt performance test environment in the lab (see Chapter 11), first, ask these questions.

- Do we have the ability to trigger certain events (interrupts) at just the "right" time?

- Do we have the ability to capture what is happening in the system and/or inside the software during interrupt processing?

- Can we emulate, simulate, or stimulate interrupt processing?

- How realistic are each of these to our project in the lab?

Many of us have chased interrupt performance bugs from the field back into the lab, because we knew that the lab had more instrumentation to assist us. Even in the lab, a tester can spend lots of time dealing with interrupt performance issues. I advise earlier analysis, attack testing in the lab, and more analysis, to perhaps minimize these bugs through architecture and design.

How to Determine If the Attack Exposes Failures

At the start of testing, consider that you may have a bug in interrupts if you see any of the following:

- Slow response to an interrupt situation

- Missed deadlines

- Missed interrupt processing

- Nearly missed deadlines (Can we push it a little harder to make it break?)

- Hardware acting unpredictably at the boundary conditions or out of bounds

Interrupt performance is hard to analyze and test for, so it is also hard to determine that you have a bug or the nature of the bug. You might suspect an interrupt bug as something else, before you realize its true source. Testers who have knowledge of the architecture can help their own testing of interrupt performance and applications of this attack in their testing efforts.

How to Conduct This Attack

To start Attack 18, you need to determine if there are interrupts, any interrupt performance considerations, and what performance analysis has already been done. This will take some previous design knowledge, as well as access to the measures of performance, which should come from requirements, system analysis, and/or engineering groups. With no interrupts in use, your job with Attack 18 is done. If the system has interrupts, but uses a deterministic architecture and is well analyzed, you may be able to do just a few attacks in this area to support timing analysis. However, lack of any interrupt timing performance analysis, when combined with a design that uses interrupts in time-based processing, indicates that a series of interrupt attacks are needed. The lack of timing information makes your job harder and maybe more critical. You may need to work with someone to create the performance analysis metrics to understand what interrupts are possible. If the team is saying "no problem," "don't worry," "we have a timing margin," and other excuses, consider running Attacks 19 and 21 while trying to break the software by stressing it with this interrupt attack.

Next, the tester needs to run a series of these attacks to gather interrupt timing data. A variety of test levels and tools can be used to gather such data. Options here can include the following:

- Emulation or simulation of interrupt timing information using models (usually an analyst does this, but it is not the "real" timing and processing information);

- Developer-based testing, where simulated time is captured during Attacks 2 or 3 (run by developers if the developer test tool environment supports gathering time data and interrupts, but substantial analysis is needed to gather and use the pieces of data [1]);

- Lab tooling such as probes or monitors to check interrupt processing and time information within the computer (can be hardware or software based, but here again, the tools must support gathering these types of data); and/or

- Internal monitoring of software to trap timing data from within the application (this is a form of built-in test where the built-in test logic is coded into the software).

Following these attacks, the test team needs to define the cases to drive the various interrupts, times, and processing that have been identified. This means defining the time/interrupt cases that need to be tested. The simpler the interrupt processing, the easier this analysis and the cases will be. However, I have seen complex multi-leveled interrupt-driven embedded systems that were hard to understand and create attack cases for, as well as to get reasonable coverage of the interrupts and time spans. This is where experience, knowledge of the system, and large amounts of creative critical thinking come into play.

At this point in the attack, I recommend the use of state modeling. In this step, you create a state model of the interrupts or parts of the system, which contain the interrupts. State modeling is supported by many tools and nicely explained in Copeland's *A Practitioner's Guide to Software Test Design* [4]. I provide a simple state model example at the end of this attack as a reference point.

To create a state model, define the states associated with the interrupts. Next, define the conditions that move the software from one state to another state. Often the transition may be a system input and other times it may be an interrupt. This information comes from the knowledge of the design of the interrupts, software processing, and system. This can be obtained using design information from development personnel (preferred) or by doing exploration of the system (which may yield incomplete state models).

Once an interrupt state model has been created containing the processing, the states, the state transitions, and associated conditions, tests cases can be generated using this model information. This is done by simply conducting a "walk" over the states and transitions. Typically, you will start in some initial or "off" state; this is the start of the attack. Next, move to the next logical state following a defined state transition. The test case continues from one state to another state following the defined state transitions, until all states and/or transitions have been covered.

There are several possible coverage measures for state transition testing, with all transitions being stronger than all states, but there are others as well—transition pairs, triples, and others. All transitions are noted as 0—switch coverage. I recommend 0—switch/transition for basic coverage. Sometimes several attack test cases will be needed to achieve this level of coverage. The "walks" will give some indication of the data or conditions needed to move from one state to another, as well as expected state results. This will become the test case input conditions. Added coverage may be needed if some triggering state transitions have multiple data values. Once basic coverage has been achieved, you may want to attack states and transitions that are "not shown" in the model by inputting data, interrupts, or trying to "force" state transitions that should not be. These may be undocumented "features" or bugs.

Setting up these cases and selecting good attack-specific points (time or data) are test design and creativity at its finest and hardest. This is a case where one attack will probably not be enough and knowing how many to create takes sound judgment and even some luck.

As the interrupt attack tests get run and evolve, you will need to gather the interrupt information and data. This is added to the state chart. The data and information can be provided to the analysts and then analyzed, often with tools such as MATLAB®. Dealing with interrupts and timing cases is tricky. The test team will likely be asked to go back and try again to gather more information. Expect to be asked questions such as "Can you cause the interrupt to happen right at some time point, a set state, or set of conditions?" The answer to these questions will give the analyst information needed to get the bug to make another appearance.

The team needs to think through various situations. Testers may find it hard to identify cases that trigger interrupt bugs. The nondeterministic behavior of timing and interrupts has caused many problems for many teams. Try not to let it drive you crazy. Following the steps in this attack will help you.

As the testing progresses, collect and organize the attack data, to allow interrupt processing and timing performance analysis to happen. This analysis can include such things as the following:

- RMA, if you have the facilities and skills for it;

- Spreadsheets or tools to summarize and analyze the data;

- Look for trends in the data, including reviewing log files, within the context of the system functionality, event activities, and hardware–software architecture;

- Look at the data graphically;

- Check for "erratic" points; and

- Review the data and help the analysts.

SIDEBAR 6.1

Interrupts: I once saw a strange hardware situation create secondary interrupts and saturation problems in the software. Now one can argue that we cannot protect against every "negative" situation that software is likely to see. But if software can handle "negative" situations with minimal processing and impacts, a much more resilient system can be the result, and this is a positive for many users. This ability is one reason why everyone is putting software into embedded systems.

For example, I once saw a system that had a hardware transistor problem, which caused a constant stream of interrupts to be sent out to software. The interrupt and system processing could not complete to determine what was going on before the next interrupt was triggered (system had interrupt saturation, resulting in processor starvation). The software processing was not designed to handle this, because no one could imagine a situation or had tested for processor saturation. A design change was determined that could "lock out" (ignore) an interrupt channel if something like this happened again. This was found on the actual (in customer's hands) fielded system—not in testing. In hindsight, it would have been much better to have found this in testing, but that wasn't our luck.

A variation of this "lock out" situation I have seen on one smartphone involved the following: (1) alarm interrupt; (2) incoming text message (another interrupt); (3) a phone call (third interrupt); and (4) a running app. The combination locked up the phone to the point where a hard reset was needed (i.e., the battery was pulled).

During these efforts, ask "what can go wrong, how can things stack up in time, and/or what changes can be made to interrupts to cause a bug (run scared)." Running other attacks should be considered too, such as the following:

- Normal case execution—see Attack 20

- Long during runs—see Attack 6

- Other time bugs—see Attack 19

Additionally, consider repeating this attack throughout the life cycle. Even with hard work and these attacks, there is still some "luck" to finding bugs in interrupts (see Sidebar 6.1).

Takeaway note: Interrupt time and priority inversion bugs are hard to find, but if you do not look, your customer will find them eventually—or for you.

STATE MODELING EXAMPLE

Table 6.2 shows a very simple embedded state model. It is not necessarily complete (see the Exercises), but shows what a person starting interrupt attacks might have. There is a picture and a table version. The picture is easier to understand at a glance; however, tables can be more efficient when dealing with software interrupt systems, where there may be many states and transitions.

This sample has 6 states, 10 state transitions, and 5 interrupts (as processes). In the table, States are the top row; Actions taken in each state are in the second row; and the

TABLE 6.2   Example of a Simple Embedded State Model for an Engine Controller for a Car

State	Engine Off	Engine Power Turn On	Engine Run Variable Power	Engine Stop	Limit Power Error	Engine Idle Power
Actions	No fuel–air flow or spark	Fuel start in choke mode and provide electronic input (spark) at start level	Fuel–air to demand level and timing of spark support fuel level	Terminate fuel–air and electronic spark	Limit fuel–air then retard spark to set levels	Fuel–air held at idle and spark set accordingly
Next state	If key on interrupt, go to "Engine power turn on" state	If started, "Engine idle power" If not started, "Engine power turn on"	Loop in "engine run variable power" until If Pedal = 0, then "Engine idle power" If "transmission error," then "Limit power error"	"Engine off" state	Go to "Engine idle power"	"Engine idle power" until peddle not "0," then go to "engine run variable power" or "Key off" go to "Engine stop" state

Notes: Interrupt conditions
User—peddle interrupt—multi-position throttle in engine system
User—key—on/off
Motor control—open valve (fuel) and open valve (air) in a coupled relationship
System sensors
  Air mass sensor
  Fuel injectors regulators
  Braking system
Errors
  Check engine
  User (transmission)—limit power to ideal because of an error in transmission

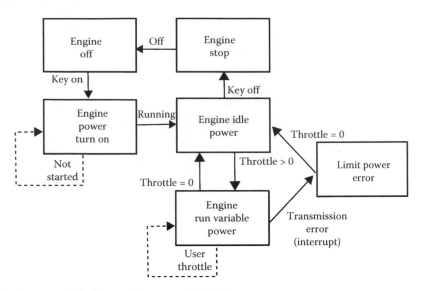

FIGURE 6.1   State model of an engine control with interrupts.

Transition actions are given in the bottom row. Likewise in Figure 6.1, boxes show the States; the arrows with the labels show the state transition events. Notes are provided at the bottom.

A test attack scenario generated from these state models might look like the following sequence:

```
Engine on -> Check Engine power on state reached -> Check engine
idle state reached -> Apply peddle -> Check engine run variable
power state reached -> Adjust peddle and check state not changed ->
trigger "transmission error" -> check limit power error state
reached -> and then check "engine idle power" state reached -> Turn
"key off" -> check engine stop state reach -> and then check
"engine off" state reached.
```

For more information on testing with state modeling, refer to Copeland's *A Practitioner's Guide to Software Test Design* [4].

ATTACK 19: FINDING TIME-RELATED BUGS

◇◇ Expert

Attack 19 targets time and timing-related bugs. Embedded systems often function within time, both long term and short term (such as nanosecond). Problems happen when time is represented differently within parts of the system: for example, two computer systems each with a different representation of time, but they are trying to coordinate actions of when something is to happen. Also errors happen in time because many mobile and embedded software systems operate with time as one of the non-functional quality factors. This attack takes some thinking to set up and again a little luck to find the bugs, since they hide in time and the interaction of factors and components.

TABLE 6.3   Timing Categories, Failures, and Notes

Category: System Has	Examples of Failures	Notes
Hour-time boundary—failure resulting from incompatible system time formats or values	International dateline Time zones Annual time changes (daylight savings-DST) Leap year Year-to-year boundary	It may seem obvious, but time has these and other boundaries, which when hit or crossed causes issues in smartphones, cars, planes, factory systems, and others
Date(s) in the systems have different date formats or values	System 1 stored value day/month/year and System 2 to expected month/day/year	Where representation formats are different, formats or dates can get out of sync between storage areas
Relative versus absolute—failure due to incompatible use of relative versus absolute time values	Launch T-minus time versus universal time code (UTC) on the first space shuttle flight	Relative time is usually a count from some point, while absolute is "wall clock" time
Race conditions	A process cannot complete because the resources it needs are locked by another process that is waiting on the first process	http://en.wikipedia.org/wiki/Race_condition

Note: Definition of race—inadequate controls of more than one mutually dependent set of software operations or resources where one process "races" to influence the system and impacts the other tasks.

When to Apply This Attack

Apply Attack 19 when the following situations exist inside or outside of the system (see Table 6.3). These have all been encountered in fielded mobile and embedded smart systems. Table 6.3 lists common time-related bugs, although it is not a comprehensive list. Your system should be analyzed for similar situations dealing with time.

Time within many mobile and embedded systems is complex, so Attack 19 must be considered and designed carefully. Because time changes within a system and is influenced by many factors, an attack for time is often well advised.

> **Takeaway note:** Look at time, how it is used, where it has boundaries, and how these all may combine in your software system. The fact that the taxonomy has shown many time errors means that testers are not thinking about time-related attacks, maybe because they are too "hard," given factors such as tester training, lack of team support, or project resources (tools, time, money, people, etc.).

What Faults Make This Attack Successful?

In addition to the example failures and associated notes in Table 6.3, testers should understand how time is used and represented in the local system and software to define "time" risk for their local context. Not every mobile or embedded device knows dates or time or uses it, but some do, and when they do, this time attack may be necessary. Time can be as simple as a count of internal CPU "clock" cycles or as complicated as linked network time across many systems such as with GPS systems. Time, how it is stored, represented, used, and calculated can result in time differences between parts of the system, subsystems, or systems. Everyone has seen instances of the clock on the wall being a different time from their watch, cell phone, or a computer's clock. Which one is correct? The answer of who or which is right is often not important directly, but when the times are different and used as coordinating or regulating system behavior, the conflict can become the source of a bug. A bug can also occur when time is not used or represented correctly.

Some smaller embedded systems do not use time, and in that case, you can ignore this attack. But be aware that even a simple counter may be used to track "time" by some programmers, so do not be too fast to exclude this attack. For example, a program may wait a second of internal CPU clock time, by counting clock ticks, before reading a sensor and a bug might lurk here. If time is stored with a different representation from the source, this can be the source of bugs. Also when time is used in decisions or calculations, bugs can result and the need for testing increases. So understanding time and how it is used becomes very important for Attack 19.

Who Conducts This Attack?

The test team, with support from the developers and systems people, needs to conduct Attack 19. The test team will need testers who know and can identify where time factors or aspects of time exist within the system. These testers should be aware of multiple places and ways that time is sourced, used, stored, and calculated. The team then uses this knowledge to define the risks, and the number of cases, and then determines how to run the attack.

Where Is This Attack Conducted?

Attack 19 is conducted on the system test bed (lab) and/or on the system itself (field testing). Testers need to know where time is calculated, sourced from (internal or external), used, and/or stored. Any test environment must support inputting time, capturing it on output, and/or creating time differences, skews, and different representations. If in the lab, testers must treat time "simply" as relative to the real world or timing faults will be missed. Testers should ask the following questions about their lab:

- Can we change system time?

- Can we input "different" times (turn time forward, move calendar dates, speed up, slow down)?

- Can we "trap" time values and representations?

- Can we analyze time representations?

Field testing means that "real" world time is used. Attack 19 is also a good one to do in the field, since time is very realistic, but this must be bound by the fact that field testing is usually limited and time is unlimited. A mix of lab and field testing is a good idea.

SIDEBAR 6.2

F22 Aircraft: An F22 fighter aircraft had to return to base, when parts of the embedded software avionics system had problems crossing the international dateline when going from Hawaii to Taiwan for the first time because of a time–date problem [5].

How to Determine If the Attack Exposes Failures

Testers should look for the following signs of time value problems:

- Multiple points in the system and software where time is sourced, used, and stored, which can result in small differences in time (find the bugs by comparing or changing the clocks).

- Time skewing and out of sequence of data or outputs (find the bug by comparing outputs with a stamp to a "universal time source").

- Representation of time information (look for small differences in time or missing information to find a bug).

- Large failures (reconfigures, shutdowns, aborts, crashes).

How to Conduct This Attack

To start Attack 19, you must understand what your system does in relation to time and the ways in which time may be represented (counts, sums, clock times, dates, etc.) in your system. The steps for Attack 19 follow:

1. Collect system timing information concepts from such things as operation or user guides, use cases, models, design-to-code, and any other information that describes in detail functions in time.

2. Define potential time "contention points" using risk analysis (Appendix G).

   Note: Contention points are where various software and hardware resources may be in demand at the same time, for instance, the network is needed by the phone, texting, and app; a software routine is called by two different apps at same time; a device gets a (human) user request at the same time as a critical hardware request; and others. These contention points may need to be worked with systems and software engineering or development personnel to understand all the cases.

3. Define attack points for test cases including

   a. Focus first on the normal situations of clock and date boundaries (think boundary value analysis).

   b. Analyze the code looking for places where "drift" can take place, including time resets, boundaries, and changes.

Note: Drift is where time or time-related information is kept in different parts of a system (e.g., two computers can become divergent).

c. Look for computer software designs that have scheduling tasks at different time periods; what is often called such things as task frequencies (routines operating at different time frequency and/or periods); multi-threading; and distribution of functionality across hardware processors. Each of these can cause time differences.

d. Review and understand industry and your company's history for time-related errors or failures (such histories can be good sources of risks and attack cases).

e. Define and understand the time-related aspects of your system, from information such as requirements, architecture, system timing flows, design, timing analyzes, and implementation, where these kinds of information are available.

f. Using these analyses, update the time-related risk list.

g. Assign a testing priority based on time-related risks, including likelihood and impact.

4. Build attack test scenarios for each targeted time or risk priority.

a. Define time point triggers.

b. Define variations around the time point.

c. Design your attacks to force time triggers and variations.

d. Attempt to force different time, dates, and other time boundary conditions.

Note: If there are a large number of time factors, consider using a decision table to help guide selections and cases for item d. Factors to go into item d or a decision table can include bogus time conditions to test the error messaging or behavior of the software; slowing clocks in various parts of the system; having clock timing signals drop off of a network; and others.

e. Create test details, data, sequencing with time, and expect results.

Note: Forcing clock time skews or race conditions can be very difficult, so special effort and/or tooling may be needed. For example, when time is used from differing mechanisms (see Sidebar 6.3).

SIDEBAR 6.3

Space Shuttle: During the first space shuttle launch attempt, the countdown had to be scrubbed because the flight and ground computers were out of sync causing the countdown to abort when the times were compared [6].

5. Run the attack.

6. Analyze test results looking for time bugs and dependencies.

7. Repeat—as schedule and budget allow, attacking other "time" risk priority items.

Pure time execution-based attacks should be supplemented with long-duration attacks (see Attacks 6, 20, 21), RMA, and other timing analyses [7,8]. They may expose information, bugs, or other areas to attack. Attack 19 may not expose a bug on the first attempt; other combinations may yield more interesting results. Time can be long and complex. Attacks involving time require diligence, discipline, tenacity, and patience.

ATTACK 20: TIME-RELATED SCENARIOS, STORIES, AND TOURS
O Novice

Testing is most realistic if you follow the way the mobile and embedded systems and software get used. Use can be defined as a sequence of activities over time, like a play, scenario, story, or tour, and not a single input or activity. Attack 20 can be thought of as grouping smaller attacks into what some authors call a story, scenario, or tour. Attack 20 targets bugs that can be found in a sequence of events and interactions over time, but not as long a time as in Attack 7. Attack 20 is related to Attacks 16 and 33. Information from Attacks 16, 20, and others can be combined and grouped to help create Attack 20.

You may have normal day-in-the-life use, "non-happy day" sequences, blizzards with ice, trips around the country—around the world or even to outer space. The Agile test world will tend to call these "user facing stories." Other testers may call this a day in the life, operations concepts, tours, or use cases, yet much of the historic test literature refers to these as test scenarios. More recently, the idea of standardized tours such as common user scenarios, identified with memorable names, has been advocated [9]. These have been around for years with different names and slight variations. For example, on one project, I had a test called Friday 13th (where everything goes wrong), main street (the normal path), normal full mission (most common expected usage), and other descriptive names for these tests. There are differences between many of the names and concepts, but the basic idea of an attack based on how the system or software gets used is the key.

For mobile and embedded software, these attacks will all have sequences of inputs, varying conditions or parameters, and plays on different sequences of events, which a user would do. Someone once said that all plays and movies are based on some work by Shakespeare, but we do not think that "old Will" would have thought about software, how it acts, or how things like smart cell phones would be used in his plays. So in Attack 20, testers must think carefully about time and sequencing of events to create stories or scenarios. If the stories are "common," can be grouped, and reused, they may become tours.

When to Apply This Attack

Apply Attacks 20, 7, 16, and 33 when time interacts with the software, events, inputs, and outputs. Time, embedded software functions, and the interplay of these are practically infinite. The likelihood of missing some sequence or combination is great, so you will need to consider risks and criticality in determining the specifics of the attack tour. The following is a checklist of things to vary or change when designing Attack 20.

- Order—software or input or output events used in a different and/or wrong order.
- Too long—software operations that took too long to complete, missing timing or tasking synchronization.
- Too fast—software operations that completed too quickly, missing timing or tasking synchronization.
- Not at right time mark or point—the story lacks synchronization.
- Late—a software operation occurs late in a sequence.
- Late or early—in parallel, something is late while something else is early.
- Early—a software operation occurred too early in a system sequence.
- Deadlocked (caused by a race condition)—more than one mutually dependent software and/or hardware operation cannot proceed.
- Extra input or output events—too many of something happening.
- Missing events—the event just does not happen.
- Wrong input/output within events—an event happens, but it is the wrong input or output.

> **Takeaway note:** Use this as a checklist to ensure that your attack does not leave out something during the series of attacks in this area. This checklist can be structured into individual one-time attacks or common tours, which are repeated with varying conditions.

What Faults Make This Attack Successful?

Time and combinations of events or functions over time are basically infinite, which means bugs can hide for many iterations, only to appear at just the right interplay of time and situations. The subtle interplay of time, events, and embedded software functions make these attacks work.

> **Takeaway note:** Time considerations grow with experience and contextual domain knowledge.

The unlikely and infrequently exercised cases are where time bugs hide. However, the same thing that makes the faults possible makes finding them in testing very difficult. This is why I recommend the idea of tours and "focused" attacks in combination with the idea of risk—or what can go wrong that will get you and your system on the nightly news. The ever shorter and more Agile development of mobile and embedded software often necessitates testers thinking more about this kind of attack and working with the team to make the time bugs happen.

Who Conducts This Attack?

Attack 20 is a test team attack. Test teams should include staff that understand the "big picture" of function and time, as well as how the system is most likely (or unlikely) to be used. This means that testers must think like users. This will likely require user understanding, hardware knowledge (what can go wrong in a sequence), software insight, and systems thinking. Testers who can ask "how will the software be used, what series of things can go wrong, and how bad can the story get" are valuable. Often you may have a former user of the system or someone with domain knowledge on the test team, but ask yourself, "How many users do we have and how can the test team 'act' like them?" This is not easy.

Note: A useful and Agile concept is for the tester to sit with the user(s) to understand what and how the software is being used. This can be applied to defining the test cases (as in Agile) and getting a tester to "think" like a user (can be easier to say than do).

Where to Conduct This Attack?

Attack 20 must be applied at a systems level with the full complement of hardware, software, and user operations. Testing can be run in the lab or on the full system in the field. Testing in the lab must be done with care so that the interplay of hardware, software, and operations is realistic. Once outside of the lab on the real system, the testing should continue.

An attack that yields a time sequence bug will need to be replicated, either in the lab or on a system in the field. Recreating failures in the field on a "real" system or even in a lab can be difficult. In this and the previous timing attack, instrumentation and test configurations can aid in testing and debugging, but can also introduce impacts into the system in the form of added time or side effects to the software you are testing. One approach is to have two configurations for testing attacks, one "pure" and one instrumented, running tests on both systems. This approach may double some efforts, so use it judiciously. However, it can remove the side effects from special test configurations during tour attacks, while allowing repeatability and testing insight.

Finally, complete control and repeatability in the attack are impossible in test environments and can result in what some call "unverified" failures. An unverified failure is where something bad "happens" but then cannot be repeated. These can become "ghosts in the machine." In some mobile and embedded environments, they may not be a big deal, while in others they may be of a higher concern, such as in man-rated, high-value, or life-critical environments. The question has been asked, "Are some of the problems seen in recent years in medical, automotive, and aerospace the result of unverified failures?" Many experts speculate that this might be the case, but the nondeterministic nature of embedded software in sequences, tours, and some systems makes the final verdict a matter of opinion. So the "where" of Attack 20 needs to be considered and structured carefully.

How to Determine If the Attack Exposes Failures

Look for bugs where behavior fails to match reasonable expectations or outright fails to meet requirements. Timing errors can first appear as rather subtle delays or other "glitches" that at first do not look or seem important. The system works but is perhaps a little slow. However, under certain conditions, the small bug becomes a big bug. So watch for the "not so obvious." Also tours or sequences of attacks expose bugs of interaction between component hardware or software and user actions. Here again, the test is looking for something that just does not seem right or does not meet timing requirements. Testers need to be careful about reporting bugs that "do not seem right" since developers will come back with a "not in the requirements" comment. They may also say this if the requirements are vague or they may just say "it is the hardware, so users must live with it." Testers must be prepared to communicate specific evidence as well as user expectations on the timing bug legitimacy.

How to Conduct This Attack

To start Attack 20, you must understand what the system does or is supposed to do in relation to time as well as some sequence of events or functions. As the attacker, you should collect system concepts of operations, user guides, use cases, models, and any other information that will detail functions and usage over time. From these, organize a sequence or set of sequences. Next, follow these steps to create a series of attacks:

1. Focus first on a normal situation.

2. Consider the off-normal nonfailure modes, including expected error conditions that should happen based on poor user behavior.

3. Look for the failure modes and effects. Does the software recover well?

4. Review and understand system errors and failure history from the field.

Additionally, it can be useful to have maps of the operations in time, which can be constructed using spreadsheets, flow charts, sequence diagrams (if they exist), state maps, existing test log files, or other methods to capture ideas. *Warning*: log files can contain large amounts of detailed data, and this can also adversely affect the performance (especially timing) of the software.

This information needs to be organized, reviewed, prioritized, and maintained for future timing attacks. Historical records can be a key component for analyzing timing attack results. Some teams in critical applications conduct review activities such as expert judgments, risk analysis or failure mode, effects, and criticality analysis (FMECA) [10] with a team staffed by hardware, systems engineering, operations and users, software, and/or test personnel. No matter the approach used, you should attack first, "whatever is most scary or of higher risk."

Using this accumulated knowledge, the team can build test scenarios for each targeted priority. Experienced teams can define more things to test with this attack than they have resources to accomplish. It is advisable to seek outside input from customers or users asking, "Do we have the right priorities?" This could mean that some priorities will change. It is also useful if the attack and/or tours have meaningful names such as "Full Mission Friday the 13th" or "flight control check."

Friday the 13th test name is an instance of a "soap opera test scenario." Soap opera tests in mobile and embedded systems have proven very useful in finding bugs, stressing timing, and providing useful information to the team. The idea of a soap opera scenario is to stack realistic activities condensed in time to stress the software system. The name was coined by Hans Buwalda [11] and comes from the TV shows. I have seen the basic idea of a soap opera attack as a specialized scenario test over and over in the embedded industry, although it was often called something else.

At this point, the creation of attack and tour details can happen. If the early information was very detailed such as in modeling, combinations, sequence tables, or FMECA, then little additional work will be needed. The attack or tour almost "flows" from these detailed analyses. If a less rigorous method was used, the attack may need to define the detailed sequence, often looking at each step and asking "What options exist now that I should do?" Once you have finalized the scenario or tour, run the attack. During and after the attacks, analyze the results looking for bugs and dependencies. Repeat with different attacks in this area as time and budget allow.

Note: Some care must be taken that this attack does not devolve into mindless scripted testing, unless that is your intention. If a script is desired for some checking function, such as requirements closure or "formal" testing for delivery, attacks from this area can serve as a good starting point. The real power of tours and attacks is their flexibility. Many embedded systems have historic libraries of attack tours in this area. They never run exactly the same "test," but they do follow the "theme" of the tour or attack. Mindlessly following a long written test script for a scenario is not a highly recommended practice.

ATTACK 21: PERFORMANCE TESTING INTRODUCTION

☐ Intermediate / ◇ Advanced

Performance testing is a fairly large topic with a large body of literature and training for information technology (IT) and personal computer (PC) systems [7,12]. I have leveraged some of Barber's and Wilson's thinking in Attack 21 and added other ideas to target the world of performance testing on mobile and embedded devices. Attack 21 is longer than most of the attacks in this book, but performance testing is quite a large and complex subject and is only a beginning.

Mobile and embedded devices often have critical performance concerns, because as covered earlier, they are typically time-processor resource constrained. Attack 21 provides a generic framework and starting point for conducting a performance assessment, as well as pointers to additional information. Testers should view performance assessment as a critical data point for the development team. The analysis should be done early and then repeated with test data points over the product cycle, as well as before delivery of a product to any customer to avoid performance surprises.

Takeaway note: Performance testing is critical. I have given references on a lot of the common thinking, some background materials, reference, and starting points, but most attackers will need to study Barber [12] and other references to become really good mobile and embedded performance testers.

My experience is that it is better to know your performance numbers and limitations from day one (or as early as possible) because these will drive architecture, functionality, design, and implementation in the embedded system. Performance testing for many mobile and embedded test teams becomes a daily focus as the project tries to fit the infamous "5 pounds of processing into a 1 pound sack." Performance analysis and attack test data can feed into the whole project to help it succeed. Often performance "curves" look like those illustrated in Figure 6.2. They start fine and continue looking good, until the system hits the performance issue (slowing-to-bad), where response times slow to a crawl. Unfortunately, waiting to discover performance issues late in the testing cycle can be costly to fix and backtrack. Further, waiting for the hardware to improve (often an option in the IT world by just buying a faster processor), but this may not be possible on your embedded release schedule. In any case, variants of Attack 21 should be run on the completed system to get the performance numbers for the as-built configuration.

In Attack 21, you are trying to avoid the following:

- Added cost of development from later rework caused by performance problems
- Delayed product launch of the software due to performance issues
- Software crashes caused by performance issues
- Slow performance with frustrated users (resulting in poor sales)
- Impacts to hardware or having to add hardware to solve performance issues

Finally, in this attack, remember that mobile app users have historic expectations of performance based on their IT, PC, and web use. Embedded devices may have less raw

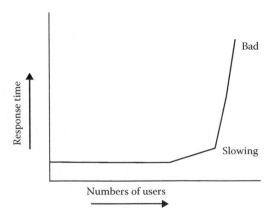

FIGURE 6.2   Performance curve example.

performance power (e.g., CPU speed) than IT systems, and users in part understand this situation, but programs should not rely on the lessened expectations of the mobile or smart user for project success. Attack performance!

When to Apply This Attack

Apply Attack 21 any time the development team determines performance is of possible concern—which may be always. The greatest functionality in a mobile smart device or game is pretty much useless if a player must wait too long for it to work. Yes, the speed of mobile and embedded processes continues to grow, but one must live within today's common performance limitations. This means that the team must know system performance capability numbers, and measures of performance for the app relative to these numbers, budgets, and future projections (see "What Faults Make This Attack Successful" in this attack) and so forth early in development. As soon as some software or even architecture and design elements become available, the team should start developing a few performance numbers at least as "goals." This may be done with budget estimate analysis, prototype assessments, models, and/or simulations.

If early analysis and testing shows a performance risk, many mobile and embedded teams continue performance testing with each new increment of the app or software, getting better data points while development does whatever they can to keep the system within allocated budgets. If no early performance issues are measured at the beginning, performance tests may be run less frequently. But watch waiting until the end to do the only performance testing since, historically, software expands to fill the time performance budgets it has and then fixes performance bugs at the end—costing money and schedule that the project may not have.

Major life cycle points to consider for performance testing include the following:

- Pre hardware and/or software development—Do analysis, set budgets, and/or run simulations or models.

- Prototype (s) or versions of hardware and software—Check and analyze software and hardware prototypes during the iterations and increments of development (do not wait).

- Qualification versions of the system—Nearly final and completed versions of the hardware and software where testing is still ongoing, but not done.

- Final version of the hardware and software—Assess the system as realistically as possible in the lab, if time allows and performance is a concern.

- New versions and maintenance releases of the system—Each upgrade of software and hardware may need performance checks, because as software ages, it tends to become bloated and slow.

Then, there is attacking performance when the product goes into the field. Here in the mobile and embedded world, we are looking for the performance items that may be hard or impossible to fully get a handle on in the lab. Not every group will have risks in performance to justify field performance testing, but many of us do—such as automotive, avionics, and factories, to name a few. When doing "in the field performance" testing, look to "stress" the following:

- Network issues such as slow, drops, different classes, or vendor networks;

- Geographic boundaries and conditions such as fringe conditions, weather, external disturbances;

- Different configurations of hardware and software such as what is loaded and used on the device;

- Variations in vendor and/or providers of service; and

- Different users (some users are fast and some are slow).

Here again, you want real, boundary, and even extreme conditions to get a variety of data points. Never say "that will never happen" (to get out of doing a test) because yes, it very well may.

Who Conducts This Attack?

Attack 21 is conducted by the full team including analysts (system engineers), testers, and, as needed, developers. The analysts and developers usually determine the estimates and budgets (theoretical numbers). Having a skilled and experienced performance tester on your team is optimal. This tester must provide the "actual" information, such as test results with measured performance over a variety of situations. The tester needs to understand time, the system, and performance target numbers, as well as how the system will get used and abused. Performance testing skills are usually developed over time since tools, analysis, and some level of creativity are all needed. A tester, who just runs the functional tests while checking CPU usage, will likely miss performance issues. The support staff will need to know timing analysis methods, how the system is likely to be used and stressed, and aspects of hardware to support the performance testing. Performance attack testers require critical thinking, specialized performance test knowledge, experience, and lots of creativity.

Where Is the Test Conducted?

Often, the performance tests (Attack 21) start in analytic efforts, continues into the lab, and then moves to the field.

The first place this effort can start is in the virtual world of computers. The team generates estimates prepared in simple tools such as spreadsheets or databases to track budget numbers and make allocations of time to system elements. Factors to consider are CPU speed, lines of code and estimation time, lags, hardware processing time, time spent in "other software" such as the OS, other apps, and dependencies (synchronized versus asynchronous behavior, user actions). The more complex a system, the more analysis and factors the budget will need. Also at this point, many systems use rules of thumb such as "be under an estimated 40% CPU usage at design time" to allow for the likely growth of software or processing. Such rules of thumb can be checkpoints at many life cycle stages. Keep in mind that there is no "right" number or point in time to make the check. But again, waiting until you are past 100% CPU usage after the final testing has started is "too late." This first step of analysis may not be done by a tester, but by an analyst.

Next more sophisticated analysis is built on the budgets. Analysis systems and tools exist to support timing studies. Tools to use in early timing analysis include the following:

- RMA [1]

- Simulation and models [13]

- Prototypes (running timing attacks on a prototype)

These tools tend to be customized, and typically, a tester is not using them exclusively, but helping the team with the analysis. There are books and instructions on how to use these timing analysis tools. These efforts form the basis of more detailed performance budgets and tracking going forward into software construction. Not every project will need every form of analysis. The critical projects may use more, but even a basic mobile app with performance concerns should probably have some budgeting of performance numbers, which someone can document in a spreadsheet and post on an informal network to remind everyone of the targets.

As the project moves its performance analysis from the "theoretical" estimates to actually running some tests, maybe on prototypes or small slices of the system, the effort moves into a test lab with performance attacks. Testers will be doing this work. If performance testing is needed, the lab needs the right support tooling, which could include the following:

- Ability to measure and monitor performance such as time—total, external, internal, and between sets of usage points;

- Ability to stress the software and system;

- Network traffic;

- Internal usage—apps that are running;

- Memory usage;

- Multiple external users and interactions;

- Network connections;

- Signal strength;

- Traffic (real or simulated);

- Numbers of users (virtual).

In the lab, we must be able to vary these items and configurations under different situations. We are looking for key factors and/or bottlenecks.

During the testing, the team may want to ask and answer the following device questions:

- What is the app doing and what can it and/or will it do?

- What is the app using (internal or external resources)?

- What else is likely to be running such as other apps, hardware, systems, and so forth, with consideration for what the minimum, normal, and maximum variations will be?

- What configuration of hardware needs to be evaluated?

- What network and/or connections or usage need to be checked (low, normal, and high)?

- For mobile devices, what are the signals and variations that drop in or out, interference or noise, and timing (most web and IT systems do not worry about these)?

- What memory use configurations need to be stressed (low, med, and high)?

- What does different hardware and network configurations do to our app's performance?

Lab performance analysis and testing may be repeated many times before a product is released to the field.

Prototypes or lab configurations usually are only parts of the system, sometimes called threads of integration. A thread can start as just the OS, generic support software, or a small part of an app with a major feature. For example, I once tested a system with only the OS, engine start, and the first 30 s of use. This thread was enough to tell the team that performance issues were going to exist.

Simple starts, using an integration thread, can be run early and find bugs, as well as provide some "real" numbers for such things as timing analysis. Threads constitute a slice of the system that can be measured and are usually just normal time flow cases at first. As features are added and threads expanded, performance numbers can be refined. The performance numbers obtained can be used as factors to project the future using the budget and estimating tools. The early performance estimates will be subject to error and debate, but the numbers are something to base engineering on—and much better than wild guesses. Measured numbers from a thread can be combined with performance analysis, modeling, or simulations to provide more complete estimates, which get refined over time.

Once a completed system is reached, more complete lab performance measures can be made. The numbers should include the normal, edge, stress, and max cases. These cases may need to be run over a variety of configurations and options. The numbers can be fed back into the earlier analysis methods to confirm and/or refine and will provide insight and clear status points before going to the field.

When possible, going into the field with the real device will be an end of project set of performance tests. Hopefully, if the performance testing run has been sufficient and budgets have been met, then there will not be any big surprises. If field numbers are surprising, added development to fix performance issues may be needed. Field testing must also be run with a variety of environments, configuration, and situations. There should not be a single user or field environment. Obtain enough data points in labs and field usage to spot trends, which may point to additional attacks. In cases where field performance testing is the first performance attack being done, be ready for "surprising" data points. When performing the testing on the completed system in the field or lab, consider the following:

- Load testing from inside the system (software)

- Load testing with the system (software and hardware)

- Load testing with the whole (system, software, user, and environment)

Load testing will be further refined in the following section. I have seen systems in the field running at 60% performance use, which had desired margins, but I have also seen systems running at 90+% CPU usage when delivered. Such a number can be scary. Again, hopefully, you have not waited until the end to know and address these kinds of issues.

What Faults Make This Attack Successful?

Attack 21 is successful because software keeps growing, and often the development environments are nice and "clean," which means things run smoothly on the programmer's computer and device, but not in the real world—meaning performance issues still exist. This happens as the developers keep adding "cool stuff" and forget that a user may be running other apps, software, or the system is otherwise "stressed." All of these things can conspire to slow the processing performance. The real world is never as nice as the development world.

The performance test team needs to really get into a "break it" mindset in performance analysis and testing. Typically, developers are focused on functionality and programming and do not worry about performance until it becomes a problem. Here are some assumptions that teams make, which create bugs.

1. We can add a faster hardware processor or more memory to solve performance (an old IT or PC leftover).

2. We can easily reduce functionality.

3. The added functionality does not take that long to run.

4. We can quickly refactor the software whenever we need to.

5. We can make the user "live with slowness" (think about the wait clock or using old values).

6. It will never take that long (stress cases).

7. The memory or network will never be used that much.

8. We will never see that many users.

9. There will never be an interdependency between apps as users will not run the device with those pieces of software.

In the embedded world, the first item (provided in this list) is often not possible within schedule, and customers hate the second item. I have seen teams that midway into the development had to revisit the software architecture or design (see items 3 and/or 4 in the list). All of these items can be costly in many ways. Projects can even get canceled because of performance issues.

Additionally, many testers want to treat mobile apps and systems the same as historic web and IT systems. Some feel that if the system (mobile–app–web–server) was sized for an additional load (say 20%), then adding mobile apps will be no big deal ("it's just a few new users"). But mobile apps add new behavior to the full overall system (an often overlooked problem) and programs must account for this aspect, or you will have bugs, including the following:

- Staying connected longer, having slower data rates, or both.

- Cell devices drop connections frequently, leaving things "hanging" (processes stalled).

- Many device configurations.

- Mobile and historic IT browsers all can be used at the same time.

- Network configurations and speeds.

- Loads and requests from many locations and configurations.

Ask these questions: "Has the testing that has been done or is being planned considered these new performance-related factors?" and "What possible performance bugs could be hiding in what places?" "Is your app and system valid under combinations of these performance attack scenarios?" If you have an existing system with previous tests and you are adding mobile apps, "What performance attack testing in the new environment is needed?"

Finally, as examples, many performance parameters can indicate problems (bugs) and so may need to be analyzed, including the following:

- CPU usage in short (milliseconds) time frame

- Peak CPU usage

- CPU usage under system load

- Systems loads can include the following

   Hardware devices being busy, hardware devices being actively used, hardware and software task being active

   User interactions, short spikes in hardware or software usage; timing during fault or error processing; memory usage/access timing

   Starvation of background or low-priority activities because of timing

   Timing of a closed-loop control cycle (how much margin is there)

   Slow and fast hardware response time

   Network timing issues caused by slow or intermittent connections; slow timing based on device location conditions

Slow timing caused by different device configuration (many kinds of mobile hardware platforms)

How long it takes for the data to be transferred from one device to another device

How long it takes for the device to conduct a search of its internal data

And many others

Testers working in mobile and embedded must keep in mind that these devices are different in many aspects of timing compared to PC or web performance testing and parameters.

How to Determine If This Attack Exposes Failures

The following errors are indications of timing bugs. Here the team is looking for the following:

- Measures of performance goals that are not met
- The system runs slow (this can have different definitions, so you need measures)
- A deadline is missed (and bad things happen or good things are missed)
- Functionality that is wrong because of timing

This is not functional testing directly, but you should be looking for nonfunctional elements, which may impact functionality. Users hate to wait (too long without information or feedback) or not have "their expected time" met. To understand performance timing failures, it is important to understand some basic timing performance concepts. And since many mobile and embedded devices are "controlling" something, the misunderstood and overloaded concept of "real time" in performance must be considered for failures in this attack. I offer some basic timing concepts in Appendix D.

After reading through Appendix D and even some of the references I have provided, you may still be unsure of timing concepts. To refine your understanding, even if you do not think you are confused, review the examples in this chapter; do the exercises; start testing in the areas of real-time mobile and embedded performance, and then come back and review these paragraphs and some of the reference materials again. Time is a funny thing since it is a dimension of the universe that we as humans live within but maybe never really understand.

Takeaway note: While this section is long, may sound repetitive at points, and may be hard to understand, it is only a beginning compared to what a real embedded performance or time tester understands. Read over this material several times. Do some performance testing and then read some more. The more you understand, the more effective you will be when performing timing attacks.

How to Conduct This Attack

I will not go into the analysis side of performance assessment, design, and allocation approaches. These are out of scope for this book and the attacks, but likely once you get into this attack and performance testing, you will want to visit other sources and consider more analysis. I will define how the test team supports some of this analysis and goes about supporting the basics of performance testing, but really it is the start of a very long journey.

First, determine the measure of performance (number) "goals" or requirements. Some mobile and embedded systems may actually have published performance requirements. If so, this is the starting point for you. Next, in most cases, the team will need to add in more performance goals. One rule of thumb I have seen is to test two or three times beyond the stated requirements or goals. This is at best a crude heuristic, but it adds to the starting point.

Often performance numbers are not given as written requirements, but must be "derived," stated as goals, or documented as best the performance team can. Here you will have to do a little digging to find numbers and goals. How do you know the meaning of too slow or too fast?

Associated with these numbers, there may be trigger thresholds used when a goal or number is exceeded (or approached) during the development and/or testing. When a threshold is met, the actions to take will be project and context unique. Triggers may lead to such things as "retry messages" or "use historic data." More than likely, these numbers will unfold and change as the performance analysis on the project progresses. While not done by the testers alone, testers should be aware of and will likely support the performance analysis. Being involved and communicating during this analysis help you to understand the numbers and the testing situations. These performance numbers serve as the basis for the attack and so must be documented at least within the testing information. In such information, you may have the details of "must meet" (requirements), the desires (goals), threshold values, extreme cut-off points, and maximum cases. These data may be the desired information within the development team.

With the performance numbers, come the conditions associated with generating the numbers. Conditions will be factors such as the following:

- CPU load or traffic

- Memory usage

- Network factors including size, traffic, priorities, and others

- Signal connection characteristics including strength, dropouts, dropdowns, and others

- Concurrent apps and programs in use

- Device conditions, batteries, CPU type, memory size, internal clocks, and external environment factors

- Data characteristics including size, type, checking, use algorithms, and so on

- Security factors including ongoing checks, screening, authentication, and others
- Hardware characteristics including limitations, size, speed, and so forth

Now, if you are getting the feeling that the performance set-up, numbers, and conditions are large and complex, you are beginning to understand how hard and how much work performance testing and analysis in mobile and embedded device can be.

Next, the team needs to determine where the attack will be done, what capabilities exist in the test environment, and if any special configurations need to be developed. If the tools and the environment exist, continue with the next steps. Often the test environment will not be ready or will not have the needed support for measuring. Here you may need to go off and do some development of your own, build hardware, get software tools, configure the lab, and even to the point of working with development to put "special" software into the application under test. Many of these approaches, which can work, have dangers of side effects, impacting the timing you are trying to measure and impacting cost and schedule. (Don't work in a vacuum—off by yourself.) For example, added tests, system, instrumentation hardware, and/ or software can impact the things you are trying to measure by adding to the clock cycles. Critical thinking on impacts and risks is required. For example, I once did unit testing to measure time of individual units on a part of a RMA, but to assess the timing and functional impacts, I did the test with the instrumentation and then reran the tests without the instrumentation. The instrumentation gave functional test data (coverage and results), and the noninstrumented run was timed. Both sets of data were important. The testing was highly automated so this was just computer time, and it was not a big deal, but it did take time and effort to set up. Once completed, the timing data was used in an RMA compilation estimate.

Further, to measure internal performance, there are tools that can probe through hardware and the internal CPU execution time, and are available from some hardware vendors. These tools can measure time from a set point to an end point, measure total time, trigger on events, and peer into memory or CPU usage. Users can also build a custom timing framework. As an example, on one project, we needed to customize the tool to the processor we were using, modifying hardware to support this, plus build some software scripting. This added time and the effort was well worthwhile, since once we were done, we could monitor time and the insides of the embedded computer. This supported both the performance attacks and debugging. The test configuration was "special," so we also had an unmodified test system for use in functional testing. However, during timing attacks, this special configuration was the way to go. It did cost time within the schedule to do this, but these efforts were justified given performance concerns.

After a test environment is determined, the tests themselves need to be considered. Do you start with normal usage of the system and stop there? Do you do performance stress cases? How many performance test points are needed to get good trend information (such as with graphs)? What kinds of models (prototype, simulation, time model) are needed? Often most systems have good performance on the good days, but on "rainy days" (high CPU usage, slow network, many users, low memory, or other conditions), performance suffers. So you should look for performance tests under load, stress, peak, beyond peak, negative, and max numbers. These "conditions" should already have been outlined during analysis and will get refined during this testing.

Consider attacking different scenarios and conditions, including the following:

- Normal usage: day-to-day use, typical operational profiles, what is likely to be seen?

- Longest path configuration: find the longest execution path in the code. Often this path is executed at unit levels and then added up to complete a system analysis.

- Load: peak expected usage points: "what are the system's goals, where is the system running apps, what is a good network connection," and so on. These data points will often be requirements or published measurements.

- Load: stress beyond peak or normal usage: number of apps, slow communications, off-normal configurations [see Scott Barber's website: http://www.perftestplus.com/], system noise, and long responses on server side.

- The breaking point: where does the performance response of the system become unacceptable?

You should be looking for a variety of numbers here in these configurations. Figure 6.3 shows an example of a small series of these scenarios and conditions, as well as how they might be stacked on each other. Often the numbers look good early on and during "normal" usage, but then bad trends start showing up in "worse" situations. Watch for the curve of Figure 6.2 and performance assumptions going into it. The nonlinear nature of performance means that the change point is hard to predict and the assumptions that drive it are even harder to know. The performance use scenarios need diversity. To define performance borders on art, so a tester will have to be a bit creative. The scripts and scenarios will need to be reviewed by the whole team, where you can make a case for doing the full gamut of testing, from positives (happy day) to negatives (rainy day) but bounding all of these factors within cost and schedule.

Also consider these scenarios when available resources (such as memory, processor, disk) are reduced to low values, to evaluate how they behave under these conditions. And at this point, a long-duration run (see Attack 6) can provide interesting performance data. Finally, volume and growth capacity performance runs may be needed for some systems. Volume testing attacks the performance of the test software or device when it is processing specified levels of incoming, network, internal, or output data. Growth capacity testing attacks the test item under conditions that may need to be supported in the future such as upgrades, new networks, changes in the cloud, hardware, software, or other factors.

Once the analysis has the needed parts of the information outlined in this section, a performance attack (really a series of attacks) can take place. As each attack is run, look for the bugs, record the performance data, and update the analysis. As you get data points, graph them. You are trying to understand the performance graph (something that will look like Figure 6.2) for your system at this point in the attack. Is your expected usage close to the knee of the curve or far from it? What assumptions did you make on the set-up? What

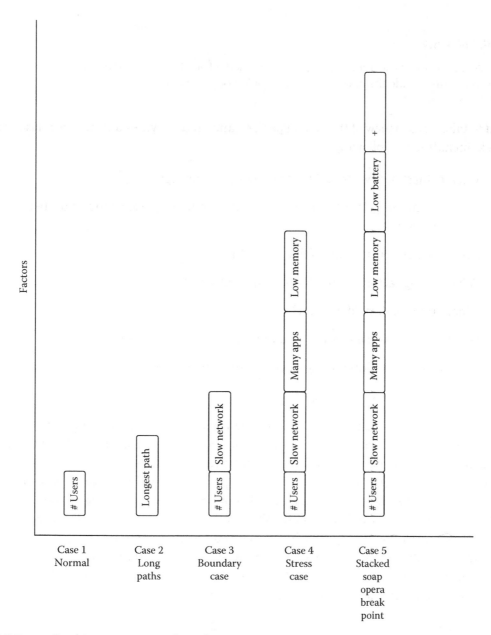

FIGURE 6.3   Stacking scenarios and conditions.

happens if your app becomes wildly successful? The performance tester and analysts must balance the "simple world" with the totally "stressed world." These graphs and assumptions will need to be reviewed with the team. There is no "right" answer here since everyone will be guessing, but the better the performance attacks and data, the more likely you will be to avoid performance failures.

The performance attacks give you baseline data points, but they will evolve as you learn more and the system matures. Learning should be a team effort, but often the performance testers must drive analysis since the development team will focus on functionality. Testers

---

SIDEBAR 6.4

In 2012, millions of smartphone users downloaded a Lady Gaga hit and caused performance problems that made the news [14]. What would your system do?

---

need to take a broad view of the system performance and activities during this phase of the attack, including the following:

- What things must work and how are they performing?

- Is there anything legally or contractually binding—performance numbers, standards, or regulations?

- Are there performance items that can kill?

- What timing factors are in play (see Appendix D)?

- What do the stakeholders want?

- What performance issues can cost the project money or sales?

- What timing conditions exist when our software is running?

- Risk—what scares you at night?

Performance testing is an art using some science and heuristics. Having a nice "clean" system to do this attack is a mistake, because the real world is messy. But putting too many items into the attack may not be realistic. A balance is needed.

Iterations, increments, refinement, and past app history can influence the performance analysis testing, but the mobile and embedded online world is changing as fast as the PC and network world once did (and does). What can a user do in parallel with the app under test? Is the user doing messaging while downloading something, listening to tunes, and checking the web at the same time? How happy or unhappy will this user be?

In addition to what we have covered during this attack, there are concepts that a project can leverage in timing attacks including the following:

- Get some outside help. Maybe hire some outside "crowd" testers to attack the system out in the field such as uTest or an IV&V team.

- Go to the cloud. Use the cloud to create a "virtual" system.

- Go into the field with the test and do a test drive or flight.

Certainly, development should care about timing performance results, but sometimes they will want to deny by saying "Well, we'd never have that many (fill in the blank) users, apps, communications, and so forth." Next management and stakeholders should probably see and understand the performance data. It may be hard for them to understand, so ask management some context-free questions to aid their understanding such as the following:

- Will unhappy users leave the app because of missed deadlines?

- What performance data do you understand (or not)?

- How long is too long?

The team, including an informed manager, needs to decide the correct answers to these questions. Many times you may ship and fix performance bugs "later." "Time to market" and "good enough" have made many companies successful even with performance issues. It is not the tester's job to make the call on performance being "good enough," but you can provide the data and reports and help management to understand the data (curves) and the implications of the data.

You will probably iterate this performance attack from the beginning of development into maintenance. Many test teams have performance curves covering many years and versions of hardware and/or software. There have also been teams who stopped performance attacks only to be rewarded with surprisingly slow performance. I have recorded embedded taxonomy performance bugs that appear years after the software went into maintenance mode.

SUPPORTING CONCEPTS

Now, if all of this is not complex enough, there are many supporting concepts of which the performance attack tester should probably be aware. I have already mentioned a few of these in examples and text thus far. Examples are not "recommendations," but are used to illustrate that this attack benefits from tools. Here, I list some concepts to consider applying when the context is right:

- State transition testing

- Use case and scenarios

- Historic data and usage from industry or similar past projects

- Industry data benchmarks—Analysis can be spreadsheet based using vendor and industry data, plus local estimates

- Performance attacks supported by hardware and software tools, both commercial and open source, including the following:

  Load testing tools—There are many vendors for both hardware and software, including tools specifically for mobile and embedded devices,

  RMA tools,

  Built-in features (timing logic) of the hardware and/or software to provide timing data or influence timing such as to stimulate a fault, which "slows" the system,

  Timing data recorders, and

  Graphical aides such as Excel.

COMPLETING AND REPORTING THE PERFORMANCE ATTACK

After you apply all the tools, run through some steps, get some data, and have information about performance, how do you present the data; whom do you report it to; and what comes next? This is the reporting phase. But first, organize the data (possibly) into charts and graphs. Pictures are usually easier for people to understand. It certainly is nice if you can see at this performance point that the system crashed, so you have bug. But often, this will not be the case. You will see a trend of slowing response, delays, warning messages, and other indications that processing is taking longer. Spending time to have good reporting and data in a clear format is worth the effort.

WRAPPING UP

This was a large and complex attack, but even so, it is only a start on this type of testing. During the attack and after, someone needs to look at the performance data. The obvious failures will likely get fixed, but many performance issues still hide. Issues can be seen in a trend or strange little "thing" in functionality. They may simply be a missed condition. Here is where some experience and skill in performance testing can yield the hunch to an expanded interesting case that exposes a really big issue. Somebody—maybe a tester— should have the charge of watching the data as it evolves.

If you are working in mobile and embedded software and/or systems as a tester and performance ends up being your task, I recommend more work and reading. The performance attacks of this chapter are only an introduction. Testers can spend their whole lives getting good with performance testing.

You want your app and/or software to make "good" history; be in demand; get lots of users; and make the blog hype. I recommend that you know the performance numbers, the assumptions, and attacks. After all, knowledge is power.

EXERCISES (ANSWERS ARE ON MY WEBSITE)

1. You are trying to test interrupts in the following situations. Define how realistic it might be to test this way for each case that follows:

   a. Interrupt time test measures are accomplished by a human for a real-time system running audio controls for a car.

   b. Interrupt time is analyzed by automated comparison to a closed-loop simulation, which is running at the same frequency as the processor running a car's braking system.

   c. Interrupt time is analyzed by automated comparison to a closed-loop simulation, which is running at two times the frequency of the processor running a car's braking system.

2. Define three possible time factor categories within a smart cell phone.

3. Define which performance criteria are important for a

- Smartphone

- Car's antilock braking system

4. You are performance testing a new smart cell phone. What time-related tools would you consider and why?

5. Answer "device questions" from Attack 21 for a medical smartphone device. This device is a smartphone that monitors via a Bluetooth device (a sensor) a patient's heart rate and sends hourly text messages to the doctor.

6. For the device in question 5, list the factors that might impact the software performance.

REFERENCES

1. Lui, S., Klein, M., and Goodenough, J.B. 1991. *Rate Monotonic Analysis for Real–Time Systems*, CMU/SEI-91-TR-006, CMU tech reports.
2. Interrupt. http://en.wikipedia.org/wiki/Interrupt (last accessed April 9, 2013).
3. Klien, M., Ralya, T., Pollak, B., Obenza, R., and Harbour, M. 1993. *A Practitioner's Handbook for Real–Time Analysis: Guide to Rate Monotonic Analysis for Real–Time Systems,* Kluwer Academic Publishers, Norwell, MA.
4. Copeland, L. 2003. *A Practitioner's Guide to Software Test Design*, Artech House Publishers, Boston, MA.
5. Defense Industry Daily. 2007. F-22 Squadron shot down by the international date line. http://www.defenseindustrydaily.com/f22-squadron-shot-down-by-the-international-date-line-03087/ (last accessed April 9, 2013).
6. STS-1. http://en.wikipedia.org/wiki/STS-1 (last accessed April 9, 2013).
7. Wilson, T. 2011. *Proformorama*, a self-published e-book, general and chapter 11.
8. Cheng, A.M.K. 2002. *Real Time Systems: Scheduling, Analysis, and Verification*, John Wiley & Sons, New York.
9. Whittaker, J. 2009. *Exploratory Software Testing*, Addison-Wesley, Boston, MA.
10. Failure mode, effects, and criticality analysis. http://en.wikipedia.org/wiki/Failure_mode_effects_and_criticality_analysis (last accessed April 9, 2013).
11. Scenario testing. http://en.wikipedia.org/wiki/Scenario_testing (last accessed April 9, 2013).
12. Barber, S. *Web Load Testing for Dummies* (free e-book). http://www.gomez.com/ebook-web-load-testing-for-dummies-generic/ (last accessed April 9, 2013).
13. Meier, J.D.,Vasireddy, S., Babbar, A., Mariani, R. and Mackman, A. 2004. Microsoft patterns and practices: "Improving .NET application performance and scalability. http://msdn.microsoft.com/en-us/library/ff647767.aspx (last accessed April 9, 2013).
14. Clark, J. May 24, 2011. Cloud, *ZDNET*, Lady Gaga download deal crushes Amazon servers. http://www.zdnet.com/lady-gaga-download-deal-crushes-amazon-servers-3040092870/ (last accessed April 9, 2013).

CHAPTER 7

Human User Interface Attacks

"The Limited (and Unlimited) User Interface"

Someone once stated that if the user does not know that the system they are using is computer driven, then it is an embedded software system. This is because, historically, mobile and embedded systems had very limited or simple (human) user interface (UI) or no UI at all. They did not have the computer box, keyboard, and screen. Although many embedded systems still have a limited UI, this is changing as embedded systems get more processing power and users have more sophisticated needs that systems must meet. The UI of mobile and smart devices have grown to the complexity of personal computer (PC) or information technology (IT) systems. To keep pace, testers in the mobile and embedded world must apply ever more sophisticated attacks to the human UI. Many mobile or embedded UIs can have some "scary impacts" such as missed alarms where equipment is lost, there are financial losses, or where there are human safety factors (Figure 7.1).

UIs are software driven, but how that software is implemented can drive errors of input or data values. There are basic UIs, such as indicator displays (cars), simple text (command lines or numeric inputs), simple selectors (buttons), alarms (lights or auditory), and other basic input streams to the software. More and more systems have moved to graphical user interface (GUI) displays, which have advanced screens displaying text, pictures, menus (pull-down or pop-up), hot links (web), and auditory (spoken words and hands free) inputs as well as advanced display and input technologies. The embedded error taxonomy identified the following areas associated with human UIs. Mobile-embedded UI attacks should consider these as focal areas.

1. False indicators such as on/off settings, hardware status, or other like actions. Failures caused by false indicators include operator action or no action based on this information.

2. User or operator interface (screen) displays incorrect information or direct inappropriate user action.

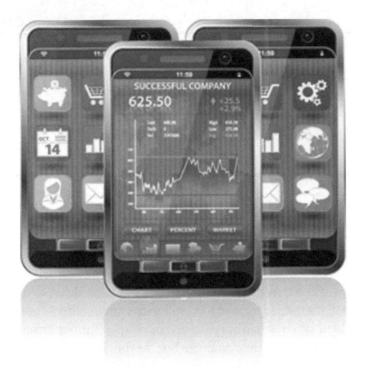

FIGURE 7.1 Example UI or GUI of mobile devices.

3. UI software did not prevent input of bad values by a human user:

a. Ultra extreme—UI did not prevent input of totally unreasonable values (see buffer overflow).

b. Data interface did not prevent unauthorized users from obtaining protected data (security or information theft).

4. Software incorrectly interprets user inputs and/or takes wrong action based on input through the UI.

5. Output overflows information and/or fails to output a value correctly because of display formats (e.g., software allows a number or text to be displayed that may not be in the correct format or correct number of characters).

6. Help files or user-related information errors—incorrect information provided to the user, causing them to make a mistake on input or slow actions resulting in time-outs or other problems.

These will be expanded as we move into the more graphically intense mobile and embedded world.

HOW TO GET STARTED—THE UI

To attack the embedded UI, start by reviewing the set of attacks well documented in Chapter 5 of Whittaker's *How to Break Software* [1]. Consider running those attacks.

TABLE 7.1 Mapping of UI Attacks from the *How to Break Software* Book to Mobile and Embedded Contexts

Attack #	Applicable Attack Name	Notes
1	Apply inputs that force all UI error messages to occur	These are error messages to a human user, which must be known and tested
4	Overflow input buffers	Identify where buffers exist and try to create inputs that exceed input buffers
5	Find inputs that may interact and test combinations of their values	Identify all human UI inputs and then use combinatorial testing (see Attack 32)
7	Force different outputs to be generated for input (variable)	Identify UI output possibilities, define the inputs that create these, and test these combinations (Note: This can be difficult for mobile and embedded if documentation is weak)
8	Force invalid outputs to be generated	Determine what the outputs are and then attempt to force invalid outputs (Note: This can be difficult in embedded systems)

Table 7.1 provides a starting map of the most applicable set of attacks from Whittaker's book. Some customization for the embedded system space is likely necessary (refer to the other tables in this chapter).

The errors of the GUI are common to both the simple and the "smart device" world. The GUI attacks in *How to Break Software* [1] apply to one degree or another here and are cross-referenced in Table 7.2.

Many of these attacks are applied as indicated in Whittaker's *How to Break Software* book. I recommend mobile and embedded testers go after the classic UI and GUI bugs first.

TABLE 7.2 Mapping of GUI Attacks from *How to Break Software* Book to Mobile and Embedded Contexts in This Book

Attack #	Applicable Attack Name	Related Attacks in This Book	Notes
10	Force the screen to refresh		Do if possible, in device
11	Apply inputs using a variety of initial conditions	Attacks 6, 7, 10, 11, and 32	Apply as in this book
12	Force a data structure to store too many or too few values		Apply per Whittaker's book but modified for mobile or embedded
13	Investigate alternate ways to modify internal data constraints		See Harty's *A Practical Guide to Testing Wireless Smartphone Applications* [2]
14	Experiment with invalid operands and operator combinations	Attacks 25 and 32	Identify and use boundaries with both valid and invalid inputs and combinations
16	Force computation results to be too large or too small	Attacks in Chapters 2, 3, and 4	Modify attacks of these chapters for a GUI
17	Find features that share data or interact poorly	Attack 32	Apply per this book

TABLE 7.3   Recommended UI Attacks Mapped to the Mobile and Embedded Space

Context → Attack ↓	Mobile and Smart Wireless Devices	Embedded (Simple) Devices	Critical Mobile Devices (Could Be Embedded, Small but Important)	Critical Large Embedded Devices (Could Be Mobile)
Attack 22	Frequent	Frequent	Sometimes	Sometimes
Attack 23	Sometimes	Frequent	Yes	Yes
Attack 24	Yes	Frequent	Yes	Yes

Notes: Seldom, seen only in specialized cases infrequently; sometimes, seen often in many different contexts; frequent, seen regularly in a variety of contexts; yes, should be considered for most contexts; no, generally not applicable.

Next, consider the security implications (see Chapter 9). Additionally, this section introduces some new attacks for consideration.

Table 7.3 expands on the attacks from Tables 7.1 and 7.2 with some new attacks from this book, and Table 7.3 provides a mapping of attacks from this book to different mobile and embedded contexts.

ATTACK 22: FINDING SUPPORTING (USER) DOCUMENTATION PROBLEMS

O Novice

We sometimes forget that the software can be "okay," but there still may be system problems. Here, the developer may say "hey, my code is just fine," so how can the system have an error? Well, consider Table 7.4 for types of support information and test cases.

For the reasons given in Table 7.4 (Source), a user inputs a value into the system or creates a use situation that creates a "problem" (Issue). The input may be by mistake or, in the case of hacking, by a malicious person's design. The situation may be something that was promised or not intended, but some supporting documentation misled the user. Attack 22 creates inputs into the system to get it to display or produce a "problem." Developers may argue that the software "works" because it meets functional requirements. However, if the user or customer is unhappy *or at risk*, the total solution (including supporting documentation) is not valid and a bug exists—albeit not in the code. This attack is only a beginning of usability testing. (For a more complete treatment of usability testing, see [3]).

Note: I treat help files separately from this one (see Attack 24), since the taxonomy indicated problems in user/product information and help files.

SIDEBAR 7.1

Ford: In 2012, Ford received bad press [4,5] for its in-dash mobile smart system. Improved user documentation and attacks on this system may or may not have improved the product. However, history from the IT world and other embedded systems indicates that tested support information (see Table 7.4) and usability testing might have made a difference in the success of Ford's product. After all, most people want "simple-to-use" or "user-friendly" devices. This attack can indicate other efforts such as usability testing (outside the scope of this book) for teams to consider.

TABLE 7.4   Information Driving UI or GUI

Source	Issue
User guides, hardcopy help files, or online guides	These may have the wrong information (see Attack 24)
Hardware manuals	A situation is not accounted for (see Attacks 4 and 13)
System operations concept documentation	Information may not account for a given situation
Sales or promotion literature	Information may be wrong, misleading, or out of date
Training	Information may be wrong or incomplete

When to Apply This Attack

Testers apply Attack 22 when there is a UI and information external to the software and UI, which could drive a user to use or misuse a device. UI or GUI testing can start during development with prototype assessments, continue into the test lab, and then go into field testing. This is a time when you have to "think" like a user, but it may be hard to "play" all the different user roles. Roles may be illegitimate (such as a person trying to hack into the system) or legitimate. This attack is a stepping stone to security attacks later in this book.

What Faults Make This Attack Successful?

Attack 22 is successful because someone makes an assumption (undocumented in any available information). Assumptions can be dangerous and can also have security risks. The system is built with assumptions. Everyone involved in development "buys into" assumptions. But the user does not know about the assumptions, does something unexpected, and then, the system behaves badly (hence a bug is born). Also an error happens if user information is wrong or incomplete. This happens often because creators and developers know too much about the system, and when writing user information, they make assumptions (again) about the user and a bug is born. Sometimes it is a technical writer creating information, yet they may not understand everything in the software or device and a bug is born. The fault actually might be between the user, supporting information, and the reality of the system or device.

Who Conducts This Attack?

Attack 22 should be done by an independent tester who is not biased by development efforts. The closer a tester (or anyone) is to the development, the more likely they will make assumptions about use, usability, and what help needs to be supplied. Fresh eyes are important with this attack. Maybe a junior tester or a newly hired tester who can really "play" the inexperienced user should be considered.

Where Is This Attack Conducted?

Attack 22 is conducted at the UI or GUI. The UI or GUI may be somewhat simple (a switch, steering wheel, touch screen) or more complex (full GUI with menus or other physical interfaces), but many questions of human factors, environments, and humans should also be considered. These can lead to many test lab configurations and system test environments. Embedded testers have labs, prototype tests, sea trials, flight tests, test drives, and other field tests. Attack 22 may be run in many of these environments using the "user information," which will be the start of assumptions that lead users to bugs. The context or domain, risks, customers, and other external factors answer the questions of where and how much. Beware of anyone who tells you that "this is all we need to do" or "the GUI is obvious." This attack starts with the UI or GUI to assess basic usability, and then, as the user documentation comes available, new testers should take this information into the test environments.

How to Determine If the Attack Exposes Failures

Issues of use, usability, being "friendly," and so forth are very subjective. And unless you are a well-trained human factors person who knows how to measure the success or failures of these system attributes, there are no simple answers. Use of history, experts, and experience can help answer "is this a bug," but always keep an open mind. Test teams would do well to have some human factors skills to help consider what is possible and what is *just not right*. Using the user information with the UI or GUI, the test team looks for the following:

- The expected or unexpected (based on user information and the reasonable person concept for this case, not requirements)

- The confused or unhappy tester (If a tester does not like it, will a user?)

- "Dumb" mistakes caused by the UI or GUI or documentation

- The strange UI or GUI input–output (this may take critical oversight and detailed analysis of results)

- Bumps and bruises (seriously, embedded devices can do harm)
  Note: Refer to Appendix F for a more detailed list of UI or GUI checks.

How to Conduct This Attack

Attack 22 is conducted by the tester playing the role of a user and depends on the development engineers and programmers missing inputs and factors that can manifest themselves in the UI. Not all input conforms to expectations. Worse, some users may attack the input such as security, performance, and other reliability issues. Testers must consider the full variety of users and their input space: valid, invalid, normal, and extreme.

Documentation information to drive the UI or GUI attack is obtained from

- Reviewing sales information, websites, and user training, if any (see Table 7.4)

- Reviewing requirements and even design documentation (do not do this exclusively)

- Defining ranges (valid, boundary, and invalid) of inputs associated with Table 7.4

- Defining the risk for these pieces of information from a UI or GUI perspective

You should design these into an attack or series of attacks while exploring the UI or GUI possibilities that the user would exercise. During these sub-effort attacks, it is good practice to complete a checklist or maybe use a survey based on Appendix F. By doing this, a tester will have a start at assessing the system device via UI or GUI from a usability viewpoint.

As this effort progresses, consider other factors including human machine interface (HMI), psychology, what can go wrong in training, less-than-adept humans, and so forth. Listen for statements such as "no one would ever do that," "why would that be input," or "it is obvious." Design this attack, usually, as a scenario (see Attack 20) or maybe as exploratory testing [6]. Once you have defined what you need for the attack, conduct it, then watch and learn (use critical thinking). After one tester has done UI documentation testing, it is good to bring in a new tester who does not know the system and, therefore, is not "jaded." Some companies rotate new testers (young or old) in, or contract with outside test groups (crowd testing). Again, the goal is to watch for things in the UI or GUI that do not seem "right."

> **Takeaway note:** Documentation can mislead, and testers that are overly familiar with the system may be biased, so consider the needed level of tester independence when this test is conducted.

SUB-ATTACK 22.1: CONFIRMING INSTALL-ABILITY
O Novice

A specialized sub-aspect of Attack 22 is to confirm the ability to install the app (software). The installation should follow associated user documentation or standards for the device(s). Therefore, after reviewing this information, testers should run Attack 22 to confirm the installation. Further, sub-Attack 22.1 should be run with numerous device configurations. I recommend using Attack 32 to identify different configuration combinations. After confirming the instructions and installation works per the "published" user installation information, I recommend some basic tests, such as Attack 33, to confirm that the software is really working. It may be useful to automate these confirmation checks.

Finally, for install-ability testing with this attack, testers must remember that there may also be de-install steps, where software is removed from the device and re-installed. In a de-install sub-attack, follow the published user instructions for the de-install and then check that the app and any associated system changes such as the removal of library and/or data files has actually been done. To check for a good de-install, you may need the help of some tools to confirm that the system is restored to a "pre-install" state (e.g., flags may have been set as part of the install but not removed on the de-install). These types of tools check the operating system, libraries, and data file status.

ATTACK 23: FINDING MISSING OR WRONG ALARMS
O Novice

Attack 23 goes after bugs in alarms or warning logic, which many mobile and embedded devices have. The error taxonomy indicated that many mobile and embedded bugs are traced to "alarm" situations. This probably happens since many embedded software systems are controlling something, and if bad events happen, alarms should be triggered. Many mobile devices also have alarms such as clocks, warnings to drivers, medical alarms, and so forth. Most of us are familiar with car warning indicator lights. The system is requesting human or user intervention, hopefully to save the day, since many times the human can handle situations that are outside of the software-system abilities. If the warning alarm is missed, a bad situation can get worse such as you "blow up" your car engine because it has too little oil.

When to Apply This Attack

Testers apply Attack 23 to the alarms and warnings and then, the systems, situation(s), condition(s), fault(s), and such that trigger them. Alarms can take many forms, for example,

- Sounds (auditory)
- Visual (colors, flashing warnings, coded strings of data)
- Actions by parts of the system (shutdown, safing, stop operations, etc.) and
- Other (vibrations or combinations of these)

> **Key point:** Many mobile or embedded software devices are controlling something or informing the user. The ability to draw quick human user attention to a situation is a key feature, and we do this with alarms, so it is important that these alarms work [7].

When running this attack, consider the user with factors such as environment, hearing or deaf, human limitations (color blindness), age, understanding, and so forth.

What Faults Make This Attack Successful?

Alarms can be minor or major. This relates to risk and consequence. The level of risk on an alarm should determine the amount and intensity of the attack testing, where major risks get the most testing. Alarm bug situations could include the following:

- False (no alarm needed, but an alarm is triggered)
- Wrong (alarm needed, but wrong warning or output)
- Missed in time (no alarm sounds at the time an alarm is needed)

These faults happen due to mistakes in the software and/or designs that miss a use case or when a requirement is missed. However, history has shown that changes to the operations, even hardware changes, can make situations occur that then result in buggy alarms. For example,

- Hardware interlocks are removed allowing alarms or warnings to be missed through a software bug that killed people [7]

- Cannot see or hear the alarm(s) or warning(s) due to a system error or human factors (such as noise on a train)

- Threshold data entered incorrectly for the use of the system (medical devices)

- A time alarm is missed (IPHONE 2010 new year bug [8])

- Wrong alarm message wording (smart phone app)

Since the situation often involves risks to humans (safety) or hazards (harm to resources), testers should attack this area and play close attention to all warnings and/or alarms.

Who Conducts This Attack?

Attack 23 should be conducted by the test team. Also it is good if the test team has "user" experience in the domain of the software use. The developers and analysts should be kept at arm's length for parts of this attack since, by the time Attack 23 is applied, these people may have convinced themselves that "it" (the alarm) is working as it should. A person knowledgeable in human factors should be consulted during this attack. Having a qualified human factors person on your team means that hardware, software, and environmental factors will be considered, such as how loud is the user's environment; what time factors are in play; and what human limitations are at play: lighting, environment, noise, and user's abilities and/or disabilities? Alarms may need to account for some of these when placed in different environments. Mobile and embedded systems are often seen in environments that are far outside of the typical home or office PC.

Where Is This Attack Conducted?

Attack 23 should be conducted on an integrated system in a test lab in a "close to real" environment and again on the full system using realistic use case scenarios for the real world. Light, sound, temperature, distractions, and other environmental issues need to be "real" to achieve valid results. Testers need to think of the possible environments the device(s) might be used in or alarm cases may be missed. The team may want to identify many environmental cases, including expected and extreme cases. A table or matrix of environments should be created and reviewed with the users and customers to make sure that Attack 23 affects appropriate situations. Watch for anyone saying, "no one would use the device there or under those conditions." This statement may be far from the truth for the device.

How to Determine If the Attack Exposes Failures

For each environment and alarm trigger case, the attack team should identify what is expected and then consider the following:

- Missing—something does not happen when it should

- Wrong alarm—incorrect codes, messages, sounds, and so forth

SIDEBAR 7.2

Factory Scenario: I know of a case where the embedded device was used in a factory, which was noisy. People used ear protection. The alarm had two factors: sound and flashing red. The alarm situation was triggered. The problem was there were other pieces of information that flashed on the screen too; the operator had ear protection on; and he was colorblind, thus the alarm was missed.

- Time issues—too short, too long, too fast, and so on

- Wrong or missed system safe action—shutdown does not complete, system does not call home, and like issues

- Alarm is triggered, but could be missed by a user (alarm is not loud enough, in a bad position and so cannot be seen, color is too dim, color blind users, etc.)

Many of these may be subjective, so likely team discussions and even involvement of actual users may be needed. It is risky to "assume" a user will see, hear, know, or have the training to understand an alarm. Finally, in looking for bug cases, consider if there are regulations or legal requirements, which must be met, since often these will document other test cases that you must run.

How to Conduct This Attack

To force alarms in an attack, a tester must identify the alarm and/or warning cases and then know and fully understand the trigger, situation, and environment for those warnings or alarms. It is reasonable to look for alarms in the user documentation information, requirements, design documentation, or even the source code. Once all of these are known, you must think like the user (remember not all users are human) to fill in the situations and environment that are likely to be encountered. The attacks consider not only the expected, which must be confirmed, but more importantly, unexpected results (could be bugs) which must be looked for, documented, and then attacked. The unexpected results are much harder, taking real focus, thought, and exploration. A decision table or some kind of diagram to capture these pieces of information would be a great tool here [9].

Note: Some alarms may be very difficult to trigger, taking special set-up and effort.

Now, using this information, define how to force the alarm. Time variations of the alarm plus environmental factors will also play into the complexity. Alarms of time may work at one boundary, say from one day to the next—but not work at others—say end of year or leap year. A variety of boundary conditions may also be needed in testing (see boundary value analysis [9]).

At this point, it is good to conduct risk analysis and define variations of the environment and situations related to the alarm condition. Your table should now be populated.

For completeness, you will want to generate an attack or test input for each entry in your table. From these analyses, you can build scenarios. Running exploratory testing using the scenarios based on the tables and prioritized by risk is next. Although scenarios—as defined in Attack 20—can also be used. As you run tests, pay close attention to what can be done together, what cannot be done, as well as the unexpected (things not in the tables or identified as risks that you learn as you test). This takes critical observation and some intuition, which is why having domain and/or user experience is valuable and why an exploration is optimal here. The alarm table will likely change and even grow as your work progresses.

As you run the attacks, determine if the alarm happened, did not happen (a bug), and/ or if something was just "not right" when it did happen (a possible bug). Repeat this for all elements in the table, as well as the important risks. However, as the repetition happens, recognize that the table, risk, and attacker knowledge will change, so be ready to go in new directions.

You may need to consider the environment, and the forcing functions for alarms and/or warnings. These are often associated with hardware, operations, and/or time-based situations, which can be difficult to create. Also once created, a fair question to ask is "how real" is the actual situation. You may also consider using boundary, edge cases, stacking alarm cases, and other combinations.

The sources for alarms come from requirements, operational concepts, user expectations, user guides, design, and the source code. Again, remember the "reasonably prudent person" concept for juries here. If a reasonable person would expect a warning in a situation and there is none, maybe it is a bug. If the warning is wrong, late, confusing, or otherwise can be missed by humans, hardware, operations, or other software, maybe it is a bug. It is better to identify as many possible situations rather than "assume" anything. The team and stakeholders should determine when a warning or alarm bug lives.

ATTACK 24: FINDING BUGS IN HELP FILES
O Novice

Bugs live in many places in embedded software systems, and Attack 24 goes after an often overlooked place—the supporting help documentation in whatever forms it occurs. An important piece of information is the helpful help documentation. This can be online via Internet, in hardcopy form, or delivered as part of the software, or some combination of these things. Embedded devices can have strange little interfaces, inputs, and command sequences. Sometimes these are difficult and not very obvious to find, particularly if the mobile or embedded device is not very smart. Many users cannot follow obtuse help information easily. This costs the vendor when the users or consumers call the Help Desk or Tech Support number, send multiple emails, or whenever they make mistakes that harm the product, which can add to warranty costs. This can also impact profit numbers or worse create a loss in sales when the user and/or consumer goes elsewhere, publishes a rant on a blog, or tells their best friend(s) not to buy this device. Attack 24 tries to "break" these helpful information sources.

Help comes in many forms such as

- Classic in-program help file, which is selected with a menu option and can be searched

- Paper manuals

- Web-based help files in the form of user guides, which are included as links within the product or downloaded as .pdf files

Each of these can be buggy resulting in "wrong" user interactions with mobile and embedded devices. Some are smart (searchable), some are indexed, and some are just information streams. The idea, of course, is to build systems with a UI or GUI in which a user does not require help. This is rare and/or hard to do.

It is fairly well known that help files have resulted in failures, confusion, frustration, and perhaps, even death. Sometimes limitations and bugs in the software are "solved" by information in the help files such as big bold letters stating "Do NOT do this." Is it better to have the help file state *this* or the software be "smart" or easy to use, thus avoiding the need for help files? In the mobile and embedded space, limitations of the software or processors often mean that the software cannot be very smart, so turning to the help documentation becomes the answer. (Many might have answered that the software should have "saved the user.") If the help information is online and the device has a wired/wireless link, then you can "access" the information from the device when you need it. When the help information exists, in whatever form, it needs to be attacked just as much as the system, software, or hardware.

When to Apply This Attack

Apply Attack 24 on the help documentation. This means that you must find the information, understand the HMI, and be able to think or "play" just like the user would. In this case, too much knowledge of the device and software can get in the way. If you understand already how the device works, it is hard to assess if the "help documentation" is really helping a typical novice user.

To assess the usability of the help information, pay close attention to the following:

- Critical or safety functions

- Complex sequences especially if time is a factor

- Nonobvious actions (is control-alt-delete obvious?)

- Interplay of hardware, time, user, and software

- User limitations (how well can an 8-year-old read)

- Actions, defined in a help file, that might not get followed

The team will want to apply Attack 24 after the development is done, once the help documentation exists, during system testing, and before delivery to a customer.

SIDEBAR 7.3

Smart Phone Game: When evaluating a smart phone game (the fastest growing area of gaming), the game functionality was found to be hard to use; the goal of the game confusing; and scoring unclear. The help file link, which was there, appeared to be inactive (or there was no help file). The game was removed from the device. As you might have surmised, this was not a successful product.

What Faults Make This Attack Successful?

Attack 24 is possible, because conveying technical information, instructions, and functions completely and succinctly is difficult. Developers or the writers writing the help documentation make assumptions and have biases about the product, which a naive user may not have. You must attack from the user's viewpoint when looking at the help documentation.

Who Conducts This Attack?

Attack 24 is run by a tester or somebody with an "independent" set of eyes from the developer who understands the system fairly well. It should also be someone who can "play" the various user roles. So, maybe, it is best if it is the "new person" or someone to outsource to.

The issue for testers here is how can you think like your user, such as your mother, a 7-year-old child, someone with a disability, or even a teenager. It is difficult but options do exist. For example, you could hire people with certain abilities or disabilities and get them to be beta testers, get "new" testers, and/or do crowd testing such as, with uTest. These people are all viable testers to assess help.

Where to Conduct This Attack?

Attack 24 should be done in the real world, with real systems, using the real help files. It can be run in the test lab, if the test lab is "realistic" and if a tester has no little or no knowledge about the device under test.

How to Determine If the Attack Exposes Failures

In Attack 24, do what the help file says exactly and blindly, meaning do not fill in any missing information or steps. If you see an attacker "making assumptions," or "filling in the blanks," there may be a bug. When the results of following a help file exactly does not produce expected or required results, you have an issue. It may not always be with the software. It could be with the help documentation itself, but it certainly is something for the project to fix. In considering a bug, there may also be a tradeoff between fixing the HMI and improving the help file. Testers should keep good notes to share with the developers, technical writers, and others. Sometimes, the team will want to change the system to make it more "user friendly."

How to Conduct This Attack

To begin, identify the format and where the help files exist. Make sure you find all help files that a user might be able to find or is expected to use.

> **Key point:** There may be multiple help files available: in printed documents, in. pdf files, on an Intranet, or on the Internet. Testers must find and attack each help file and elements of the help file or otherwise assure that they are the same since sometimes they are not.

You next create use case scenarios based on what the help file says. Watch for any point where you need to "fill in," add missing information, or assume what to do. These may be bugs. As a help file scenario is built, exercise the scenario as the attack. The tester must practice close observation watching for where the help file says X will happen and y happens. This may be a bug too. Also you should watch for where an action just does not work. This likely is a bug. Repeat this attack for all of the help files and file options that exist. Here a decision table to track any complexity and options may be useful. Take good notes and feed them back to the person who writes or works on the help files, so that the files can be corrected and/or improved. Also consider that maybe the device or HMI may need improvement, allowing the help file to be "simpler." This would be a bug to report to development. Finally, consider repeating your steps with a different tester once you become too familiar with the help files. A new tester can play different roles (new user, operator, manager, etc.).

EXERCISES (ANSWERS ARE ON MY WEBSITE)

1. For the system you are testing, identify all user guidance information and then conduct a risk analysis of what could be wrong with it in the software implementation.

    a. Or do the same for the "Winter Park Ski Area" app, which you load on to your smart phone.

    b. Then conduct Attack 22 for 1 or 1a.

2. Locate the "clock" app on a smart phone and then define the alarms it uses in your system or do the same for a system you are responsible for testing that has alarms.

    a. What kind of alarms are they (short term, long term, sound, message, warning, or what)?

    b. Define the environments and conditions that might be important to test for the alarm.

    c. Define attacks for these alarms (input triggers, how to create, and expected results).

    d. Execute the attack.

3. For a smart phone "tank" game (simulates a battle tank), identify the different possible users of the game's help file.

REFERENCES

1. Whittaker, J. 2003. *How to Break Software*, Addison-Wesley, Boston, MA.
2. Harty, J. 2010. *A Practical Guide to Testing Wireless Smartphone Applications*, Morgan & Claypool Publishers, San Rafael, CA.
3. Rubin, J. 1994. *Handbook of Usability Testing: How to Plan, Design, and Conduct Effective Tests*, John Wiley & Sons, New York.
4. Ayapana, E. March 6, 2012. There, I fixed it: MyFord touch update released, *Motor Trend*. http://wot.motortrend.com/there-i-fixed-it-myford-touch-update-released-176907.html
5. Hill, B. June 23, 2011. MyFord touch drop kicks ford from 5th to 23rd in J.D. Power IQS rankings, *DailyTech*. http://www.dailytech.com/MyFord+Touch+Drop+Kicks+Ford+from+5th+to+23rd+in+JD+Power+IQS+Rankings/article21994.htm
6. Whittaker, J. 2009. *Exploratory Software Testing*, Addison-Wesley, Boston, MA.
7. Leveson, N. 1995. *Safeware: System Safety and Computers*, Addison-Wesley, Boston, MA, Appendix A.
8. Chen, A. January 11, 2011. The i-phone bug that made you late for work, *Gawker*. http://gawker.com/5722444/the-iphone-alarm-bug-that-made-you-late-for-work (last accessed April 11, 2013).
9. Copeland, L. 2003. *A Practitioner's Guide to Software Test Design*, Artech House Publishers, Boston, MA, Chapter 4.

CHAPTER 8

Smart and/or Mobile Phone Attacks

FOLLOWING UP FROM THE USER INTERFACE (UI) SECTION, the world of mobile–embedded software has been changing and expanding to include devices with more capabilities and features. This moves mobile–embedded closer to the personal computer (PC) and information technology (IT) world. There are now no clear boundaries between mobile, embedded, and IT software. The evolution is seen in control systems with graphical user interfaces (GUIs), phones, tablets, and other devices once thought of as " purely embedded," but are now taking on many characteristics, and hence the potential problems of the PC and IT world, due to cheap and powerful processors as well as massive memory. While these devices take on features of the IT or PC world, many of them still have classic embedded bug types.

Table 8.1 offers a mapping of the attacks in this chapter to general mobile and embedded contexts as a starting point guide.

Table 8.2 defines a few of the more common "web" attacks from the *How to Break Web Software* [1] book that embedded mobile smart device testers should consider using.

GENERAL NOTES AND ATTACK CONCEPTS APPLICABLE TO MOST MOBILE–EMBEDDED DEVICES

There are a variety of factors to check and attack in the mobile and embedded device world. These factors include screen size, screen types (touch, nontouch), colors, input devices (screen, tracker ball, key pads—hard, software, and other types), input and output ports, memory cards (subscriber identification module [SIM] or secure digital [SD] cards, etc.), batteries (different types and differing power profiles), and many others. I recommend considering these during this chapter as well as Attacks 32 and 33. Besides this chapter, an excellent reference to be familiar with is Kohl's [2] "Isliceupfun."

For many mobile and embedded devices in this attack area, there can be legal issues to consider. We see young children with cell devices. What restrictions should be in place (think adult websites and predators)? Also many apps now request the user or buyer to

TABLE 8.1   Smart Attacks to Mobile and Embedded Contexts

Context → Attack ↓	Mobile and Smart Wireless Devices	Embedded (Simple) Devices	Critical Mobile Devices (Could Be Embedded, Small but Important)	Critical Large Embedded Devices (Could Be Mobile)
Attack 25	Yes	No	Sometimes	No
Attack 26	Yes	No	No	No
Attack 27	Yes	Sometimes	Sometimes	Sometimes

Notes: Seldom, seen only in specialized cases infrequently; sometimes, seen often in many different contexts; frequent, seen regularly in a variety of contexts; yes, should be considered for most contexts; no, generally not applicable.

TABLE 8.2   Map of Attacks from *How to Break Web Software* to Mobile and Embedded Domains

No.	Attack Name	Relates to This Book	How to Apply	Notes
1	Panning for gold	Attack within the implementation or developer area	Learn about your code that is customized	Is it possible to obtain the source code and, if so, check to see what kinds of developer testing and verification have been performed, then "design" attacks for weaker areas
2	Holes left by other people	Security	Look for source code reuse, libraries in applications, back doors, etc.	Security "holes" are constantly being uncovered in the OS, applications, and support structures of smart mobile devices. Just as in web and IT systems, testers must watch for these holes and monitor industry news reports, because these "holes" are constantly evolving (see Chapter 9)
3	Bypass restrictions on input choices	Modify the attack for local mobile contexts	Extend attack to client side testing, not server or web side, to determine if any app might bypass the server side	See *How to Break Software Security* for more notes. Classic tests such as, buffer overflows, SQL injection, and others apply here
4	Bypass client-side validation	Modify the attack for mobile contexts	Try to use the app to bypass client-side security checks. This attack is beyond the App	See *How to Break Web Software* for more notes

accept legal terms (think a contract), but many people, again children, cannot sign contacts. How does the app you are testing deal with this? Then, there is also the side of legal called regulations (think Food and Drug Administration) or tort law (think getting sued). Does your device and software have such legal factors that the attacks and tests should address, and if so how? These are all factors a development team and testers may need to consider. You want to stand on solid legal ground in many cases with the attack, results, and information retained. There can be and are legal implications of many of the things we test—or do not test. Again, ask yourself, "What if I am on the nightly news, in court, or my mother is impacted by this?"

Finally, in planning these and other attacks, it may be useful to have "data" about your mobile device. This brings us to mobile analytics, which studies the behavior of mobile site visitors and use information [3]. From a testing viewpoint, this analytics tool can help determine which phones, features, and stories may need to be tested or test more frequently. Data collected in mobile analytics can be of use by testers in defining test information including number of page views, sites visits, device information, country user information, and other data that can help to design the tests that follow.

ATTACK 25: FINDING BUGS IN APPS
☐ Intermediate

Attack 25 is specifically aimed at breaking applications (apps)—the software or code that makes smart devices—particularly cell phones, "smart." There are stores to supply apps and websites to download apps, and we have all heard the new cliché, "there's an app for that." Many apps are free. Many are "selling" something. Some come at a cost, and some are "updated" as time goes on. Apps use the system to provide a service to the user. They interface with the operating system (OS), hardware, other software (apps), a network, the web, a user, and many processes. Often in the app world of small, smart devices, one app uses another app to create a fully functioning feature. For example, a translation program uses the text-to-speech (TTS) program, which uses parts of the OS and read only memory (ROM) to type, identify, or translate words, and then provides the voice as well for a word or phrase. These interacting programs become part of the testing space. So one cannot just "test" the app alone.

When to Apply This Attack

Apply Attack 25 when you have one or more apps, which is probably all of the time when dealing with mobile smart devices. Besides this specific attack, consider the other attacks and concepts found within this book when you are planning and setting up to test apps.

> **Key point:** Attack 25 should be part of a series of attacks including Attacks 20, 33, and other attacks run by developers (see the attacks of Chapter 2).

What Faults Make This Attack Successful?

All the problems that exist in any software can exist in any application. Of particular concern are the following:

- Communications external to a network

- Data exchanges within the device

- Interference to/from other apps and/or the OS

- Small size of the apps
- Limited UI
- Programmer mistakes (classic coding mistakes)
- User needs (unmet requirements)
- Design issues (classic logic and programming bugs)

As testers on apps, we all have the same problems of classic testing, embedded testing, and a few new challenges introduced by the topology of the mobile and embedded world. Additionally, most "official app stores" have considerations (rules) on how apps are "readied" and approved to be made available. Testers must know these rules and ensure that an app meets those requirements, so that the app stands a better chance of being made available in an app store. Rules will likely tighten over time, possibly through store regulations. Of course, there will always be "unofficial" apps that some devices try to lock out. However, many users and now even malicious apps (malware) will find a way to get those apps onto the device.

Who Conducts This Attack?

To start with, developers and testers should consider testing apps with the attacks of Chapter 2 and using the implementation test ideas in references such as *A Practical Guide to Testing Wireless Smartphone Applications* [4] to drive testing on their software before proceeding with this attack. Since apps are relatively "small" (few lines of code), some test teams may be small or even just a couple of people. Someone needs to be tasked as being the official tester, since testers can provide independent information and a different viewpoint from that of the developer. This strategy helps to find errors that hide in developer's blind spots.

Where Is This Attack Conducted?

Testing apps will likely start on simulators and/or emulators during developer tests. This environment can also be used by the "official" testers. Once tests pass integration and interoperability in simulated and emulated environments, test facilities with hardware in the loop should be used. Here the team may want to consider "open box" architecture [5] or special lab hardware, which can better support some testing that can be difficult at a full system level with "closed" boxes. Open box architecture can include various forms of automated testing. The real environment is where this app attack finishes. The attack here is as "realistic" as can be possible.

How to Determine If This Attack Exposes Failures

Test the app against advertised or expected functionality

Check against standards and/or regulations

Check that the app works online

Check that the app works offline

End-to-end functional tests (perhaps as an addition)

SIDEBAR 8.1

Apps: In the last 10–15 years, new areas of software have "exploded" and new devices have come to market (sometimes prematurely), and there is an increase in the numbers of companies creating new apps to meet consumer demand. This has happened in computers, small PCs, IT systems, and web systems and now cellular devices. Apps are the latest in this line. (Remember the IPHONE 5 and its problems with APPLE Maps?) History shows us that testing takes a backseat to the rush to market in the early days of an explosion, but eventually the market demands better software, which means testing attacks for many apps will become of increasing importance, just as it has in other domains. Understanding bugs helps testers to find them as does using a variety of attacks, techniques, and approaches in combination during testing.

There may be other nonfunctional areas, which some app testers may want to consider, such as performance or security testing. When one of these functional or nonfunctional areas fails to meet expectations, you have the start of a bug.

How to Conduct This Attack

Here are the steps for Attack 25.

1. Gather the functions and linkages by exploring (playing) with the app. These can be listed on a piece of paper or by constructing a mind map of the app (see Attack 33).

2. Identify and construct a test use scenario (see Attack 20) or storyboard [6] for each element (functions or links) in the app.
   Note: For multifeature apps, or ones that use web, more than likely there will be a series of use scenarios or storyboards based on different user types, common, and uncommon usages. These will come from operational concepts, marketing information, agile stories, and so forth. Also, if Attack 33 has been done already, you can reuse the mind map, expanding the map with the scenario information.

3. Define valid data or input for each user selection to drive the data into each display or entries to the stories (capture these in the list or mind map).

4. Define any invalid data or input options for each display or entry option.

5. Construct a matrix of valid and, where possible, invalid data inputs (see Table 8.3).
   Note: Valid/invalid selections can be done using equivalent classes, boundary value, or other classic test techniques [7].

6. Construct a minimum set of test coverage based on the matrix.

7. Create specific data points or values to cover as much of the matrix as is reasonable within factors such as cost, schedule, and/or risk.

Knowing the functions, linkages, and valid/invalid inputs can be a challenge, particularly in small shops where everyone just "knows" the app and where there may be infamous

TABLE 8.3 Table Example Valid–Invalid Test Matrix Table or Map for Attack 25

Page ID	Entry Item	Valid Data Entry Value	Invalid Value	Test Case Assignment	Specific Test Value	Notes
Splash page	Open App	Click to open		Test 1	5 s for display of page	Check timing
Landing first page	LinktoFlight Res	Click on link	None possible	Test 1	Entry click	Travel to sub page
Landing first page	User ID login	Enter valid email		Test 1	jon@company	Accepted on entry
Landing first page	User ID login		Enter invalid email	Test stress 1	Enter user ID not in the system	Any invalid user ID
Landing first page	Use ID password	Enter valid password		Test 1	My password	Accepted on entry
Landing first page	Use ID password	Valid email		Test stress 2	jon@company	Accepted on entry
Landing first page	Use ID password		Enter invalid password	Test stress 2	Try case hacking with a really big password entry value	Good user ID but bad password
Landing first page	User login submit	Click on link		Test 1, Test stress 1, and Test stress 2	Entry click	Test 1 = works; Test stress 2 and 3 = error message

Notes:
1. You must test a bad password with a good user ID.
   Test case names (Test 1, Test stress 1, etc.) and so forth are picked by the tester to help track into list, mind map, or storyboards.
2. Other test matrix table or maps and columns are possible.

undocumented features. Finding rich resources is a good idea at this point such as talk to developers, review requirements, work with users/customers, look in the code (if possible), and other information sources.

Checking basic functions with Attack 33 with only valid or typical inputs is a start in the world of attacks, but it can miss many bugs that this attack may find. In Attack 25, using such a valid/invalid matrix, a tester spends more attack time in the invalid domain trying to find bugs. However, the list matrix can get large, and as its size grows, a spreadsheet or "subtables" may be useful in a "divide and conquer" approach to manage a series of attacks.

Once a series of these attacks are known, the tester identifies the test environment and then can construct full end-to-end test scenarios. A scenario tells a story of how the device will be used, combining a series of attacks both in the expected (valid) and in the unexpected (invalid) situations. These situations can be strung together to create longer or more complex stories. In fact, some teams may be able to expand these attacks into standard patterns of "reuse" that some call tours [8], which are classic stories that occur over and over.

Finally, testers will run the attack (attack, scenario, story, and/or tour), and as it progresses, the coverage in the matrix should be assessed. As higher levels of coverage in the matrix are reached, completion of this attack can be considered. Likely, no single attack scenario or tour can hit all of the valid and invalid situations at once, so a series of efforts may be necessary. Good testers learn as they explore, so expect to find new testing objectives in this attack area [9].

> **Key point:** It is hard to know when to stop testing, but in a competitive world like apps, you may have less time to test. So why not make your attacks count? Do them in parallel with development and be ready to ship in days or hours, not weeks or months, but still within project context. In the world of apps, minutes even hours can count.

ATTACK 26: TESTING MOBILE AND EMBEDDED GAMES
○ Novice

Attack 26 is aimed at finding bugs within games on mobile, embedded, or smart devices.

Recent data show that some 80% of the apps loaded and bought for smart phone devices are games, and mobile game apps are the fastest area of the software gaming industry. The game production industry and testers have been around for many years now, and it is big business. It is not surprising that such high percentages of apps are games and that embedded and mobile devices are loaded with games. Game testers, including for "toys" like the X–BOX or WII, have years of experience, and it is no surprise that these border on embedded devices. Many game testers do not understand the embedded mobile world, and testers that are familiar with the embedded world often do not have experience with games. Both sets of testers have things to learn so that they can conduct Attack 26 [10].

> **Key point:** I expect many app testers to be game testers too. You need this attack, but many other attacks should be considered too.

When to Apply This Attack

When you are testing a gaming app, consider applying Attack 26, or when the game is going to be interacting with other parts of the system, consider applying this attack. There was once a joke that software games did not need much testing or evaluation. Companies produced games with bugs, and therefore, many of those gaming companies failed. As the software gaming industry progressed, the criticality and levels of testing increased to levels equal to many critical software domains. We may not be talking life and death, but billions of dollars are being spent on software games, and the game app world continues to grow and be valued [11].

As in many other attacks, when attacking the game app, start with the first version of code and continue incrementally until the end of the project. Design and screen concepts need to be explored, and while more of a development domain, the test team benefits from being in the loop during this up-front play planning. As the code comes into existence, the developers should conduct their attacks (see Attacks 2 and 3), followed by the on-staff testers running attacks. The on-project developers and testers should be the first "gamers (players)." As the game matures and the release deadline nears, the test team should continue to add new game tester roles as defined in "who conducts this attack" section. Also, there are a variety of game references testers should consider [10,12–14].

What Faults Make This Attack Successful?

A game app is just a piece of software. The faults that can make most other software fail can make game apps fail, too. Plus, in the mobile and embedded world, there are the limitations of the device to factor into software projects.

The particular complicating factors to watch for include network communications, memory limitations, screen size, app interaction, power usage, disk (if used), and CPU speed. Many players will expect graphics and features of full-size gaming systems. These may not be possible, and developers will "cheat" or make compromises. These may or may not be bugs, so testers must think carefully "is this really a bug or just a feature." Finally, there are the download sites and/or app store regulations with which to comply. All of these factors combine to either make bugs or not.

The last thing, and most important factor, that makes Attack 26 successful is fun, or more to the point—the lack of fun. The game app must be fun. Fun is like art and beauty: in the eye of the beholder. This means that you must have a clear focus on "who" is doing the attack and if they understand the app's potential users. If you can understand your users a little better, maybe you can understand when a game is fun or when it is not fun. When it is not fun, then perhaps there is a bug.

Who Conducts This Attack?

The selection of "who" in game app testing is perhaps the most critical aspect of Attack 26, so let us spend a bit of time on some details. Traditional embedded testers and staff will need to stretch a long way here to understand the gaming world, just as the traditional game tester may need to pick up a few new ideas on what a mobile and embedded bug might be. There will be a management push to keep the test staff numbers small, and this will limit the number of user roles that can be played. Many game teams will have everyone testing. The more game testing eyes there are the better, with a couple of new twists, but similar to the earlier lists of test staffing. I would recommend the following tester (gamer [10]) roles:

1. The developer(s)—See Attacks 1, 2, and 3.

2. The app game architect or director—This person (or these people) will be the people with the vision of what the game should be. Think of them as the master painter or director of the movie. They have the idea and vision and so should be very "demanding" about the art of the game. If the architect or director is not happy, then there may be a bug.

3. On team game tester(s)—These testers will know the games, know the project, think testing, and enjoy creative destruction. They should be experienced in engineering, gaming, industry history (what other vendors are doing), development, art, and "sick–o" attack testing (testing that no one else thinks about). This tester should also know the mobile and embedded test domain space, as well as classic testing techniques.

4. In company "dog food" testers—These people will be team members from outside of the immediate game app team and should run game attacks. These people test with prototypes (partial versions) and "eat the dog food the company creates" (so you want the game food to be good). Bringing in these outside resources for a weekend of play can be good data points particularly while the game is maturing (during iterations of development).

5. Independent test player—After this basic internal to team or company testing, you may want to get a tester with fresh "eyes," representing the people likely to download the app and play it for the first time. A starting option here is to have the hardcore gamers and testers representing likely game players. These people's feedback is critical, if you expect to sell the app. However, care is needed because not every game is liked by every "newbie" on the first play, or at all. If one newbie does not like the game, look to the next person. This means that the team may need several newbie testers. Crowd source testing becomes an attractive option at this point, but watch for legal and "product" leak issues as the gaming world is competitive in more than just those who play games.

    Note on roles: During testing effort, and as it progresses, do not forget that there are many different roles for newbies, some of whom are not "hardcore gamers." These people buy and play games. Example candidates may include business executives, children, teenagers, junior gamers, experienced gamers, even moms, and dads. Each role is a customer that the game attack may need to address. These roles are feedback points, and if you want the "big picture," a team may need to get feedback from many of them. It is hard to tell what the next viral app game will be, and getting data from these roles may help in making a viral explosion when crafting the game.

6. Mass beta trial testers—In some games, there may be a series of game "trial" sessions both internal to the company and external where testing (game playing) is done using

such tools as the checklist in Appendix F or a survey on a nearly complete game. The film industry does this with trial showings, test audiences, and private showings. The survey can be used before this, but here the team is trying for "customer reaction."

7. Not a tester—Finally, consider who should *not* be playing the game in this attack and then exclude them from the "who" list since they can give you bogus data, and Attack 26 is about getting valid user data (see Exercise 6 at the end of this chapter).

So, you may want a core test team of developers, testers, and insiders followed by a sampling of friends and other groups in the "newbie" category. The end of project testing should likely be a variety of the "newbie" testers and outsourced test teams, if you can keep "control" over them. The gaming of the game attack may be the largest challenge for the test manager because fielding a game that lacks fun may be the kiss of death for your game app.

> **Key point:** The "who" in game testing is important—often many people and players. You should get inside and outside test inputs. Finding the right testers, mix of people, and phasing is difficult. It takes the experience of a good game test manager.

In staffing, ask yourself these questions: "Do I have skeptics, hardcore gamers, newbies, pro supporters, skilled testers, developers, and others all contributing meaning?" "Do I have the time and budget?" Definitely find people who want to have fun (that is what games are about) and are not just people doing a job for a paycheck.

Where Is This Attack Conducted?

The gaming may start on device emulators (see Chapter 11) and, hopefully, make use of automation. Quickly, as the emulator yields good results, the testing should shift to a lab, where you will have some of the actual device hardware—and maybe a variety of configurations available to test. Finally, you will shift out into the real world (the field) with the official testers and new test team members. In each environment, exercise care to record and track the results of the game and player impressions with all observations.

How to Determine If This Attack Exposes Failures

Testers should start by classifying bugs: performance, nonfunctional, features not working, and so on. You can use other attacks from this book against the game app. Also watch for the bored tester or the "I hate this" comment. Attack 26 focuses on the art and fun of the game. Not everyone likes every piece of music, so watch that you do not have a jazz person listening to rap. The current taxonomy of errors and factors for game apps yielded Appendix F as a checklist of things for testers.

To determine if the game is "working" from the art side, the test manger gathers information from as many tester sources as possible. You should watch for misfits and miss directions from team testers, as well as the positive feedback and useful comments. As the test team gets data, careful filtering of the feedback is needed. You do not need to act on every negative statement and report, but you should be recording these statements to pass along

SIDEBAR 8.2

Game Apps: I was testing an app game for fun (both the game and the testing). I downloaded it, accepted the terms and conditions, and started to play. I could not figure out the following: how the basic movement on the field was done, how scoring worked, what the objective of the game was, or even how the help information was to work. Within 10 min, I abandoned the game and deleted it. Needless to say, this is not the success story the app creator or company wanted.

to management, developers, and others. This is art after all. A good development team and test manager must mix and match the gaming data to shape the game and find a winner. Bugs that are found to be critical should be considered for a fix. However, some bugs will persist, and in gaming, there is usually more concern for "fun" and the "wow factor." Look for the art and follow the team's muse.

How to Conduct This Attack

This testing starts with the developer, using selected attacks from Chapter 2. After some of the traditional attacks, the gaming attack picks up with "how do you like the game" and "is it really fun." You should be less interested in failure bugs but more interested in bugs of unhappiness. The team will "cycle" through the roles of game testers.

Attack 26 is done by basically "playing" the game app. This is really a form of exploratory testing (see [8] and Attack 33). During this attack, a tester or test manager must somehow "watch" the game in play and/or gather feedback. Feedback could be surveys, question sessions, debriefs, and asking for reports from the various testers. Watching can be in person or via video. Further, most mobile environments now support various configurations of capture and playback test tooling. These tools can be used to "capture" the attack and, when needed, play it back to see what the tester was doing.

Game playing can take two forms: guided explorations or ad hoc. Guided exploratory testing follows some rigor and rules: using expected use stories, checking help files, exploratory test charters, and other attacks in this book. Guided exploratory cases include combinatorial testing (Attack 32), stress cases, and security attacks (Chapter 9). Exploratory can be done in parallel with or in addition to more informal ad hoc attacking, which is free form, unstructured game playing (just go play).

The ad hoc side of Attack 26 uses hardcore gamers, newbies, and people who just "play it." During the observation of these testers, do not interfere with comments such as "let me show you," "try this idea," or other unhelpful comments. There will be a tendency to interfere but *do not*, because this will introduce more jading to the information coming from the attack. The idea of this informal testing is to understand how the app will be played by real buyers, any problems in play, and how much fun the gamers are having. Fun may be the most important and is difficult to quantify, so do not try. Just ask, "Is this fun or what?"

In both approaches, it is good if the tester in this attack has something to be checking for, for example, I have provided a starting point in Appendix F with a gaming checklist based on bugs. You should tailor the checklist to the local project, game, or context. The checklist can be used by testers during game attacks. It can be used as a starting point for a "game user" survey. Or, it can be used during game observations. During the observation of either exploration or ad hoc, the attack test team's observer is watching for body language, newbie frustration, looking at or for help files, having fun, wanting more, and/or just giving up. Some of these are positives and some are negatives. The negatives indicate a UI or GUI that must change, features that do not work, or bugs that must be fixed. The positives point to things that are going right, games that might sell, and are fun.

If you are the independent or newbie tester, hired or volunteered to do Attack 26, you should be using the checklist too. You can fill out your report or survey (based on Appendix F), which may be how you get paid or get your name in lights. The more information you can provide, the more pay you should receive. The testing may or may not be "watched" by the development team, but is still important, so take good notes. It may mean a return chance with other game apps. You may need to become an advocate for the things you see or do not like by writing clear reports, reporting good data, as well as unbiased bug statements. There is a job market in crowd source testing companies, and the good "newbie" testers can become quite in demand.

Finally, during the testing, the game testers should watch for bias. Developers and development teams will say "that is not a bug," "this feature is really cool," or "this tester does not get our great vision." And while these statements can be true sometimes, if patterns of "no fun," "hard to understand," "why would anyone do this," and so forth appear over and over, consider that biased views may be at play. Bias is where teams can lie to themselves and get into the market with a game app that fails. Remember that the team's goal is a killer app that will sell. This goal can be very hard given the tens of thousands of apps. Your job during the attack is providing information that tries to "break" the fun.

ATTACK 27: ATTACKING APP–CLOUD DEPENDENCIES
☐ Intermediate

The use of remote computers for processing and storage of information is increasing daily. This encompasses topics and items such as cloud computing, remote processing, web computing, and extension of classic client/server computing. Given the limited size, storage, and computing power of mobile or embedded devices, the attraction of using the cloud or other remote processing is obvious and likely to increase. There are unique aspects of cloud smart devices that need consideration, so whenever you are testing the cloud side, refer to the test techniques in *How to Break Web Software* [1].

> **Key point:** Many things are moving to the cloud and giving mobile and embedded devices limitations. The cloud is an obvious place to move data, processing, and interfaces too. As a friend once said, "Why download the picture locally. Look at it online." App testers will need to "deal" with the cloud.

When to Apply This Attack
Apply Attack 27 when the mobile and embedded application is dependent upon a server-side connection for computation cycles and/or information. When the app has dependencies, attacks can be done as soon as all components start to be functional. Assuming incremental

and iterative development, it is best to start with preliminary versions and interfaces. While Attack 27 focuses on the app side, as the testing progresses, expect to refine and find bugs on both sides of the interface in parallel because of the interdependency on both sides. Additionally, over the test time, the iterations will cause regression issues as new versions are available. A final set of attacks can be done when everything is complete, but this can impact cost and schedule if "late" integration–interface bugs are found.

Who Conducts This Attack?

Attack 27 is run by a test team specializing in integration even before testing the app has started. The test team should include knowledgeable people for both sides—the app and the cloud. This test integration team has a united responsibility where the developers on each side may be making assumptions that are creating the bugs. Attack 27 can also be done by independent testers such as labs or crowd testers. This may be a little more problematic, since the independence may mean lack of knowledge about the app or cloud–server side. Here, the test team must make up for the deficit by working hard with integration items (design, interface control document (ICD), basic understanding of how the cloud works, and other information) when accessible. Ask questions such as "What can go wrong at the interface to impact the app?" "What information can the app access?" "Where does the data come from?" and "What references (help files, literature, and so on) are available?"

Where Is This Attack Conducted?

This test can be run in system test labs or in the field, where both sides (app and cloud) exist. Often it is run in the lab where special tooling, data files, or configuration support for the app and interface testing exist. The labs may not need special tooling if other attacks and test efforts are being practiced on the system. Additionally, it may be of benefit if the lab or field can create or simulate a variety of input and environment conditions, which can influence the app. These input conditions can include things such as partial or incomplete data file, transmission dropouts, weak signals, out of date data, server side problems (busy, overloaded, and other web/cloud problems), and other interesting scenarios.

A large potential value of where (and when) the testing is done in a lab can be using the servers or cloud to create virtualization [15]. In virtualization, automated scenarios can be played into the app quickly, allow hundreds or even thousands of attacks to be quickly, though these can be artificial and risk unrealistic testing. Note: if virtualization is used, the real devices with the app running into cloud and/or servers need to be included at some point.

What Faults Make This Attack Successful?

Attack 27 is successful because of the app–cloud–server dependency. The more dependency and data exchange, the more areas this attack will need to address. What makes a bug here is when each side of the interface has assumed something about the other side but failed to communicate and account for the assumption during development. For example, the cloud side may assume trust of the mobile and embedded app for some data. When this trust fails, information and security holes (bugs) can open up. Finally, the embedded error taxonomy (Appendix A) indicates many problems occur at the interface.

SIDEBAR 8.3

More Apps: In testing a mobile app that displays ski area information (see example in Figure 8.1) including ski area name, new snow, snow depth, and date information, a request to update the data was made and the update apparently completed, but nothing was displayed (no information at all). And there was no indication of a problem such that the network or update was not available. In other cases, the information was displayed with date information, but parts of the data were known to be outdated and false. Unhappy users (skiers) might then move to a new app or go to a different ski area.

How to Determine If the Attack Exposes Failures

Here you are looking for some breakdown in the information flow, use, or display. Many times the break is not obvious, as the app will not crash or give any warning that anything is amiss. Close attention to the data, processing flow, timing, and display is needed to find the bug.

Bugs in the app and between the app and cloud can cost business in more ways than just the app developer losing sales. Apps identify and display information users depend on. For example, the app may display sales information which, if it is not right, a sale can be lost, costing the vendor using the app. Here the tester is looking for bugs such as an app that displays wrong information, a slow network, missing indicators, or out-of-date displays. Business analysts depend on accurate data, so keep their role in mind when testing apps also.

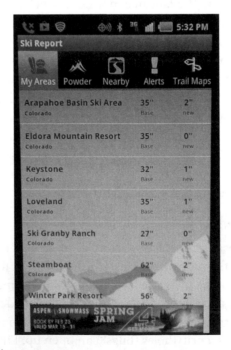

FIGURE 8.1   Example of ski area app.

To know there has been a failure in the processing or interface, you must understand what the server does and does not do well, as much as what are the correct data. You will have expectations about how the app should work, based on experience, documentation, conversations with developers, and knowledge of the architecture. For example, the tester should assess if checks and balances such as date/time stamp missing and/or if the update indicators exist on the data, and if they do not exist—you may have a bug.

How to Conduct This Attack

Testers will start Attack 27 by understanding what information the app uses and where the information comes from then bring up the app. This analysis and understanding may take some time when it is not documented or when such information is not provided to the tester. Some of the information may be obtained from the developer in such documents as an interface control document (most times called ICD), design notes, data exchange specifications, and so forth. Creating a map of the data or an app display table can be a good way to organize the knowledge gained from understanding the app. It will be easier on the tester if the information comes from developers, but testers can create such a table or map by hand over a series of iterations, if necessary.

If development's design information is lacking, start generating the needed information using the following approach. Within the device, disable network communication such as enable airplane mode, which takes the device off the mobile network. While in this mode, run through this test scenario:

1. Run the attack or otherwise explore the app to determine if it is not functioning or has reduced functionality or information.

2. Make notes of the dependencies and missing data elements, because these are the foundation of this attack.

3. Watch for old and missing data or app functionality that is not working. These data can be on both input and output data elements.

4. Continue until the table or map has been initially generated. As Attack 27 continues, expect the table or map to be updated. This approach will work in many cases, but some apps will only respond that the network or cloud is not available.

Once you know each type of data element coming from the server or cloud and have these elements identified in a table or map, you have the basic information to be tested. From this basic information, the tester determines the data interchange approaches (internal, cloud, mixed) and adds this to the table or map of the interfaces. Each type of data and interface needs to be tested. For example, ask if there is any offline use of the app. If there is, then any offline functions should be tested in an offline mode. The app may be largely dependent on being online, which is common. Should this be the case, then the majority of functions should be tested while the device is online and a few "features" such as error messages tested while offline.

Here is a sample table for a ski area app (Table 8.4).

TABLE 8.4   Example Ski Area App Mapping Table

Data Field	Use	Data Display Interface	Notes/Check for
Ski area name	App Display 1	From cloud	Static database item
Date	App Display 2	From cloud	Today's date
Snowfall today	App Display 3	From cloud	Receive data from ski area, other app, or database
Base snow to date	App Display 4	From cloud	Receive from ski area, other app, or database
Ad notes	Pop up	Preloaded ad	Pays for site (important), so determine when loaded and how updated and does it require any other app(s)

During the testing, you would use the table or map to answer the following types of questions:

- Can you test each item and do you know what constitutes success or failure of the item?

- Are there dependencies between table elements and should you test the combinations of inputs?

- Should you test the combinations of outputs?

- Is there some data that is more important to get right than others?

- Where is the data to be displayed?

Based on the answers to these questions, you can start running attacks on the app at the interface. This attack will follow a use case scenario or approach, see Attack 20, exploratory testing, or function testing of Attack 33, each driven by information in the table or map. Each attack should address and show coverage of table elements (cross things in the table off as you go). As you run these tests, watch for signs of bad, old, missing data coming from the server or cloud. Often the app will not know or report anything amiss even when it displays bad information. This may be a bug. Besides not displaying bad or old information, what warnings or error messages should be displayed? Lack of error messages when error information should be displayed to the user can be bugs to be watched for and reported.

Sometimes the app may take wrong actions, perform incorrect computations, or display bad data. You should watch for these bugs too. For example, I have seen map displays fail to update picture elements or missing pixels. Such display bugs may be temporary, related to connection speed, noise, and/or user actions. Is it a bug or not? What impact can bad or temporary bad data have on the app or user or worse—the company and its bottom line profits? You may or may not have a bug. Still, you will need to report the results to the team so that a determination can be made and implications examined. Questions to keep in the forefront are the following:

- How soon should you conduct this attack during iterative life cycle development?

- How much effort should be done to test the app?

- Where (lab, field, simulated, and so on) to attack the app?

- Who decides if an update is not correct or a display is too slow, or has some other issue worth fixing?

There is no universal right or wrong answer to these questions, so again careful thought about the answers to these questions by testers and developers alike is needed. Bottom line is this: "What will the user do, say, or think?"

If iterations on Attack 27 continue, consider the following:

- Ways to implement different attacks on the list or the table as the attack evolves

- Interface coverage, if known

- Risk-based testing (what is critical on the displays and processing?)

- Additional exploratory testing

It is good to start Attack 27 "early" and, then, repeat it. This is an attack that may be revisited as the interface and both sides of the system evolve, mature, and get maintained.

EXERCISES (ANSWERS ARE ON MY WEBSITE)

1. Take out your smart phone and test your favorite app (or your project's app). Options to use for students include a mobile travel app such as Orbitz or Travelocity.

    a. Construct a storyboard

    b. Define the app's inputs

    c. Define the app's valid classes

    d. Define the app's invalid classes

    e. Construct a matrix of these (see Table 8.3 as an example)

    f. Determine minimum coverage for a set of tests

    g. Create data to cover these tests

    h. Run Attack 27

2. For your smart phone, find the "angry birds game" or pick your company's game. Load the app on to your device.

    a. Select and define your gamer role

    b. Run Attack 26

    c. Complete the checklists (see the checklist in Appendix F)

3. Conduct Attack 25 on

   a. Your project app or

   b. The Orbitz travel app

4. Obtain the Orbitz app for your smart phone and conduct Attack 27.

5. You are testing a violent war game for adult players. Who should *not* be included in the set of "user" players and why?

REFERENCES

1. Andrews, M. and Whittaker, J. 2006. *How to Break Web Software*, Pearson Addison Wesley, Upper Saddle River, NJ.
2. Kohl, J. Oct. 7, 2010. *"Test Mobile Applications with I SLICED UP FUN!"* Kohl Concepts http://www.kohl.ca/2010/test-mobile-apps-with-i-sliced-up-fun/ (Last accessed April 11, 2013).
3. Mobile web analytics http://en.wikipedia.org/wiki/Mobile_web_analytics (last accessed April 11, 2013).
4. Harty, J. 2010. *A Practical Guide to Testing Wireless Smartphone Applications*, Morgan & Claypool Publishers, San Rafael, CA.
5. Open architecture http://en.wikipedia.org/wiki/Open_architecture (last accessed April 11, 2013).
6. Storyboard http://en.wikipedia.org/wiki/Storyboard (see the interactive media section for the detail of a storyboard) (last accessed April 11, 2013).
7. Copeland, L. 2003. *A Practitioner's Guide to Software Test Design*, Artech House Publishers, Boston, MA, London, U.K.
8. Whittaker, J. 2009. *Exploratory Software Testing*, Addison–Wesley Professional, Upper Saddle River, NJ.
9. Harrison, J.A. 2012. *Mobile Testing to Boldly Go*, CAST 2012.
10. Schultz, C.P., Bryant, R., and Langdell, T. 2005. *Game Testing All in One*, Thomson Course Technology PTR, Boston, MA.
11. Patrick Martin April 13, 2009 "Pentagon holds war game simulating world financial crisis" World Socialist Web Site http://www.wsws.org/articles/2009/apr2009/game-a13.shtml (last accessed April 11, 2013).
12. Van der Spuy, R. 2009. *Foundation Game Design with Flash*, Friends of Ed, Berkeley, CA, ISBN:9781430218210.
13. Rogers, R. 2011. *Learning Android Game Programming: A Hands-On Guide to Building Your First Android Game*, Addison-Wesley Professional, Upper Saddle River, NJ.
14. Khode, A. 2011. *Designing Test Cases for Mobile Applications: Things to Consider*, Tea-time with testers, November 2011, p. 41.
15. Virtualization http://en.wikipedia.org/wiki/Virtualization (last accessed April 11, 2013).

CHAPTER 9

Mobile/Embedded Security

LARGE LEVELS OF SOFTWARE NOW EXIST in mobile and embedded devices such as smart phones, airplanes, factories, medical devices, robots, and many other consumer products. The communication between these devices and the outside world gives rise to security concerns. Until recently, hackers and the security world did not worry too much about such devices. They were too few, not well connected to the outside world, and did not have much interesting information. Now, all of that has changed. Security hacking can cause substantial financial and human losses as well as loss of integrity and system functionality. Security constitutes a risk for many mobile–embedded systems; therefore, testers of at-risk systems should focus on security testing with a bit more intensity than information technology (IT) and personal computer (PC) systems since embedded and mobile devices appear poised to be used in many ways, which malicious people will likely exploit. Many security attacks are common to the existing security testing world, but here I introduce a few more.

It is surprising how many mobile and embedded project testers seem to ignore or not worry about penetration, spoofing, and other security testing. When I talk to groups, they seem to suffer the psychological block of denial, reasoning that it has not happened to them yet, so therefore, it will not happen. This approach has affected people with occurrences of Stuxnet, phone hacks, and many others. Malicious people rely on developers and testers being lazy, which in turn makes hacking attacks viable. It is, in part, a game of one-upmanship, so you must balance risk with cost and schedule to determine how much of this kind of attack is reasonable. Again, look at your local context.

This book focuses mostly on software and data attacks. The following list covers important aspects of security and vulnerabilities for mobile and embedded systems; however, the items appearing in italics, although important, are out of scope for this book.

- *Theft or loss of devices (encryption, IT controls, memory wipe programs, and so on)*;

- *Physical control (locked door and limited access to facilities)*;

- *IT operations (virtual private network [VPN], network control, access monitors, registry logon)*;

- Development and test processes such as developers leaving back doors in the code, testers doing something they should not when they should not (addressed in Attack 28 and Chapter 2), and so forth;

- Software bugs (Attack 28);

- *Operating systems (OS) and COTS bugs not created by our development (secure OS, encrypted files, authenticated files, etc.)*;

- Regulatory and legal constraints (all applicable, including Sarbanes–Oxley Act [SOX] and Health Insurance Portability and Accountability Act [HIPPA], do an Internet search for these terms);

- Data/file input and output (Penetration Attack 28); and

- Impersonation (spoofing Attack 30).

THE CURRENT SITUATION

The addition of large amounts of software and network connections to the outside world makes mobile–embedded devices subject to security testing considerations no matter if they are called smart, cellular, embedded, or otherwise. Hacking or social engineering [1] is a continuing industry problem and will continue to be a concern in the future.

Interest from hackers and publicized hacks has prompted interest in embedded device security. The Coverity Scan 2010 Open Source Integrity Report [2] revealed quite a number of high security risks in open source code primarily focused on the ANDROID kernel but also covering LINUX, SAMBA, APACHE, PHP, and others. This report covered a static analysis test attack and found 0.47 defects per 1000 lines of code (LOC), many of which might be exploited by hackers. While this error rate is 1/10 of some reported industry averages of 5 defects per 1000 LOC on older PC systems, the effort still found 359 defects in total, 88 of which were considered "high risk" in the security domain. And while other embedded devices may be more or less secure, nobody should think that the problem is contained or even well understood.

Mobile and embedded systems are highly integrated hardware–software–system solutions that perform dedicated functions that often require real-time requirements. Often, such systems must be highly trustworthy since they handle sensitive data and perform critical tasks, interface with banks, keep people alive, and control airplanes, to cite just a few examples.

REUSING SECURITY ATTACKS

Security attacks were discussed in *How to Break Software Security* [3], which targeted traditional PC and IT systems. Table 9.1 provides a mapping of these security attacks to consider when attacking embedded or mobile devices. Mobile–embedded software testers can use Table 9.1, adapt these security attacks to mobile–embedded, and then use the attacks for their local context. Additionally, I have defined a few more embedded security attacks for you to consider.

Table 9.2 addresses the attacks of this chapter (excluding the attacks from Whittaker/ Thompson) within different mobile and embedded contexts.

TABLE 9.1   Map of Whittaker/Thompson Attacks Mapped to Embedded Contexts

Attack No.	Attack	How to Adapt (Start)… Then Follow Book	Notes
1	Block access to libraries and/or OS internals	Determine internal files of device	May require insight into the OS
2	Manipulate the application's registry values	Determine device/apps registry locations…	If present…follow the book
3	Force the application to use corrupt files	Determine internal files of embedded device…	Internal embedded files may take special tooling or access to determine
4	Manipulate and replace files that the application creates, reads from, writes to, or executes	Determine internal files of device	Tooling and/or mobile and embedded simulators may need to be used to aid this attack
5	Force the application to operate in low memory, disk–device, and network availability conditions	Force network connect weakness, slow, low RAM, low SD memory, and/or hijack network…	Special tooling and access may be needed to implement this
6	Overflow input buffers	Follow the book	
7	Examine all common switches and options	Determine internal switch/ options of device…	Special tooling and access may be needed to implement this
8	Explore escape characters, character sets, and commands	Follow the book	
9	Try common default and test account names and passwords	Follow the book	
10	Use a tool to expose unprotected application program interfaces (APIs)	Follow the book	Tooling and/or mobile and embedded simulators may need to be used
11	Connect to all ports	Check device's "ports"— USB, Wi-Fi, or network	Determine what can be hijacked
12	Fake the source of data	Follow the book	
13	Create loop conditions in any App that interprets script, code, or other user-supplied logic	Follow the book	Implement using simulator
14	Use alternate routes (in the app) to accomplish the same task	Limited applicability as some devices have limited access paths, but identify them anyway and…	
15	Force the system to reset values	Follow the book	
17	Create files with the same name as files protected with a higher classification	Determine if internal files of device exist…	
18	Force all error messages	Follow the book	
19	Use a tool to look for temporary files and screen their contents for sensitive info	Determine internal files or storage of device…	

Source: Whittaker, J.A. and Thompson, H.H., *How to Break Software Security*, Addison-Wesley, Boston, MA, 2003.

Note: "Follow the book" here means follow the attack as it appears in Whittaker/Thompson's book.

TABLE 9.2   Security Attacks Mapped to Mobile and Embedded Contexts

Context → Attack ↓	Mobile and Smart Wireless Devices	Embedded (Simple) Devices	Critical Mobile Devices (Could Be Embedded, Small but Important)	Critical Large Embedded Devices (Could Be Mobile)
Attack 28	Yes	Seldom	Sometimes	Yes
Attack 29	Yes	No	No	No
Attack 30	Yes	Seldom	Sometimes	Sometimes
Attack 31	No	No	Sometimes	Sometimes

Notes: Seldom, seen only in specialized cases infrequently; sometimes, seen often in many different contexts; frequent, seen regularly in a variety of contexts; yes, should be considered for most contexts; no, generally not applicable.

ATTACK 28: PENETRATION ATTACK TEST

◇ Advanced

Penetration hacking is one of the most common security threats today. In this type of hack, the "bad guy" attempts to penetrate the software–system without proper authorization. Many organizations now do the same kind of thing with penetration tests to assess how secure their systems are. This attack is used in some at-risk IT groups and companies. Attack 28 defines a mobile and embedded software device penetration test. Attack 28 is really a basic attack with a series of related sub-attacks going after several common bugs or holes, which occur all too often in software–systems. The "bad guys" (malicious people) attempt to exploit such vulnerabilities. Attack 28 follows the same patterns of hacking that the bad guys follow. The basic idea and process is introduced here, but to become a good mobile and embedded security attack tester, much more work is needed. So view Attack 28 as merely a starting point, for those in the mobile and embedded domain who, because of risk to their software, need to conduct penetration testing.

When to Apply This Attack

You should apply Attack 28 when your software and/or device is found to be at risk of penetration vulnerabilities, because it is integrated with the outside world (such as in network communications); has information that needs restricted access (such as medical or personal information); or if the data is compromised in any fashion that can allow bad things to happen (such as crashing a control system in a factory or accessing personal data including phone numbers). If your app has a registry feature (e.g., user and password), you may need Attack 28. Finally, if the app under test might pose a risk as a gateway into these areas, consider running some penetration attacks.

Who Conducts This Attack?

Attack 28 should be conducted by an independent, and maybe scared, test team. The team can start with the attacks in this chapter. It is a benefit if this team has some skill in the security domain, such as with software vulnerabilities [5,6], and can think like the bad guys. The team can also benefit from knowing security tools. Another good reference for the team is a book called *Hacking Exposed* [7].

SIDEBAR 9.1

Know Your Enemy: Cyber security specialists are asking the wrong question when it comes to securing computer networks against external threats, said Richard Bejtlich, chief security officer for private security contractor Mandiant of Alexandria, Va. Testing for network vulnerabilities gives Air Force cyber officials no idea of the enemy they are up against, said Bejtlich on March 23 at AFA's Cyber Futures Conference in National Harbor, Md. Instead, cyber defenders should develop requirements for an "Are you compromised" assessment to provide real intelligence on the enemy, he suggested. "I think that could be a real game changer, because right now we're going onto a football field"—nobody knows what the score is, we have a sense that we're getting killed, but the only metrics we have "are on our own defenses and not the forces the enemy is bringing to bear," he explained. "You just paid a lot of money for a test [for which] you know what the answers are going to be," he added.

—From Arie Church [4]

(Reprinted by permission from *Air Force Magazine,* published by the Air Force Association.)

A team composed of security testers, regular testers, security personnel, systems engineers, developers, and/or someone representing the user constitutes a good composite team. This special team still can benefit from the knowledge coming from the developers (code structures), users (who will use the system), and classic software testing concepts and approaches.

Where Is This Attack Conducted?

Attack 28 should be run in a test lab and maybe the field too, but be careful with field testing. Make sure you have the safeguards and permissions to do field security penetration testing. As much as driving a car over the speed limit is illegal, trying to hack software without permission is far worse, and you can get into *big* trouble with the law. Officials (law enforcement and companies) have dedicated units looking for bad guys attempting hacking or trying to do malicious things with their software or systems. You do *not* want to be perceived as a "bad guy." However, just like a race track, you can establish controlled environments and places where penetration testing can be performed "with approval."

Key point: Do penetration testing but make sure you have approval.

The lab test environment, while not as "realistic" as field testing is safer for the security test team and may allow more attacks to be run in a shorter period of time. The lab "sand box" environments should have the following features.

1. Have at least the basic architecture (see sample in Figure 9.1)

2. Have the complete system or as much as possible of it to be tested such as hardware, software, and interfaces or links to external environments

3. Be regarded as a "safe" sand box to attack and run tests in

4. Be well tooled (have appropriate tools)

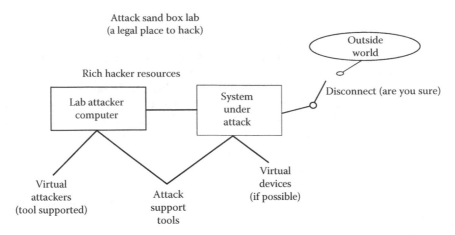

FIGURE 9.1   Attack sand box lab.

The basic attack lab architecture needs to support the hacking attack but still keep your software–system "safe," such as in a "sand box." As an example, students in a lab were able to hack a pacemaker [8], something I do not think anyone wants to see in the real world. As this example demonstrates, the sand box concept allows the attack to happen without escapements into the outside world. If you are trying to hack into a cell phone in the field, you might break into someone else's phone. This is illegal. The configuration must allow a firm and complete disconnect from the outside world, then you are a little freer to play the "evil hacker."

Note: Even in the lab, there may be issues of license agreements, intellectual property, and the ability to reset devices to known states to contend with in security testing.

The ability to create virtual victims and/or attackers can be another nice lab feature. Virtualization in a private isolated cloud can allow penetration through a variety of sub-attacks and attack automation as well as having many attacks ongoing simultaneously. Such lab configurations can be costly and time consuming to set up. To help in this, I recommend you start with these references: *Hacking Exposed* [7], and *The Basics of Hacking and Penetration Testing* [9].

If you are called on to do penetration testing in the real world, *be careful*. You will need to obtain official (legal) permits and document agreements from all stakeholders of the system(s) you are attacking. They need to understand what you are likely to be doing as well as the risks involved. A written understanding of your "risks" to outside parties (e.g., those contracting you or stakeholders) is a good idea. Again, *be careful*. You really do not want to go to jail!

Penetration testing can be run in the lab, in the field, or in the real world. In these scenarios, penetration testing should be conducted as if you were the hacker, which means you use the same tools, information, and skills a hacker is likely to employ. You may also practice social engineering and reconnaissance [1] as part of this type of testing. In the field, you will be trying to determine the ease of penetration or level of risk for your device, software, or app. All of this information is important to pass along to the developers and stakeholders.

One big reason to stay in the lab—besides being on the right side of the law—is that security attacks can damage things. Since in the embedded world, you will be dealing with hardware—sometimes one of a kind hardware, and it is possible to physically damage that hardware. When attacking in the real world, be *very careful* (Have I said that before? It deserves repeating.). You can also damage the software and, via corruption, the data. A simple reload may solve the problem, if it is really a full reload, which may mean the OS, supporting apps, databases, and so forth. There are cost and schedule factors associated with reloading all of these items. This is where the lab and tooling can make life much easier by supporting fast reloads and getting back to a "clean" test configuration.

What Faults Make This Attack Successful?

Attack 28 is going after weaknesses in access points, such as the following:

- Registry—passwords and authentication

- Data input streams—checking source or validity

- Connections—checking validity

- Controls—where has the data been or gone within the system

These weaknesses are only a start and will change over time, so the diligent attacker must stay on top of the literature and trends. The sub-attacks that I have included tend to focus on aspects of each of them.

How to Determine If This Attack Exposes Failures

Ultimately, when you are conducting the penetration attack, you are looking to gain control over the system or obtain information that you should not have access to. Some call this exploitation of vulnerabilities. When you have the information on the vulnerability, you have something that you can report to stakeholders. Whether a project team decides to fix the problem and how is unique to the situation, but you still have to provide the test information to them.

How to Conduct This Attack

First, understanding and knowledge by what some hackers call reconnaissance or information gathering will aid in performing this attack and will be your starting point. This may be built on information mining from development sources; public information from the Internet; covert information gained from social engineering; plus your basic tester knowledge built up over years. What you do and how you do it are largely determined by the context. If you are an internal tester, many informational avenues are open such as code internals, planned usage, internal checks, bug reports, architectures, languages, and so on. If you are an external independent tester, you may be relegated to public information, your knowledge of security systems (and where bugs hide), and social engineering (again, *please be legal*). Your target(s) may be one or more of the following.

- How is the device to be used that a penetration hacker might be able to leverage?

- What are the test objectives and legal concerns of your security testing?

- Where are the inputs coming from (to the software or app or directly to the device)?

- How are the inputs being used (target names, addresses, internet protocols [IPs], etc.)?

- Where are the (software or app's) outputs going or being used?

- What, how, and where is the data flow in the system or device between input, usage, and output (if you have such visibility)?

- Who can access the system or data (users, hackers, systems, networks, not just humans)?

- What is the device or software doing and how (functional and nonfunctional requirements)?

- How will users, including hackers, act (this is where use cases or operational profiles come in handy)?

- What information can tools provide you (see tools in Chapter 11)?

- Have you identified common and unusual threats (data exposed, client information, money losses, plane crashes, etc.)?

From this, you create a starting list of weaknesses or threats for your software or app. This list will be made up of the kinds of things that hackers might exploit. Depending on the context, these lists may be long or short, simple or complex, but in any case, you are trying to gather and organize the information to improve your understanding. You should also expect to come back to this activity and revise your list over and over as you learn about your system and its vulnerabilities.

From this starting point, you then create a specific risk list (see Appendix G). To add to the risk list, the penetration attack team should ask—on an ongoing basis—the following types of questions.

- What item(s) are at risk (data, control, information, access, etc.)?

- What is the value of the system, data, or processing if lost or compromised (lost money, lost life, lost equipment, power outages, etc.)?

- How is the information or risk processed in the system and software (where do vulnerabilities provide access (see [1])?

- What protections are there (or not) and how can they be circumvented (passwords, trusted systems, identification checks, back doors, trusted systems with lesser security, etc.)?

In this ongoing set of activities, you want to identify all points of penetration (such as network, registry, input file, etc.) to populate a security risk list. This is done during the whole cycle of security attacks [7,9,10].

Next, as the system comes online, you will need to analyze both the hardware and the software. Analysis approaches include inspection, modeling, and checking. In inspections, you visually examine the system, hardware, software, and development information looking for penetration points. This may be the easiest place to start but the hardest to do well. I suggest doing a brainstorming session [11] (see Appendix G). Try to capture the following information:

- An understanding of the system and its functionality or nonfunctionality;

- Key points and/or known concerns;

- End-to-end uses (likely a series of these) done in use cases, stories, or operational concept scenarios (examples only);

- From these, identify key cases and/or scenarios (higher risks); and

- Known in use security mechanisms with any pattern or ideas on how to break them (see sub-attacks).

This information gathering should be agile and informal.

The next technique in your "bag of skills" is modeling. In modeling, you create a more formal representation of some types of information for the system in an attempt to understand weaknesses. Models can be diagrams in semiformal tools such as PowerPoint or other graphical programs; in a mind map; or using a more formal language such as unified modeling language (UML) and/or systems modeling language (SysML). The model attempts to capture and identify the weak points. In modeling, you will capture your understanding of the security of the system, which can then serve as a "map" of the attack on the system.

Once you have inspected and modeled your situation, then start some detailed check testing. In checking, you will use visual scanning and/or automated scanning tools (see my website for some examples) to look for the high-risk penetration points of control. The scans identify candidates. Once you have completed them, check to see which ones are real high priorities, real but lesser priority, and which ones are not worthy of a penetration attack.

During the checking and scanning efforts, you will likely be moving into a lab using tools to get information on the system and software. In this early analysis and testing, you are trying to determine first if you can detect a live system (think *ping*) when you should not be able to do this. Next, can you detect an entry point, such as an open communication port, and if so, you will then want to see if the tools you have can detect vulnerabilities in the system, such as a port that is not protected.

The exact scan, tooling, and vulnerability assessment is very system dependent. This is detailed information mining. *You will learn.* Update your vulnerabilities risk list as you do these things. Also, be forewarned because at this point you may start exposing actual holes and vulnerabilities. For example, you may find unlocked or unprotected devices, ports, or files that should be locked. Look for rich resources, tools, code, and other information to help you understand the vulnerabilities.

After these initial probing activities, update your risk matrix.

You cannot test everything on a penetration risk list, so use your matrix to prioritize and determine a level of testing rigor. This will drive how many attack tests to run and when, which is detailed test planning. Now, with the risk list and plan in hand, do some test design [12] and implementation using other attacks and sub-attacks.

> **SIDEBAR 9.2**
>
> Pacemaker Hack: In 2012, college students with just a PC and software radios were able to hack into a medical pacemaker with minimal effort [8]. What if they had done this while the pacemaker was inside of a patient?

Next, implement and run the penetration attacks based on other items in the risk matrix. The specifics of the implementation will leverage the vulnerabilities you have defined. The penetration attack tests should be fast, likely unscripted (informal), exploratory in nature, and wide ranging (covering as much as you can but with depth too). You may be able to justify more of either cost or schedule for added attacks, depending on what you learn during this testing.

> **Key point:** Penetration testing is a specialized form of exploratory testing, which targets vulnerabilities. A good penetration tester is both a security–vulnerability wizard and an exploratory guru. It takes much practice over years to develop the skill and mental models to do this well (and, again, to remain legal). See my website for more information on this attack.

ATTACK 28.1: PENETRATION SUB-ATTACKS: AUTHENTICATION—PASSWORD ATTACK

☐ Intermediate

This is classic cracking into the software system. You are trying to penetrate the app or embedded device via the app's registry information, by thwarting the authentication. The registry includes user identities (IDs) and passwords. Password and registry penetration testing is detailed in books such as *How to Break Software Security* [3], *Hacking Exposed* [7], and others. Additionally, there are specialized sub-attacks on registry authentication such as buffer overflow, Structured Query Language (SQL) injection, and others, which should be considered at the registry point, but these sub-attacks are not covered by this book (see my website [13]).

Authentication is where a computer system positively identifies a user. For the basics—you probably all know and love it—the user name (or ID) and password combination. It is also the favorite weakness or attack point in computer security today. Multi-factor authentication adds a critical second layer of security to user logins and transactions. It works by requiring any two or more of the following:

- A thing you create or change that is "secret" (typically a password)

- A thing the system gives you (user ID and/or account number on a mobile or embedded device)

- Something you are (biometrics)

For the current sub-attack, I target the first two, since biometrics are not in general use, yet. Additionally, I will not get into how to hijack a device based on internal device identification, but this is a concern area for many security testers.

How to Conduct This Attack

In this sub-penetration attack, you are trying to break the user ID and/or password. First, you must know the user account identification. For testers totally external to the development group, obtaining user IDs that are not known can follow social engineering "tricks," such as ask someone, look on desktops, peer over someone's shoulder, search people's trash, and others. Common user IDs can be guessed based on things like user email IDs, last names, last name–first name, first names, company names, account numbers, and other publicly available information. For testers internal to development team, the tester can use a user ID that is known in the system, since the password is often seen as the main gatekeeper.

Once you have the user name or ID, then you must break the password. Industry data points indicate that many passwords are poorly constructed. There are tools that will "guess" or crack passwords. Wikipedia lists over 10 such programs including Cracker, DSniff, and (my favorite) John the Ripper. Quick internet searches will yield a variety of programs of which many will be specific to a context.

Lacking such a tool, you can use common passwords such as reversing the user IDs, user phone numbers, pet and children names, and other passwords (see Table 9.3). I have hacked into accounts in just a few minutes using passwords from this list. I do recommend a tool, but the common password table work too.

Once you have a tool, generate the candidate passwords into a list or table. The exact mechanics of generating passwords from a tool will be determined by following the tool's user information. In the case of manual generation using Table 9.3, the tester takes their "best guess." I recommend generating several hundred passwords or more.

If you are doing manual testing, you will enter each user name–password pair, one at a time, until you break in. This can be time consuming, so with a little bit of programming, using a language like Perl or Ruby and with an interface to your device, or using a capture–playback style input tool (see Chapter 11), you can create and input many name–password pairings, which can then be entered automatically into the registry until logon is achieved.

Note: Depending on the nature of the registry, more or less programming may be needed to account for system actions and responses.

If you have now gotten in (hacked) an account with minimal effort, you have the beginning of a bug to report. And you certainly have information that the stakeholders may find actionable (such as improving passwords). Many organizations now require "hard" passwords (a mix of letters, numbers, and special characters). Still, many do not. Getting into an account may or may not be a bug.

SIDEBAR 9.3

Smart Meters: In 2012, researchers found security flaws in "smart" meters, which are designed to replace the electric meter on your house. These meters allow your house to be connected to the power grid. The flaws would allow access to user data and worse, turn power on or off to the consumer [14].

TABLE 9.3   Most Commonly Used Passwords

Most Commonly Used Passwords (in the Order of Use)
Password
123456
1234567
12345678
Qwerty
abc123
Monkey
Letmein
trustno1
dragon
baseball
111111
Iloveyou
Master
Sunshine
Ashley
Bailey
Dilbert
passw0rd
shadow
123123
654321
A repeated keyboard sequence, repeated five or six times (e.g., asdfasdfasdfasdfasdfasdf)
Superman
Qazwsx
Michael
Football
A repeated number or multiple numbers repeated five or six times (e.g., *77777777-77777777-77777777-77777777-77777777* or 12345678-12345678-12345678-12345678-12345678)
Baseball
Sport123
idontcare
A first or middle name often with 123 at the end
Names of spouses, children, or family pets often with 123 at the end
Home address info

ATTACK 28.2: SUB-ATTACK FUZZ TEST

◇◇ Expert

Fuzz testing is a software test/attack/approach that provides large amounts of valid and/or invalid inputs to the device. These inputs are in unexpected orders or are random in nature. Given the volumes and nature of the data, there are usually tools that generate the data for automated or manual input. During the attack, the intent is to crash the program or get highly unexpected results, such as memory leaks, file access, warning messages, overflows, or hanging systems, which perhaps can be exploited. The bugs are exposed by pushing the software into code paths and fault recovery logic, which were not properly designed or tested in earlier attacks. Additionally, if large data sets are input, some kind of log program is used to monitor the software under test and record the unexpected, so it can be replayed and debugged later. This approach has been used for general software testing, but most recently, it has grown in use for security testing during penetration or other hacking attempts at the registry. This penetration hack has some advantages over Attack 28.1, since you can use this sub-attack even on the user name and password without knowing a user ID.

How to Conduct This Attack

There are several approaches to generate the test inputs. One approach creates inputs from a representation of the input space based on combinatorial input modeling (see Attack 32). Another approach uses totally randomly generated inputs (think random number generators). A final approach randomly mutates (changes the inputs of) actual inputs from earlier testing. Even the very basic random generation of inputs can often find defects, which may represent security flaws.

In the fuzz attack, the tester starts by selecting one of the approaches and then selects and configures the tools. There are now generic "fuzz testing" tools for many types of systems. Tool names change rapidly here, so I suggest an Internet search (for fuzz testing tools) refined for your project context. As password sub-attack, you will need at least two tools, one to generate the inputs and one to input or monitor the execution (a capture playback or logging tool as in Chapter 11), unless you want to manually enter huge amounts of data (possible but not fun). Once you know how to set up and use the tools, you can run the attack (series of inputs) by generating large numbers of inputs and executing them.

Here again, iteration and increments will be needed. In some cases, you may need to refine your input approaches, by changing things like tools used, different data points, volume of inputs, and other variations of input. Run the attack again. During and after the runs, it is advisable to review the recording (log) tool. Upon reviewing the resulting logs, you may see interesting results to follow up on, learn from, or even a bug (e.g., a short error message), in which case you have items to follow up on. More often than not, additional fuzz attacks or other security attacks may be needed to isolate a "real bug." This is where tester skill, hacker knowledge, and creativity will come into play. To get good at these, you will likely need to do some added reading and research on this topic. I recommend the following sources: http://en.wikipedia.org/wiki/Fuzz_testing, and then Art Manion and Orlando Michael's presentation, *Fuzz Testing for Dummies* ICSJWG May 2011 [16]. Fuzz away.

ATTACK 29: INFORMATION THEFT—STEALING DEVICE DATA
☐ Intermediate

Mobile and embedded devices have security issues and are new enough that the mobile and embedded security bug history is still being written. At the same time, smart devices are used to do shopping, pay bills, order stocks, check finances, transfer money, airport security checks, keep people alive, store personal data, emails, and even register truck and trailer loads with ports of entry. The mobile nature of embedded smart devices adds new security concerns for attack targets.

Mobile and embedded device information theft and detecting it will be a source of security bugs and attack testing for quite some time. For example, I am out shopping using an e-commerce-enabled smart device. I am in my favorite hardware store scanning bar codes, comparing prices, and looking for good deals. The phone may identify "I am Jon's phone, and that I am located at x.xx position with the following personal data…" How does anyone know that to be true, and what if it is not me? What is possible/impossible, and what will the hackers do?

Do not think that this is a big issue and common situation? It is estimated that over 14% of e-commerce transactions will involve smart devices by 2014, and that figure continues to grow every year [17].

This attack provides a framework to assess the information theft risk that mobile and embedded devices face. The attack can expose weaknesses or bugs that a team may want to fix. Again, the attack follows the cracking patterns reported in books and websites, which many of the bad guys use.

When to Apply This Attack

Consider this attack when the device or application can be used for mobile e-commerce and contains personal information including, but not limited to

- Account numbers
- User-IDs and passwords
- Location tags
- Other personal data (bank account numbers, credit card information, emails, medical information, company intellectual data, etc.)
- eMoney (used to buy and/or pay for things, move money around, etc.)

All of this data needs to be protected from information theft on the user side (smart device) as well as on the institution or server side that the user is going through for transactions. Your testing may be limited to the mobile software–app side, but you cannot forget the server side, as in some cases, you may need to attack test that too. You are dealing with systems now where testing an app in isolation may lead to big security holes that may actually allow hackers access.

Who Conducts This Attack?

The tester and/or a security test team will conduct this attack. Some specialized knowledge, tooling, and skills are required here. It also helps to have some consulting engineers or senior knowledge experts who are up to date on the latest hacks and approaches. Testers should pay attention to the hacker sites, books, and latest news articles. Multiple alerts on security issues are published daily. Thus, the attack details should change as quickly, and the team needs to keep up with these details by keeping current on what the bad guys are doing.

"Quality must be built in," so the saying goes (Appendix C). And, as a quality charac-teristic, security must also be built in. However, the tester's independent eyes and mindset should be brought to bear to provide security information, as well as data points on what is working or not. Specialized knowledge and skills are required here. Finally, some testers are just more devious and will make good security testers. Hire them! Also companies special-izing in security testing have some experience in handheld devices and are good candidates to consider too.

Where to Conduct This Attack?

A specialized test lab with the right tooling is the starting point here. A security test lab with the full system, security test tools, and access to both sides of the equation (server and mobile client), as well as access to the software under test, is called for here. This means that you are going to do white box, system black box, and in between testing. A lab that has tools to access the internals of the hardware and software (such as source code, execution traces, internal data, timing points, etc.) can provide useful data points, which may feed into secu-rity. The bad guys have access to lots of tools. The lab should also have these tools, and more. (Follow OWASP, events like SANS Appsec events, and standards published by OMTP, WAC, BONDI, and details on ARM TrustZone. Research these items on the Internet.)

Further, the level of risk the app faces determines criticality and level of security testing. A new networked game app, where you are in the development space (and source code), may need less security attacks and focus, but consider this: should issues with "hijacking" be worried about? For more critical apps with access to banking records, personal infor-mation, medical information, financial information, other sensitive data will need more of this attack and its subparts likely in different locations and facilities.

I suggest lab test time, running both white and black box tests. In the lab, apply test tools such as static security analysis and multiple attacks. Follow this with field testing time, where devious testers attack. It may be necessary to have several different testers in different locations and/or situations, since after a while, an attacker may lose some of their creativity.

How to Determine If This Attack Exposes Failures

Here the tester is looking for a social engineering hack. Did the attacker gain access to private or personalized data? Were they able to make an illegitimate charge to an account? Did they get access to information that they should not have access to such as medical or banking infor-mation? How can you find information that would help a hacker to commit fraud by stealing personal information? Can you pick up someone else's phone and do "bad things" before the phone is "wiped"? Look outside of the phone itself to the social environment in general.

Specifically, you should be looking for bypass points, access to registers, internal file information, or other sensitive data, in whatever form it might appear. Understanding the entry points and protections of the system, software, and/or network will help determine errors that may be seen and which attacks to run. During the attack, you should be watch-ing for subtle clues such as file access and indicators that something is not quite right. Many security bugs do not announce themselves. Instead they may hide in system crashes. For example, during a buffer overflow, strange error messages like error-402, or subtle holes such as apps that upload or download unexpectedly. When subtle clues are found, it will be necessary to attack further or gain team understanding to expose the real security bug. Each stage and substage in the attack defines specifics, which will need to be checked. Hackers spend hours (days) exploiting small vulnerabilities into larger ones. Test teams may need to do this too, depending on project context.

What Faults Make This Attack Successful?

Just as in IT-based e-commerce and real-world commerce, information theft can happen anytime. People assume our identities, steal our accounts and passwords, or in other ways commit information theft. This is nothing new, but smart mobile and embedded devices offer new avenues for the bad guys. Currently, the approaches on embedded authentication for detecting and preventing information theft include the following:

- App–software to server-based registry (device ID, user ID/password)

- Location–device based (device ID)

- Profile based

Server-based apps capture information about a user's browser and embedded device at logon. If the application is browser-based, a script captures a unique ID and other data to identify the embedded device. If the application is fully native on the embedded device, hardware serial numbers and network information can be provided back to the server as an "identity." In some cases, either with this information or in place of it, a registry entry (user ID or password) may be used to access and log on to a system or account server such as web pages, bank accounts, email, and so on. These identifiers create the "identity." As with IT systems, this is a major entry point for hackers. The use of smart cards, tokens, and other "hard" identifiers is also possible.

The second method uses the embedded device's location, internal identification codes, and the fact that the embedded device is on (broadcasting itself). Location can help authenticate the user through correlation of information such as a user's home address, work, city, or other specified location data. The device has an identity as hardware and/or software based, as mentioned in the earlier paragraph. Additionally, a user may opt in and must "approve" location authentication information, like "yes, I am Jon (his phone) located in Hardware ZYX store right now." The "opt in" remote locations are usually temporary, while the home locations tend to be more permanent. Again, these identifiers create an "identity" and location of a user or device.

The third approach is similar in concept to the risk scoring done by credit card companies, where they look for bank card transaction fraud. However, this approach is created for mobile applications. Here, the server looks at device, user, location, device identifications, user characteristics, and other factors to decide if the device and/or connection has "identity" and is legitimate or not. As these systems evolve, it may be possible to check that a user or location matches logically. For example, if I used my device in Arizona this morning, it is not likely that I will be in Russia this afternoon. These types of systems can refine identity.

Once the system determines the device is legitimate, an approved "identity" is created, and certain actions become possible.

These systems will evolve, have hybrids, and blend over time. Risky situations involving loss of money or critical information may require more levels of authentication. The addition of concepts like biometrics and other advanced identification methods will likely come to the market place soon.

> **Key point:** Information theft in mobile devices will be a key security risk for many mobile and even embedded software apps.

How to Conduct This Attack

First, determine if information theft possibilities exist within the scope of your mobile or embedded app testing. Then, use the following steps and sub-attacks.

Determine the identity checking method being used such as one of the standard methods or something new. If no method is being used, be concerned because you may have a bug or at least a hole—as well as information that the stakeholders should consider.

This initial information is used to determine the nature and amounts of testing and what to do next. Create an information theft risk list. This classification should be coordinated with stakeholders. Since usually security test efforts have limited time, budget, or both, the items classified as "lower" risks will get fewer (or no) attacks. Higher risks will get many more attacks. The balance between low and high risks needs to be considered and reviewed as the attack efforts provide information (risk plans are likely to change).

This is a composite attack. You use one or more of the following attacks, based on risk and products needed to conduct Attack 29:

- Attack 28.1

- Attack 30

- Social engineering Sub-Attack 29.1

- Fuzz Attack 28.2

- Attacks from Table 9.1

In these efforts, you are trying to take over the target device. You will not make the information theft happen on the first attack, so you should iterate the attacks to the level of risk of the product learning about weaknesses and vulnerabilities as you go.

ATTACK 29.1: SUB-ATTACK—IDENTITY SOCIAL ENGINEERING

◇ Advanced

Is testing of the server in your testing charter? If you are chartered to only test the app, you may need to stop and/or politely inform the server side testers "this is the information that I have or can provide from my attacks so far to attack the server software." The app test team may be told to continue and not worry about the server testing, but you have now provided information to the stakeholders. And in the world of clouds and distributed processing teams, you should be aware of all sides of the system, because these types of "holes" and excuses such as "don't worry it is someone else's problem" are the places where the bad guys will take advantage. Good security attack testers consider the whole system.

If you are chartered to test in whole or in part into the server or are looking for complete system-level app testing, you should provide different spoofing identity attacks to the server. Ask these questions: "What happens?" "What accesses can you gain?" "How much mayhem can you create?"

After gathering some initial early test data from these attacks, the team should review the social impacts of the test data on app and/or server systems. Minor social impacts may mean minor variations of these information datasets are needed. Major social impacts such as loss of company data on an ongoing basis, large money losses, somebody could die, risk of lawsuits, and so on may mean more data sets, and even continuing attacks over the life of the app's use are required. Ongoing diligence is increasingly common in organizations with apps and devices at risk. Many companies have dedicated vulnerability detection and avoidance groups based on the risks of information loss.

ATTACK 30: SPOOFING ATTACKS

◇◇ Expert

Attack 30 outlines how to start and conduct a mobile or embedded device spoof. Spoofing, in general, is fooling somebody or, in this case, software into "thinking or believing that one thing is something else." This is done to gain access to information or control over the system. In the mobile and embedded world, spoofing is alive and well. This attack presents concepts to allow testers to spoof the app or system with a goal of providing information to the stakeholders on the ease and nature of the spoofing vulnerability. This spoofing attack should be done when the risk justifies the attack.

A rumored spoof is given in Sidebar 9.4. The details are and likely will remain classified, but even if this story is partially true, mobile and embedded attack testers should take notice.

When to Apply This Attack

Testers should consider running Attack 30 when information in the mobile and embedded app is coming from somewhere else; when the information is used by the app for something

---

SIDEBAR 9.4

U.S. Drone Spoof: It has been widely reported in the media [18] that a U.S. spy stealth drone was subject of a global positioning system (GPS) spoof, which resulted in the systems capture by Iran. The details and exact nature of the attack and even if there was an attack are subject of much speculation, but if this story is even partly true, the use of GPS spoofing represents a concern to many mobile and embedded systems that use GPS. I suggest you read and learn more about this GPS spoof.

that is critical or otherwise important to the user; and when that information can be compromised before coming into the device. Spoofs that have been seen or reported in mobile and embedded systems include the following:

- Caller ID spoofing
- Website spoofing (not specific to the device or app)
- Email spoofing
- IP address spoofing
- Location spoofing including spoofing signals, GPS, or other
- Network spoofing
- SMS spoofing
- Referrer spoofing, a type of spoofing attack (http://en.wikipedia.org/wiki/Spoofing_attack)

Note: Most of these spoofing attacks can be found on the Internet using Google or Wikipedia.

Having or using these features is a first indication of spoofing risks, and as a test "attacker," you should give each respective spoofing risk its due attention.

Who Conducts This Attack?

As in other security testing attacks, Attack 30 is conducted by an independent test team and those with special skills. This team should

1. Have the capability to play the role of the bad guy and be very devious
2. Have knowledge that the bad guy would have including bugs, holes, tools, techniques, environments of software and systems, as well as knowledge of specific security and technical books
3. Have great knowledge of the product to be spoofed, how it will be used, and the implications of it being spoofed and
4. Have an understanding of the motives of bad guys

As mentioned in the earlier attacks, persons involved in performing this testing may be an exceptional single "hacker" person or more likely a composite team of security testers, regular testers, security people, systems engineers, developers, and/or user representatives.

Where Is This Attack Conducted?

As I stated in the other attacks in this chapter, Attack 30 is likely not conducted in the field—at least not without great care, permissions, and resources, but in a lab where special safeguards can be in place and the right resources exist. For the basics on the lab set-up, refer to Attack 28, as a beginning.

To refine the lab to support spoofing, consider adding the following:

- Specialized tools

- For location spoofing, special signal and transmitter generators may be needed

- For other spoofing attacks, "peak and poke" tool features to peak or poke into files and/or memory may be useful

- Skills and knowledge to use each of these

What Faults Make This Attack Successful?

Attack 30 is successful because the app or software trusts some of the incoming information. This trusted information can be displayed or used in other ways in one app, multiple apps, or on the device. The trust is often because there is no checking, authentication, verification, or validation of the information and/or the source of the information. In this case, the data may have been hacked prior to delivery to the device or app by someone outside of the software or system. The level of checking, authentication, verification or validation, as well as trust can vary, but the more critical the information is in the overall success of the software or app, the more checks may be needed, which means more testing is required. Checks can include but should not be limited to the following:

- File signatures (internal file identifiers)

- Asking the user for authentication

- Asking the sender for their credentials with verification and validation of credential data

- Sender trust (how the systems are configured where one system "trust" another)

If these mechanisms are not in place or misused, then the software is weak and will accept the information as truth without checks in place. This should be considered a fault and possibly a high-risk situation.

The fault can also come from "outside" of the software from hardware or the environment. Here you are concerned with signal overloads, false connections, fraudulent systems or sites, and so on. In this case, your software or app may need to determine the spoof by various crosscheck means, which can be very difficult or impossible to do. Here you should be just looking to demonstrate the *possibility of a spoof*, not how to fix it.

In all of this, the question you should be trying to answer is "What is possible?" This question must be asked with a mindset of cost, schedule, and the risk to the software or app. Things cannot be totally "spoof proof," but the risks should be considered for most software and apps. *There is no single right or wrong answer here.*

Chapter 11 (security section) lists some tools to get you started, but keep in mind this is not a recommended list, just a starting point.

How to Determine If This Attack Exposes Failures

You will know you are successful with Attack 30 because the app or program will do the following kinds of things:

- Display false data

- Accept and/or act on false data

- Provide no warning or indication of concern about "trusted" data when it is "accepted" without verification

To determine that you have a spoofing vulnerability, you must see if you can "trick" the software or app into accepting the information. If the information is used, then the criticality and likelihood of the spoof must be determined, before development action can be determined. Finally, not all unverified and trusted data constitutes a risk if spoofed, so spoof risk is project context dependent.

How to Conduct This Attack

As in Attack 28 (you should be familiar with that attack too), you need a solid understanding and knowledge of your system gathered with reconnaissance or information gathering. I will not repeat all of the information of that attack here. Instead, I encourage you to follow its basic outline with tailoring for spoofing the data and inputs. Your targets may include the following:

1. How the device is to be used, particularly where there is a chance of spoofing incoming data.

2. What are the objectives and legal concerns of your security testing (this helps determine the level of spoofing)?

3. How are the inputs coming in (network, land line, radio signals, hard device [universal serial bus or USB], and nature of these inputs)?

4. How are the inputs being used (target names, addresses, IP addresses, etc.) in the processing?

5. Where are the apps or software outputs going or being used (other software or apps, which might be at risk of spoofing)?

6. Identifying and understanding all the interfaces and inputs (look for back doors?).

7. Understand the data flow in the system or device between input, usage, and output (if you have such visibility).

You are trying to understand the spoofing weaknesses and threats for your software or app that might be exploited. This data will be used to feed a risk analysis, which, in turn, will produce a risk list (or matrix) as in Attack 28. However, this list will be different since it will be based on the spoofing. This list will also grow and change over time. The spoofing risk matrix will prioritize the testing attacks. This early risk analysis activity can be started during concept definition and early system analysis, but can also be done later on.

As the system (hardware and software) is developed and comes available, you will need to continue analysis, updating the risk matrix as needed. As outlined in Attack 28, the analysis approaches can include inspection, modeling, and checking of development information and products. You are looking for interface points and data inputs where spoofing can happen. Each of these points or inputs can follow the concepts outlined in Attack 28, but in Attack 30, you should try to capture the following information:

- A good understanding of the system and its functionality and/or nonfunctionality.

- Refinement of the actual interface and data points open to spoofing including file information, standards being used, and data specifications.

- End-to-end uses (likely a series of these) captured, for example, in use cases or operational concept scenarios.

If input and output data interface models and information models already exist and are complete, you can use them during this analysis. You can also generate your own models and information to supplement any existing data. From all of these, you will identify key spoofing cases and/or scenarios, prioritized by the risk matrix.

Once you have inspected and modeled your situation, I recommend you start some detailed exploratory testing with peaking and poking into as much of the software as you can. Here you are trying to find the interfaces and data details that are not known from the inspection or modeling. You are also trying to understand what to spoof and how easy or hard the spoof will be. Easy spoofs might indicate a bug to follow. Hard to do spoofs may require more work and could still indicate a bug. However, both of these spoofs must be bound within cost and schedule factors.

At some point during the analysis, checking, and scanning activities, you are likely to be moving into an attack lab. You may have security attack tools (listed in Chapter 11) to improve your understanding. Try first iterations of your spoofing attack concepts on the system and software. In this early prototype testing, you are trying to determine first, if you can access the system data (can you see the file, data, and/or is it encrypted?). If you can access the data, next you want to see that you have and can "change" and/or "fool" with the data and what the effects to the system or software might be. Maybe you should or shouldn't be able to spoof data. Often "fooling" with the data will require the use of other tools. This is a trial and error process where you will get smarter as you go, so document your efforts, as much as possible. You should be creative. You will learn. You may not get the data spoofed on the first try, or if you do, you may have a pretty good start at finding a bug. Update your vulnerabilities risk list as you work. If you spoof things, the team should ask, "Is the spoof of concern and what if anything should be done about it." Also, as the spoofing attack continues, look for rich resources, tools, code, and other information, to help you understand the vulnerabilities. You may have to add tools and search for new ways to conduct the spoof.

These early attack activities are not complete, but they form the basis for the rest of Attack 30 based on knowledge and history gained so far. After these activities and with an updated risk matrix, prioritize the rest of the spoofing attacks that you want to run. You may have found some bugs and specific weaknesses or other important information that can be provided to team.

Using the risk list, do some test design and implementation. The specifics of the implementation will leverage the spoofing vulnerabilities you have defined. The risk matrix drives the attack. You will design the attack tests in the form of spoofing scenarios to leverage the identified weaknesses including sub-attacks.

After designing the attack, run the test. As you run spoof attack tests, LEARN. As you learn, REPEAT the activities provided earlier—as time and money allow. The spoof attacks should be fast, likely unscripted and informal, exploratory in nature, and wide ranging (covering as much as you can but with depth too).

> **Key point:** Spoofing attacks are a "learn as you go" advanced concept. This "how to" section is a basic starting point only.

In spoofing of data, some hackers will try to hide that the data or system has been spoofed, so that false information is used. This may or may not be important to try to do in the attack at this point. It may be enough just to know that the data has vulnerability and that may be good enough for a bug report. Other times, you may need to show how far the spoof can go, and what measures are in place to detect and/or correct the spoof. How far to take this effort is risk and situation dependent.

Testers that get good at spoofing will expand beyond the basic patterns in this book. A larger amount of creativity is possible here. Also, the tools keep changing, so efforts to stay current are needed. Likewise, the checks and safeguards in the software or apps will keep improving, but nothing is complete. So, while the basics of this attack may fail in initial spoofing attempts, that does not mean that a spoof is not possible, only that more work and thinking are needed.

ATTACK 30.1: LOCATION AND/OR USER PROFILE SPOOF SUB-ATTACK
◇◇ Expert

In this sub-attack, you are trying to fool or spoof the device or app identity based on use and/or location where the device is being used. For starters, you should see if the identity can be "hijacked" and data accessed or sniffed. This can be done using a tool like Wireshark [19], which can sniff and decode data being broadcast. Wireshark analyzes UNIX and WINDOWS network protocols and gives visibility into the network internals and messages. Tools like this are available in the public domain and even have an active user community who will help in their use and modification (see Chapter 11, security tools for more information).

As this sub-attack continues, you should go after the mobile and embedded devices network using such a tool. Monitor the traffic of the developers or other testers. What data is sniffed out? What data can be "ferreted out" by digging a little with other tools? If you are finding user IDs, passwords, or other useful information, the identity is spoofed and you have a bug.

Next, if location is used in determining identity, check to see how the location is used and if authorization is temporary or permanent. If it is temporary, the attack should check for remnant data files in temporary storage areas, which might contain sensitive data that might be subject to data mining. Use development tools and/or the OS to poke around in the file system of your device. *Warning*: The file may be encrypted, in which case you may need a file encryption cracker for that type of file encryption (such as pkcrack). Your tools should be specific to your file or system. There are many sites that will tell you how to use tools—even YouTube may have step-by-step instructions. When sensitive data is obtained, you have a bug.

If the "location and identity" file is not temporary, you next need to determine if any of the permanent information can be accessed, abused, or corrupted to aid in hijacking or information theft spoofing. Look for storage files in the device's short- or long-term memory. In many devices and apps, this data should be encrypted and here again apply the cracking encryption tools. If sensitive data is accessed, you have a possible bug.

Once you have the location–identity file information, ask yourself "Can I spoof the location on either inside of the device or what it is broadcasting?" Each system or app will be structured differently, and the security checks will be growing. This attack hack will be a little harder than the first two and may not be possible at all. Of course, if a hack is possible, you have a bug.

Closely related to location identify spoofing is the user profile spoof sub-attack. Here you will attempt to take over an identity by understanding how user profile checks work or do not work. This will require understanding the internal data points of what your system is checking. This will take some cooperation from the developers or a lot of hacking time. Many data point factors are closely held secrets, but if too few are used or not used properly, you may be able to "assume" an identity. And if a tester can do it, people with malicious intent can too. User profile factors to look for are location, time, where transactions are occurring, type of transaction, money amounts in transaction, provider or store, product, and biometric data. These combinations of factors might be a good place to consider combinatorial testing or the fuzz test sub-attack. Once these factors are determined, you may want to input them into the system, then determine if the server gets confused or gives the wrong or sensitive data. This borders into a server test, so be careful with this step. (Is it within your scope?)

ATTACK 30.2: GPS SPOOF SUB-ATTACK

◇◇ Expert

In GPS spoofing, the attack tries to fool the GPS system by providing a more powerful signal, which overpowers and fakes one of the on-orbit GPS satellites. The GPS device and the system being spoofed will then use incorrect location coordinates in making decisions and calculations. The spoofing information must be determined by the attacker ahead and during the attack. The difficulty is that a spoofing signal must be generated to resemble the normal GPS data, but modified in such a way as to cause the receiver to determine its position to be somewhere other than where it actually is. GPS systems are based on time–distance differentials between multiple satellite signals (each satellite outputs a time signal). GPS spoofing requires that the attacker know where the target device or system is and then adjusts the higher power spoofing signal with the appropriate time changes to "fake" the system under attack to a false position. This sounds easy but usually takes multiple passes, since most GPS systems may detect signal "problems." Substeps in the attack are thus:

1. Determine the need for the attack based on the risk.

2. Establish lab environments necessary to conduct the spoof such as labs with transmitters that can vary power levels while adjusting the time stamp signal.

3. Determine the system's location exactly before the attack.

4. Determine where you want the spoof location to be and calculate time differences.

5. Broadcast the spoofing signal at the same time–location as that of the satellite, which is going to be "replaced" in the spoof, to make sure the device under attack "accepts" the new signal and stays in the same location.

6. Slowly adjust the time signal to change the location.
   (*Note*: If done incorrectly (moving too quickly, or wrong signals, and so forth), the device being attacked may "drop" or lose signal and "lock out" the spoofing high power signal; if so, you will need to try again).

7. If the location appears to change and the system under attack does not detect this, your system has been spoofed (you now have information on a possible bug).

Attack 30.2 will likely be done in a lab to determine the level of impact and susceptibility for systems where GPS spoofing is a risk. This attack is not at all easy (for a novice). If it is desired to have information on a system impact, or how or if backup location systems are functioning, this attack would be needed.

Note: Some fault-tolerant systems may use GPS and inertial systems with cross checks to determine location such as in aircraft. If the system has such fault tolerance, this attack may be used to assess the system's ability to know location and not be "fooled."

ATTACK 31: ATTACKING VIRUSES ON THE RUN IN FACTORIES OR PLCS

☐ Intermediate

In the embedded world, the gray zone area of factory controllers and/or industrial systems has enjoyed minimal concern from security and hacking, since many teams viewed viruses as an exclusive problem of the Internet, and IT or PC world. This changed in 2010 with the Stuxnet virus [20,21].

Stuxnet represents a first of its kind of virus evolution. Stuxnet and its offspring (e.g., Duqu) will continue to present a threat that testers need to be aware of into the future. Embedded industrial equipment manufacturers and operators of factory controllers must now consider security attacks in addition to traditional testing. To do this, one must understand the malware and conditions that allow these devices to be vulnerable.

Stuxnet used programmable logic controllers (PLCs) to target supervisory control and data acquisition (SCADA) systems [22], which are generally referred to as industrial control systems that control and monitor industrial processes, infrastructure, or facility-based processes. Industrial processes under threat include manufacturing, power generation, production, refining, or fabrication. Infrastructure processes under threat can include but are not limited to electrical transmission systems, civil defense systems, water treatment facilities, oil and gas pipelines, and large communication systems. Facility-based processes include building, airports, ships, space stations, which monitor and control heating, ventilation, and air conditioning (HVAC) systems, power and energy consumption, and control devices. Basically, these things make modern life convenient for us all.

PLCs are computers used to automate electromechanical processes such as lighting fixtures, amusement rides, and factory machinery. The programs that control PLCs are usually stored in nonvolatile memory. PLCs are hard real-time systems whose output results are produced in response to input conditions within a bounded time. (Refer to Wikipedia for more detailed definitions and examples.)

Most software test attacks in the *How to Break* book series are generic, some with specific examples. Attack 31 is specific with generic "add on" parts. I do this because Stuxnet represents a major milestone in viruses and malware constituting an entry into a new embedded computer subdomain. Reported noteworthy Stuxnet characteristics include thus:

1. A targeted attack, which has the ability to disrupt modern industrial (not IT) control systems with possibly "devastating" results such as severe schedule impacts, losses of jobs, poor product quality, massive destruction of property, losses of millions of dollars, even illness or death to humanity, and so on.

2. While the original virus was targeted at one specific system, the original virus has been retargeted to other systems. Worse, the offspring of Stuxnet will continue to evolve, since the code has become available.

3. It used some 20 or more vulnerability exploits with cross platform targeting and transmission.

4. It has been described as a "master work" and a "weapon."

5. Stuxnet also has other sophisticated features, such as stolen certifications, hardcode passwords, hides itself, hides its code, multiple platform infection, and appears to have "insider" knowledge (meaning someone who really understands these systems and software).

As time goes on, some PLC or SCADA holes will be closed, virus and malware protection will improve, and operators will think they are safe. But if history from the IT world shows us anything, the hacking will continue since malicious people now have a new avenue for viruses, malware, and hacking. *The genie is out of the bottle*. Worse, since these viruses can be used as a weapon, perhaps they will be used as such, but by whom? Will governments use it? Will your competition use it? Will organized crime use it? Will the hackers change it? The answer to many of these questions is likely "yes."

So, as a tester of these types of systems, you must run attacks to find the holes (vulnerabilities) before the bad guys do, while also checking for the unique threats from Stuxnet—and now its offspring like Duqu. The industrial control systems world now faces a scary future, and given society's and the many business' dependencies on these types of systems, increased diligence is in order. This attack introduces how to start attacking systems to assess if they might be vulnerable to viruses or malware. Once your testing is complete, provide that information to stakeholders so that determinations can be made on the strategies to correct the vulnerabilities. Then, be prepared to test some more.

When to Apply This Attack

If you are testing a factory embedded device, for example, PLC and/or SCADA, forget thinking of just basic attacks and historic test techniques. You need to include security attacks in your strategy. Attack 31 should be considered during the development, deployment, and later operations and maintenance of controller or industrial control systems. The levels and specifics of attacks will be different for each of the life cycle points (as well as the products), but just because one phase completes successfully, the bad guys will not go away during later phases. Since Stuxnet (and like viruses) are now on the loose, the team must work this effort based on the risks to the system or implicated and/or adjoining systems. More risk equals more attacks at more life cycle points. PLCs are controlling the power grid, nuclear plants, big factories, oil rigs, water delivery, and wastewater treatment and distribution systems, and many other industrial systems. What is the impact of your system going down or causing damage? What is the impact of your system connected to another system going down, taking down both or other adjoining systems? The local context, as well as the risk level, should direct your testing efforts.

> **Key point:** Embedded software testers need to think about viruses on such devices (as PLCs and SCADAs) since they are getting more "connected," and society is more dependent on them. We are no longer "safe."

Who Conducts This Attack?

Attack 31 should be conducted by staff, including the developers, testers, security experts, and perhaps an independent testing organization. This is a specialized group. Searching for Stuxnet or like viruses or malware requires some totally different critical thinking! Software development teams should consider the threat(s) and then consider how to deal with them in the architecture and design. Defensive programming, developer attack testing, and virus scans provide the data to the team to make good choices. PLC or SCADA testers must expand their traditional testing approaches to include security attacks—maybe for the first time ever—along with the lessons learned from Stuxnet's characteristics. Developers and development-based test teams in some cases may not be enough. Independent test teams, often called IV&V teams or security test groups, can be an option to add testing skills to the mix. Independent test teams have a long history of high-risk software systems, and now these teams can expand into security testing of factory and industrial systems. Consider the levels of independence in the test teams as well as more in-depth testing when the risk is high such as the controller is controlling million-dollar elements (hazard costs), humans can be hurt or killed (safety factors), or taking down a large system impacts a schedule (and time is BIG money).

Certainly, the test teams must arm themselves with the knowledge of Stuxnet and its offspring, as well as other security threats. For many embedded testers, this will mean learning new things. Good! Testers who have security and/or hacking test experience from the IT and PC worlds can become great additions to any test team, although these kinds of testers will need to learn the nature of embedded and factory systems. A composite test team starts to take shape here. Yes, this costs, but how much will losing the industrial system cost? Testers will likely need to be advocates for better and more security testing.

Where Is This Attack Conducted?

Security attacks should start in the developer's test labs. Developer-based attacks defined in Chapter 2 are the first line of defense. Each of these attacks should be run. Developer test environments may have different variations, tools, and usages, each with their pros and cons when dealing with viruses like Stuxnet.

After the developer's labs and testing come the project test labs. Project labs can be added to the developer's facilities, using additional test lab configurations, or it may be a separate independent test environment. If the test lab is built on top of development, have some concern for lab deficiencies due to commonality, which must be a consideration for possible risks and updates to the full test lab. Lab independence offers some advantages, but in any case, lab deficiency risks must still be defined and offset, as needed. Specific security lab testing features to have can include the following:

- Code scanners

- Automated execution

- Analytic or combinatorial tooling

- Monitoring and instrumentation tooling

- Realistic network, usage, and hardware–software configurations

The lab test environment must address how the system's components (PLC or SCADA software, user, and operations) interplay with each other. The tester must understand the local context or what "normal" usage of the system will be; how infections might take place during normal and abnormal use; and how site or factory customization might impact the security of the overall system. Many systems may have a product line approach, where a generic framework of hardware and software is created, so that they can be reused in many places. Other factory systems are totally customized by vendors and/or internal teams.

Whatever approach is used, security concerns and risks need to be documented, addressed, and security attacked in the lab environments.

Once the product (hardware and software) is in the field or factory, field or site testing should be considered. Questions to consider here include the following:

- What attacks should be run based on the local context such as system size, factory usage, test lab limitations, security holes from industry, regulations, legal concerns, and so forth?

- How do the user and factory environments impact "social engineering"?

- What are the industry trends and news that might impact the attack?

Finally, as the operations and maintenance happen, periodic regression security attacks and virus scans may be need to be considered and recommended. Remember: *Run scared*.

How to Determine If the Attack Exposes Failures

Here, the team is looking for the virus or malware or its offspring, as well as holes that might let hackers in. Scanner and virus detection programs are a start, but the challenge is more than just finding the virus. You want to find the holes, too. Stuxnet uses multiple vulnerabilities. Many of these vulnerabilities have been plugged by now, but the bad guys will be looking for "new holes" and so must testers. Some example holes to watch for include the following:

- Hardcoded passwords

- File configurations and signature checks (is it a trusted file or program)

- "Zero day" vulnerability

- Ease of stealing certifications

- Security cross checks or safeguards that do not exist

How to Conduct This Attack

Attack 31 is a hybrid attack. For starters, I recommend that factory development and test teams apply and understand other attacks throughout this book. There is no best or "only" set of attacks. Testers should know the local context of the system (or controller) such as how it is built, used, and maintained, as well as all of its risks need to be considered. Consider the following as a starting point:

- Scan for Stuxnet (or its offspring) with a virus checker as part of your system deployment.

- Apply static analysis, developer attacks, and system testing to identify (and remove) vulnerabilities.

- Research the latest Stuxnet variants and exploits on the Internet and then use this knowledge to tailor the attacks from this book.

A variety of vendors are now providing virus protection tools tailored to this world and this virus. Consider using them. This may not directly be the test team's responsibility to perform, but a good tester will find out what is being done and recommend improvements, as necessary. Virus protection software should be part of a complete system, and testers should be providing information about how "complete" a system is or is not.

Next, investigate how "secure" the system under test will be. This is part of risk analysis. For example,

- Is the factory or is the user environment controlling physical (security) access to system?

- What kind of network and device interface protocols exists?

- Who can modify and/or access the systems?

- How much access does anyone have to the system and are there levels of access?

- When can the access and changes happen?

The risk analysis should continue including Stuxnet taxonomy information, current security vulnerabilities, and impacts to the system from security failures. If you are dealing with a nuclear plant, *be very scared*. If your system is controlling the factory for your company's main product line, *be scared*. If a failure of your system can put your company on the nightly news, *be scared*. If your system is going to be controlling the local wastewater treatment system, how deep can you be in it? Management and other stakeholders need to evaluate these levels of risk from Stuxnet and its relatives. The risk may be high or low. Higher risk means more attacks should be done, but in any case, a balanced set of attacks must be determined and results reported to stakeholders.

Next, the test team must consider "who" does the testing. Is the attack just done by developers? Is this reasonable? What are local development testers doing? Are there independent testers? How much planning and risk management are being done by these teams? What attacks have we done to date and what have we learned? The answers to these questions may drive more or less security testing.

Answers to these questions should be supplied to development personnel, management, users, and customer teams—maybe even compliance or governance authorities. These items and context considerations are dynamic over time. The answer during initial development may be one thing. Maintenance efforts will get a different answer. Once your system is in the field being used, it may require more or less attacks. Decisions are not the security attack team's alone. The attacks provide major data and information toward test planning of what to attack and how much to attack. Plans unfold over time with agreement from the stakeholders.

This attack will end up being a series of other attacks from this and other books. Consider creating a security test attack plan. For example, an initial plan might have these attack patterns:

1. Code: Implementation test attacks (Attacks 2 and 3)

2. Code: Static analysis attack, specifically aimed at and against Stuxnet vulnerabilities (Attack 1)

3. Integration: Combinatorial test configuration attacks (Attack 32)

4. System: Security attacks (Attack 29, and/or 30)

5. Site: Regression test attacks and penetration (see Attack 28)

Your real attacks would be different, often much more extensive, and will change over time.

EXERCISES (ANSWERS ARE ON MY WEBSITE)

1. Without doing actual attacks, identify which attacks from this book and Table 9.1 that you might conduct on a mobile banking app (create an attack plan) and cite why you would run each attack.

2. Define how you would build a security test lab "sand box."

3. Define the spoofing vulnerabilities that a mobile map app might have.

4. Build a risk matrix for Exercise 3.

5. Research the offspring of Stuxnet and the kinds of systems (what types of embedded devices) they are targeting now. Make some notes to yourself about the characteristics of each one and note the tests that might find that "bug."

6. For a mobile banking app, define the "users" of the app, particularly identifying which ones might be or can become the "bad guy" hacker- or cracker-type users.

REFERENCES

1. Social engineering. http://en.wikipedia.org/wiki/Social_engineering_(security) (last accessed April 9, 2013).
2. Coverity scan 2010 open source integrity report. http://www.coverity.com/html/press/coverity-scan-2010-report-reveals-high-risk-software-flaws-in-android.html (last accessed April 9, 2013).
3. Whittaker, J.A. and Thompson, H.H. 2003. *How to Break Software Security*, Addison-Wesley, Boston, MA.
4. Church, A. 2012. Know your enemy. http://www.airforce-magazine.com/DRArchive/Pages/2012/March%202012/March%2023%202012/KnowYourEnemy.aspx (last accessed April 9, 2013).
5. Du, W. and Mathur, A. 1998. *Categorization of Software Errors that led to Security Breaks*, Purdue University, West Lafayette, IN.
6. OMTP, 2009. Security threats on embedded consumer devices, OMTP tech report.
7. Scambray, J. 2001. *Hacking Exposed: Network Security Secrets and Solutions*, 2nd edn., McClure, S. and Kurtz, G. (eds.), Osborne/McGraw-Hill, Berkeley, CA.

8. Monaco, A. July 16, 2012. Keeping hackers out of medical implanted devices: Researchers find way to prevent attacks on wireless medical equipment, The Institute and IEEE News. http:// theinstitute.ieee.org/technology-focus/technology-topic/keeping-hackers-out-of-implanted-medical-devices (last accessed April 9, 2013).
9. Engebretson, P. 2011. *The Basics of Hacking and Penetration Testing: Ethical Hacking and Penetration Testing Made Easy*, Syngress Basics Series, Waltham, MA.
10. Thompson, H. and Chase, S. 2005. *The Software Vulnerability Guide*, Charles River Media, Hingham, MA.
11. Brainstorming. http://en.wikipedia.org/wiki/Brainstorming (last accessed April 9, 2013).
12. Copeland, L. 2003. *A Practitioner's Guide to Software Test Design*, Artech House Publishers, Boston, MA.
13. Jon D. Hagar's web site. http://breakingembeddedsoftware.com (last accessed April 9, 2013).
14. Protalinski, E. July 22, 2012. Smart meter hacking tool released, *ZDNET*. http://www.zdnet. com/smart-meter-hacking-tool-released-7000001338/ (last accessed April 9, 2013).
15. Rathaus, N. and Evron, G. 2007. *Open Source Fuzzing Tools*, Syngress, Burlington, MA.
16. Manion, A. and Orlando, M. May 2011. Fuzz testing for dummies, in *ICS Joint Working Group Conference*. http://www.us-cert.gov/control_systems/icsjwg/presentations/spring2011/ ag_16b_ICSJWG_Spring_2011_Conf_Manion_Orlando.pdf (last accessed April 9, 2013).
17. e-commerce transaction growth, *SD Times*, October 15, 2010, Issue 256.
18. Schwartz, M. 2011. Exploit of well–known bug in a drone's software which made it "think" it was landing at an American airfield—not 140 miles inside Iran! *Information Week*, December 16, 2011.
19. Wireshark tools. http://www.wireshark.org (last accessed April 9, 2013).
20. Clayton, M. January 6, 2012. Stuxnet cyberweapon looks to be one on a production line, researchers say. http://www.csmonitor.com/USA/2012/0106/Stuxnet-cyberweapon-looks-to-be-one-on-a-production-line-researchers-say (last accessed April 9, 2013).
21. Broad, W.J., Markoff, J., and Sanger, D.E. January 15, 2011. Israeli test on worm called crucial in Iran nuclear delay. http://www.nytimes.com/2011/01/16/world/middleeast/16stuxnet. html?_r=1 (last accessed April 9, 2013).
22. SCADA. http://en.wikipedia.org/wiki/SCADA (last accessed April 9, 2013).

Generic Attacks

M OST OF THE ATTACKS IN THIS BOOK have been targeted at specific bugs or situations found in the error taxonomy (Appendix A). This is not totally true for the following attacks: Attack 1 Static Code Analysis (SCA), which targets many bug types that this analysis could find; Attack 17 Simulation and Stimulation, which can be used in combination with other attacks as well as against the lab; and Attack 20 Scenario, Story and Tours, which included the aspect of time sequences (stories) of events, which could be used in many situations.

The final two attacks of this book are even more "generic" in that their patterns can be used standalone or in combination with other attacks. These attacks are important additions to any attack pattern inventory and can also be used with other testing concepts and techniques, but all of this takes creative thinking. As I have stated previously, attack testing is based on the idea of patterns, and patterns can be used in many different ways and combinations. As you begin thinking about attacking mobile and/or embedded software, know the target bugs for the context, and this will help you to select your attacks. Keep in mind, however, that you can combine, mix and match, and be creative to maximize your bug-finding capabilities within cost and schedule constraints.

I present the last two attacks in this chapter because they are an error-finding technique that crosses the boundaries of the attacks found in previous chapters.

Table 10.1 defines the final attacks for this book, which are universal in that they can support many other attacks and testing efforts.

ATTACK 32: USING COMBINATORIAL TESTS

☐ Intermediate / ◇ Advanced

A limitation that testers quickly realize is that you cannot test everything much less test everything exhaustively. There are infinite combinations of inputs, conditions, hardware, apps, or software, as well as output configurations. Most test techniques and many of the attacks in this book are aimed at controlling and/or reducing this complexity and vastness. Some of the attacks in this book are performed standalone and might not be combined.

TABLE 10.1   Final Attacks Mapped to Mobile and Embedded Contexts

Context → Attack ↓	Mobile and Smart Wireless Devices	Embedded (Simple) Devices	Critical Mobile Devices (Could Be Embedded, Small but Important)	Critical Large Embedded Devices (Could Be Mobile)
Attack 32	Yes	Yes	Yes	Yes
Attack 33	Yes	Yes	Yes	Yes

Notes: Yes, should be considered for most contexts; no, generally not applicable.

The more powerful techniques and attacks can be combined as in the story of *killing two birds with a single stone*, but this requires a lot more critical thinking. Attack 32 is aimed at bugs, which occur in association with complex combinations of things: hardware, software, variables, data, operations, and other items, which when taken together result in a bug that single attacks or techniques will often miss.

Attack 32 presents concepts and techniques that can be used to deal with the complexity. This attack can be combined with techniques and other attacks in this book, since it is a realization of a math-based test technique called combinatorial testing (CT) [1]. The technique has a history going back to the early days of computers (1970s), yet many testers do not make sufficient use of it, either by itself or in combination with other attacks.

The basic concept of CT is to use math to identify a set of inputs (or outputs) from a function's total input, where there are interacting variables. If variables do not interact, they can be tested individually. In software, due to the concept of coupling, which causes side effects, variables are often related to each other. Testing a variable in isolation misses many interacting errors. Consider five variables each with five choices, all of which interact such as five menu fields on a screen with five pull downs. This will give you the following formula for possible test cases:

$$5*5*5*5*5=3125 \text{ combinations of test cases}$$

Ouch! Most testers would pass on this many tests. And this test does not have that many variables or choices! CT provides some math to help us with scenarios like this. The variable-choice CT set (for this sample test) can be done in just 42 tests (versus 3125 tests), which is a much more manageable number of tests. Would you agree?

> **Key point:** In the mobile and embedded software domains, there are usually many combinations of things that impact the software and cause side effect bugs. Using CT offers testers an approach to deal with the sheer numbers of testing combinations. For advanced testers, CT can be used with other attacks to achieve better testing for the same cost and schedule. Spend some time thinking about how to combine the CT attack with other attacks to take advantage of combinatorial benefits.

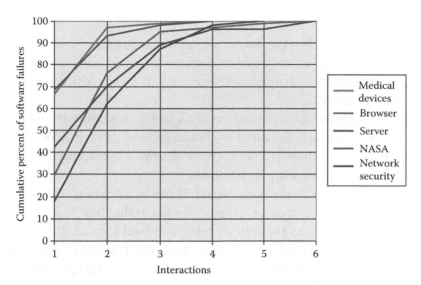

FIGURE 10.1 NIST study on interactions versus cumulative percent of software errors. (Reprinted from Kuhn, D.R., Kacker, R.N., and Lei, Y., *Practical Combinatorial Testing*, NIST SP 800–142, U.S. Department of Commerce, Technology Administration, National Institute of Standards and Technology, Gaithersburg, MD, 2010. With permission from National Institute of Standards.)

This is just a sampling, meaning that you are not testing all combinations. National Institute of Standards and Technology (NIST) studies (see Figure 10.1 [2]) have shown that the error-finding ability of these reduced numbers of test sets is still good. The NIST study found that for less than 2%–10% of the effort, 60% of the interacting variable-related bugs can be found. This is a good tradeoff, when the situation justifies it.

You can run Attack 32 without knowing the code, but you have to know the variables and choices in the code. And while the CT attack is often done on the "valid" (positive) cases, it can also be used for testing invalid inputs (negative). CT is math based and a close cousin to concepts such as Design of Experiments (DOE), Taguchi Methods, and statistical testing. For a decent familiarity with these and many other related concepts, spend some time searching for the terms on the Internet. Also, for good additional information on CT, spend some time reviewing the materials found in the references [2] for this chapter. The CT math details such as Orthogonal Arrays, All Pairs, and All Triples are out of scope for this book. To apply this attack concept, you do not need to understand the details, but you will need a tool (see Chapter 11 on tools or these references [4–10]) and a basic understanding of the CT attack.

And while an ideal would be to test everything, as I have already stated, usually exhaustive testing is not practical. Further, the studies have shown that when testers "guess" in large numbers of combinations, due to human bias, they create duplicate or missing test sets. The tests are inefficient. Human guessing approaches do not scale up when combinatorial "explosions" of input and state values are involved. I recommend that testers add Attack 32 to their toolbox.

When to Apply This Attack

A tester applies Attack 32 when the data space, inputs or outputs, become very large, has interactions between variables, and a selection scheme becomes attractive. When you have a list of dependent variables with options, choices, or parameters, this specific attack technique should be considered.

Apply CT when the influencing parameters or variables are known. It is also a good time to apply this attack, if you have the variable–choices and you can run a lot of tests quickly. This might be within an automated system, or if during a test scenario attack, you can introduce a large number of test input conditions.

Finally, as I have mentioned previously, Attack 32 can be combined with other attacks or techniques. I will cite some basic concepts and examples in the "how to part" of this attack. I also provide an extensive list of references, tools, and books for additional information. I am not saying this attack should be used all the time, but I do feel that it is vastly underused in favor of testers just guessing when they have "too many things" to test or just testing one thing at a time and ignoring interactions. The taxonomy of Appendix A indicated many interaction bugs between software, hardware, and the real world.

Who Conducts This Attack?

Attack 32 should be conducted by the test team with the right tooling. However, if the developers or analysts have testing or assessments that they want to make over a combination of variable input spaces, this attack can be good for them too. A tester can quickly set up the combinations for these other groups, once the variables and choices are defined.

Where Is This Attack Conducted?

Attack 32 can be conducted in almost any environment, including simulators and emulators, test labs for hardware, software, and systems, and field testing. The environment merely needs to support the testing, which means you need access to the CT tools.

What Faults Make This Attack Successful?

Attack 32 works because in software, the hardware, system, and the components and variables can all interact. This can be because of many of the things we have talked about already such as software–software communication, coupling, hardware–software interaction, system combinations, and so on. Many testers tend to think of things in isolation (and one at a time). You may have already read that much of the embedded world has dependencies, but the testing of combinations can be daunting. Attack 32 goes after the bugs of interaction by testing a reasonable set of them.

How to Determine If This Attack Exposes Failures

Basically, if you are running Attack 32 within or supported by another attack, you will use the determination criteria of the other attack. If you are doing this attack as a standalone effort, which is possible, you are looking for the classic wrong answer or response, given the inputs. To determine right and wrong, you will need some kind of success criteria based on those inputs. If your answer does not match the expected answer, you have a bug. Success criteria can come from human judgment, a prototype, simulation, model, and so forth.

How to Conduct This Attack

To get started, you will need to have a CT tool and to be familiar with configuring and running the tool, as well as understanding what data the tool needs. Most of the tools are pretty easy and can be learned in just a few hours. For the attacks in this book, I use the NIST tool advanced combinatorial testing system (ACTS) [3]. It is well supported, documented, and easy to use. There are others, but this one works and is free.

Next, you need to recognize that the problem you are looking at has dependent parameter variables and choices. While this sounds simple, experience has shown that this is the first hurdle and where many testers fail. Here is your starting point:

- Analyze your problem to identify the dependent variables and

- Determine the number of choices for each variable.

To understand this a little better, look at a very simple example. I am testing a cell phone app that will run on ANDROID and IPHONE. I also want to test on a few different carrier networks such as AT&T, Verizon, One, and Cheapo. Finally, maybe I have three hardware platforms to test on: APPLE, SAMSUNG, and Motorola.

My starting points would be

$$2 \times 4 \times 3 \text{ combinations or } 24 \text{ total test cases}$$

This is where I have variables or choices of parameters:

- Operation system (OS) = IPHONE, ANDROID
    (*Note*: See invalid test case section)

- Hardware = APPLE, SAMSUNG, Motorola

- Carrier = AT&T, Verizon, One, and Cheapo

These parameters are shown as inputs in Table 10.2 and outputs from the ACTS tool in Table 10.3.

TABLE 10.2   Sample Variables and Choices as Input Parameters

Parameter	Inputs
OS	IPHONE®, ANDROID
Hardware	APPLE, SAMSUNG, Motorola
Carrier	AT&T, Verizon, One, Cheapo

TABLE 10.3   Output Test Case Set

Item	Test OS	Hardware	Carrier
1	IPHONE®	APPLE	AT&T
2	ANDROID	APPLE	AT&T
3	IPHONE®	SAMSUNG	Verizon
4	ANDROID	SAMSUNG	Verizon
5	IPHONE®	Motorola	One
6	ANDROID	Motorola	One
7	IPHONE®	*	Cheapo
8	ANDROID	*	Cheapo
9	*	SAMSUNG	AT&T
10	*	Motorola	AT&T
11	*	APPLE	Verizon
12	*	Motorola	Verizon
13	*	APPLE	One
14	*	SAMSUNG	One
15	*	APPLE	Cheapo
16	*	SAMSUNG	Cheapo
17		Motorola	Cheapo

Note: *means any value can be used.

The tool produced 12 tests with the combinations listed, as shown in Table 10.3. No big deal, you may say. I have reduced the test sets from 24 down to 12 and, yes, that is not a big deal in a small case. However, what if we expand our cases a bit?

Suppose you are testing an airplane and the airplane had

- 11 country or language configurations

- 2 different computer processor options

- 4 different landing altitude brackets

- 20 different instrument option packages

- 4 different engine configurations

This would be $(11 * 2 * 4 * 20 * 4) = 7040$ test case sets in the software to test the different combinations. That is a real ouch! But if you put this into ACTS, you would only need 220 tests, which is a much more manageable number of tests, particularly if upward of 60% of the interaction errors will be found [1].

Once identified, the variables and options are entered into the CT tool. The tool does all the hard work. It will generate test cases based on various algorithms.

Next, the tester will determine an expected result answer for each test case from the CT tool. This is the oracle problem.

Once you have the tool test cases with expected results, you can run an attack for each test case. The attack can follow a basic pattern of inputting the test case into the software via whatever input options exist and then checking that the returned result matches the expended result, while watching for the bugs. There is also the option of combining this attack with another attack. Consider using Attacks: 2, 3, 6, 9, 10, 15, 20, 25, 28, 29, and 33. If these attacks are used, Attack 32 will supply input test cases that can be used in these other attacks.

Finally, there is an option with Attack 32 of doing configuration testing, compatibility checks, or the integration of Attack 12, which are important for many mobile apps. In these tests/attacks, you are trying to determine if the software under evaluation works with different configurations of hardware and/or software. In this variation, the same basic Attack 32 pattern is followed, which is identifying variables/options based on configuration of hardware or software, generating CT cases, and executing tests. However, this version of the attack is looking for bugs such as mismatched configurations or incompatibilities, which can be found by executing different configurations.

A note on invalid test cases—Consider Table 10.3 again. Some of you may have noticed that several of the tests are "impossible" such as ANDROID, APPLE, and AT&T. They do not need to be tested. CT tools have a way to deal with this called "constraints," which are rules that are basically Boolean logic where the tester can specify "this combination is invalid, so do not generate." I do not address exactly how to specify the constraint, since this configuration is tool specific (refer to the tool documentation), but after identifying parameters or choices, the next hardest thing testers must find in CT will be the defining constraints. The tools and supporting documentation elaborate on this topic. Once you define the constraints, you will want to rerun the tool to generate a new set.

CT can be used in many ways with other attacks and techniques. CT can be used with hardware, software, system, security testing, and others. I explore these topics on my website. Also, my website maintains a list of CT tools.

ATTACK 33: ATTACKING FUNCTIONAL BUGS
○ Novice

Earlier, I stated that I was not going to extensively cover traditional, functional domain, and requirements-based testing. A few attacks have touched on aspects of functions, concepts of use, and requirements or stories. There are many books on testing that cover these aspects of testing, but in the mobile space, many testers may be coming in without access to a lot of these books or may have scant knowledge in this area. Therefore, as a wrap up, I address a basic functional attack approach. Attack 33 addresses basic functional coverage (basic functions of the software or app), which could be used with or without requirements or stories. Attack 33 can be used on functions internal to mobile or embedded devices and has interfaces to the external logic, although it does not fully target testing external systems.

When to Apply This Attack

You can apply Attack 33 as your first attack or at almost any point in testing. You can do this attack when you are trying to understand and assess your software. This attack can be useful in providing the understanding for other attacks, but this attack is where too many testers spend their test lives, so do not apply it exclusively as your only attack.

> **Key point:** You must understand your system, and Attack 33 can help in that regard. This attack can be a beginning with exploratory concepts to quickly test the software of interest. It can be a follow-up with other attacks depending on what it is that you need to attack.

Who Conducts This Attack?

Attack 33 can be conducted by an individual, the entire team, or anyone in between. It can be done inside or outside of the development efforts.

Where Is This Test Conducted?

Attack 33 can be done pretty much anywhere and at any level of testing, though probably most often it will be done in the lab or field. When it is done in a lab, you do not need lots of lab tools, but it cannot hurt to have tools—provided they supply useful information. If you are doing field testing (as an independent tester), all you really need is the software and the device to execute it.

What Faults Make This Attack Successful?

This attack is successful because most software has a ton of features, and it is easy to get one or more of them wrong when coding. Take a look at your typical web or network-enabled software app for a smart phone such as a travel booking app. There are many buttons, options, even subpages, features for logging in, feedback points, and links to other sites. These are all "functions" or services, which will need to be checked. In some software, these are tested over and over by many different levels of testing. If these items are tested, they are more likely to be working correctly. However, in some mobile and embedded domains such as the app world, time to market and limited budgets result in software that has had less functional testing run, resulting in bugs escaping and users finding them. The more complex the software, the easier it is for a developer or tester to miss the bugs that happen due to the software's complexity. The functional attacks need to be balanced with the complexity and context of the software.

How to Determine If This Attack Exposes Failures

In Attack 33, you are looking for a function that is not implemented or not implemented correctly, as defined by such things as requirements, concepts of operations, or in a reasonable user's expectation. We all expect to be able to log in, and we expect buttons to work.

> **SIDEBAR 10.1**
>
> Citibank's Double Charging App: In 2012, Citibank released a mobile app that charged some users twice for a single Bill Pay selection. This was wrong (unhappy users) and resulted in much bad press (probably worse). The cost of fixing the problem and undoing the charges probably would have paid for a little more functional testing of the app before releasing it to the world. The bad public image impact is maybe worse than the cost in the longer term.

We expect a web page to load and error messages to be meaningful. So you can determine errors from both formal information (requirements) and based on your own expectations. Once you see potential issues, you should capture them and maybe even explore them further to determine if there really is a bug and how bad the bug might be. All of this is information to report to the stakeholders through some reporting mechanism such as a bug report.

How to Conduct This Attack

In Attack 33, you first want to access or load the software under test on the device. You need to understand the software version; configuration of the device; the hardware and software involved in the testing; and the ideals of what the software is supposed to do. The ideals come from places such as user guides, help files, requirements, user stories, basic human expectations, and so forth. For example, if I tell you the app to be tested is a system that lets users access store product information for comparison shopping or multiplayer games—even without the app, you start having expectations given these pieces of information. You want as much information as possible about the app. I have heard testers say, "I cannot test without requirements or test cases (stories)." This is not true! It is true that requirements help for a certain kind of testing such as in requirements verification. Yes, you do need requirements, but anyone can attack software and find bugs without them. For example, you can base your testing on your historic understanding of types of software, your knowledge of the OS, and so on, without knowing the specific requirements.

There are many ways to start Attack 33. You can do exploratory testing [11] by creating a list of features as you find them. I prefer to start by building a mind map while doing some exploration [12,13]. In exploration, load the app and open it up. As you explore or "play" with the app, take note of things such as

- Pages you land on

- Functions and links on each page

- Overall color scheme and/or theme of the page

- Wording on the page (Does the wording make sense? Is the spelling correct?)

- Placement of objects and icons on the page (What is visible on the screen you are using?)

- Icon style and nature (Do they make sense for your device?)
  (*Note*: Some devices have guidelines for these display items.)

- Are there indicators (progress or time bars) to give users feedback and let them know something is processing or happening?

- Do menus appear (visible), make sense, and work (try selecting all items even if they should not be selectable)?

- Does the app's logic (i.e., pages) "flow" or make sense? (Would your mother or a 5-year-old be able to use it?)

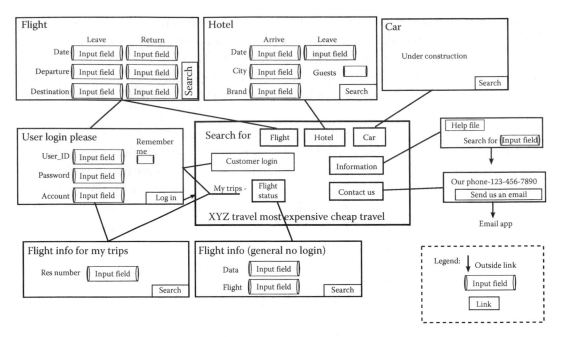

FIGURE 10.2   Mind map example. (*Note*: Boxes are used here instead of circles for readability in this example.)

- For more items to check on a first pass, see the list in Appendix F
- Build these things into a map as pictures and notes (see Figure 10.2)

Build a mind map by simply drawing boxes (or circles) as you go for each item listed earlier, topics/areas/items that you find noteworthy. Mind maps can be very free form (minimal rules). Links between elements are shown in the map with a line (see [13] for more information). Take note of "coverage" of each mind map item as you go. Coverage in a mind map is achieved when each noted mind map element (box or circle) is addressed (or touched). Make note of any observations or bugs as you find them. This can be done on the mind map (if there is room) or separately.

After gathering as much information as possible, stop and refine the mind map and consider what is next. Factors to consider include thus:

- What is the landing page?
- What can and cannot be done on each page?
- What do you need to use to access the system?
- Should you continue this attack or another to reach coverage of mind map elements?
- What do you see or not see that perhaps you should?

After this first session and organization, you will have a better idea of what you are testing. You may now have an idea of added information you need to continue attacking such as a user account, data, or other information to "feed" into the test. This data should be obtained before starting your test. Next, you might want to get a little more formal. Consider doing one or more of the following activities:

- Risk analysis

- Expanded mind map—maybe using a tool to formalize the test

- Modeling with a state map

- A table listing functions from the information you have such as requirements, stories, operation concepts, and so on

You may have some of these already, and they can be reviewed or expanded. Basic versions of these first items can be generated in less than an hour and allow you to find bugs.

Once you have some of this information, it can serve as a framework for a more organized set of activities for this attack. You should be looking at some level of systematic attacking, maybe reaching coverage of the mind map, risks, or state model. This is not "required," but if you want to make any "claims" about the software, it is nice to be able to say what has and has not been tested—at least for yourself and with some form of coverage. How many risks have been tested and how? How much of the mind map have you covered? What states are not covered? The number of times to apply this attack and others is governed by time, cost, and what you are trying to achieve. It is good to set some limits or "time box" the attacks to govern time and cost. I also recommend completing the user interface/graphic user interface (UI/GUI) checklist of Appendix F during these functional attacks.

EXERCISES (ANSWERS ARE ON MY WEBSITE)

1. Do one of the following:

   a. Load your favorite app or the app you work on. Apply Attack 33. Draw a mind map, showing the app's activities and links.

   b. Load a travel website app onto your smart phone, such as the Orbitz app. Apply Attack 33. Draw a mind map, showing app activities and links.

2. Produce a risk list for the software of Exercise 1. Define how you apply this to improve your testing/attack. (Refer to Appendix G for more on risk analysis.)

3. Produce a state chart for the software of Exercise 1. Define how you apply this to improve your testing/attack.

4. Get a CT tool (e.g., NIST or other tool) and load it on your computer. Define and apply CT to the software of Exercise 1.

REFERENCES

1. Kuhn, D.R., Senior Member, Wallace, D.R., and Gallo, A.M. Jr. 2004. Software fault interactions and implications for software testing, *IEEE Transactions on Software Engineering* 30(6):418–421.
2. National Institute of Standards and Technology, Combinatorial methods in software testing. http://csrc.nist.gov/groups/SNS/acts/index.html (last accessed April 11, 2013).

3. National Institute of Standards and Technology ACTS download site http://csrc.nist.gov/groups/SNS/acts/documents/comparison-report.html#acts (last accessed April 11, 2013).

4. James Bach's AllPairs (Open Source) tool. http://www.satisfice.com/tools.shtml (last accessed April 11, 2013).

5. Bob Jenkins' jenny (Open Source). http://burtleburtle.net/bob/math/jenny.html (last accessed April 11, 2013).

6. National Institute of Standards and Technology tools download site http://csrc.nist.gov/groups/SNS/acts/documents/comparison-report.html (last accessed April 11, 2013).

7. Hexwisetool. http://hexawise.com/ (last accessed April 11, 2013).

8. rdExpert tool. http://www.phadkeassociates.com/rdexpert.htm (last accessed April 11, 2013).

9. IBM's Combinatorial Testing Service. http://www.alphaworks.ibm.com/tech/cts (last accessed April 11, 2013).

10. Kuhn, D.R., Kacker, R.N., and Lei, Y. 2010. *Practical Combinatorial Testing*, NIST SP 800-142, U.S. Department of Commerce, Technology Administration, National Institute of Standards and Technology, Gaithersburg, MD.

11 Whittaker, J. 2009. *Exploratory Software Testing*, Addison-Wesley, Boston, MA.

12. Johnson, K. The mindmap of Web domain was inspired by conversations with Karen Johnson, 2012. *Note*: The mind map pictures in this attack do not follow standard mindmap notation used by many tools, since we were trying to "replicate" the look of a screen device. Most mind map tools use circles, not squares.

13. Mindmap. http://en.wikipedia.org/wiki/Mind_map (last accessed April 11, 2013).

Mobile and Embedded System Labs

I OFFER THIS CHAPTER AS ADVICE based on my own experience firsthand and a culmination of reading. I hope that you find specific items or sections within this chapter that you can take directly into your local project context and apply those concepts to you lab in addition to the attacks of previous chapters. There are many kinds of labs, but for this book, the focus will be on the software test labs.

> **Takeaway note:** To test many mobile and embedded devices, a specialized supporting test lab is needed.

INTRODUCTION TO LABS

An important feature of any test program is the environment in which testing is conducted. In the personal computer (PC) or information technology (IT) world, it may be as simple as having a generic computer sitting on a desktop in an office. In the mobile and embedded test world, having a dedicated test facility and/or lab with tooling, software, and models can be an important part of successful and cost-effective testing (Figure 11.1).

Mobile and embedded test environments range from small to large and can include the following:

- A facility or buildings, rooms, furniture, and so forth

- The actual mobile or embedded hardware supporting the device such as processor(s), hardware, many different cell devices, and communications networks, and so forth

- The software, which is going to be tested, such as developed software, off-the-shelf software, or other software

- Test lab support hardware such as racks, test support computers, electronics, and so on

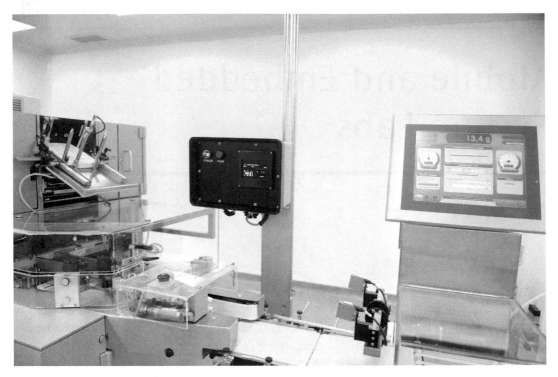

FIGURE 11.1 Example embedded test facility.

- Test lab support software such as tools, models, analysis support, and so on

- Test operations support resources such as people, technicians, standard operating procedures, and so on

- Field support versions incorporating special support equipment (optional)

This chapter talks about the basics of these items covering a conceptual test facility, elements, and its life cycle. I have seen facilities ranging from the mobile phone you have in your pocket to large environments, which took tens of millions of dollars to set up and run. The makeup and nature of your test environment are completely project and/or context dependent.

TO START

The test environment and support tools must be developed—just like the software. This is vital for success and has to happen prior to the beginning of a test cycle; otherwise, testing can hold up the progress of the project. Here, we examine how to set up facilities, the processes, and a lab to achieve good attacks. Given that one main goal of attack testing is to expose errors, the areas of software risk and types of errors must be weighed carefully when selecting tools, modeling, and test environments, as well as actually setting up the lab.

Finally, facilities or labs will have tools. Tools provide useful support features for testing. Labs and tools have limitations in an integrated test environment, and understanding these limitations must be factored into the planning and test development effort.

To be successful, the test environment construction (facility or lab) effort includes the hardware, software, support techniques, time, management, and people and must also factor in any limitations of the environment. Each of these elements must be planned, developed, implemented, and controlled.

TEST FACILITIES

Many companies have for years developed software and hardware test environments. It is important to distinguish the test environments between facilities and the labs or equipment they contain, although many places often tend to use the two interchangeably. For example, a test facility can include the following:

- A building or room(s) in a building housing a test lab containing tools (hardware, software and so on) and people.

- A mock-up of the mobile and embedded system such as smart device, ground station, aircraft cockpit, factory floor, automobile compartment, and so on.

  Note: The product versions found in the labs may not necessarily be a final production version, although final testing typically uses a production level of software before its release.

- Test cells or labs such as environmental, electromagnetic, acoustic, electro-mechanical hardware configurations, radio wave generation, software test bed, and so on.

  Note: While some of these labs can be more for hardware testing, sometimes the system–software attack team may use them too. Such labs are where a tester may serve many roles such as software tester, hardware evaluator, performance analyst, and others.

- Outside pads or structures such as cell towers, radar systems, hardware stands, test tracks, support equipment, and so on.

Generally speaking, a facility is a room, a complex, or a building with a mock-up or prototype of the system–device(s). Inside of these larger facilities, there may be one or more test labs. A test lab could be a subset of the capabilities in the larger facility, depending on what is being tested. For example, a facility may have a hardware prototype test lab, a formal software test lab, a system test lab, and the equipment and/or set-up to support field testing. A facility and lab might also just be a cell device hooked to a monitoring PC sitting on a desktop in a crowd tester's basement. A facility may have labs, which are more hardware focused, but in close proximity to the software test labs. Again, my focus in this book will be on the software test labs.

A software test lab in our definition, here, is a collection of computers supporting interfaced hardware and software that allows realistic testing of the software under production. The test lab(s) can include special equipment that aides in the definition, execution, and analysis of test results. This can include specialized computers with test points, probes, oscilloscopes, voltmeters, input devices, and data-gathering tools and/or recorders. Many test labs typically have specially trained staff and test procedures to aid the test process, or testers will need access and communication with subject matter experts (more commonly called SMEs). The labs are used to run the testing, having both the inputs to drive the software and the facilities to record the results. Labs may even become "mobile" for testing in the field. Examples of a field test lab could be

- A series of cell phones you can carry to the outside

- An automobile test vehicle used for demonstration in the field (to travel perhaps across states and/or countries)

- A specially configured airplane, craft, or other vehicle that can perform in the "real world"

- Prototype systems that can be taken into the field

- Medical devices used in clinical trials or in "experimental" surgeries

Such field test environments come with special equipment and features to allow "in field" assessments. The simplest field lab may just be the smart phone you hold in your hand, but with the right tools (e.g., your brain).

Now, while not every mobile and embedded attacker needs a large test facility or lab, some will. Generally, when creating and operating facilities or labs, you (or a member of the team) should coordinate the development, quality(ies), security, operations, safety, and things like facility service or logistics (i.e., cleaning, trash removal, maintenance), all of which need to be taken care of in a timely manner.

Further, as I will cover in detail later in this chapter, the creation and life cycle of facilities and labs are like "a project inside of a project." Many organizations have test facilities and labs that are reused from project to project with some modifications in between projects. In fact, many organizations (or companies) doing mobile and embedded work have large investments and teams to maintain and run test facilities and labs. As covered in this chapter, the project inside of a project means that many of the same development life cycle stages (planning, requirements, design, implementation, and testing) exist.

Takeaway note: Software test labs are usually part of a larger test facility

WHY SHOULD A TESTER CARE?

I cover facilities and labs and their supporting elements as a special chapter because I have seen many projects rely heavily on the facilities and labs in their attacks, while others have forgotten these details until late in development only have large cost and schedule impacts. Management should not forget about the test lab and/or facility. As with much of testing, the team should start thinking and planning the facility or lab up front, during project concept or definition. Start by asking a few simple questions: "How many labs (or facilities)?"

and "What types of labs do we need to test effectively on the project?" Here are some other detailed questions to ask:

- Should there be a separate lab or environment or facility for the developers?

- When do the testers (and which testing teams) need the lab (will sharing or reuse work)?

- Should there be a hardware–software integration lab, environment, and/or facility?

- How many lab test beds are needed? (Remember, you tend to expand to fill whatever space you have.)

- Can we "buy" third-party vendor support for labs and virtualization?

- Can we hire crowd testers, where they bring their own equipment, or must we supply to them our equipment?

- What tools, equipment, and people with what specialized skills are needed?

- Can we expand the facilities and labs, as needed or when we are successful?

- What tools, equipment, and features does the lab need?

A person doing test planning should understand the answers to these questions and can even treat them as a checklist for the lab. Considerations and expectations for these topics are addressed in this chapter. However, the focus will be on software test labs, tools, and operations, including the following:

- Types of labs

- Developing labs

- Lab lessons learned and limitations

- Automation, tool, and modeling support in labs, including

  Test tool concepts covering the complete test effort, not just test case execution

- Modeling can increase the effectiveness of test labs by providing "missing" data in the form of inputs, supporting simulation, or output analysis

- Important concepts for labs, staffing, and test operations

- Reference pointers to other materials

WHAT PROBLEM DOES A TEST LAB SOLVE?

Most projects accept testing as a needed and valued activity; however, to do testing, testers need a place to actually do the testing. Typically, this is called a test lab. The lab provides the "playground" to conduct attacks, often in a "safe" environment (meaning errors, if found, are contained in the lab and are not let loose in the world). The lab also provides attack support in the form of tools, models, and abilities that the real world may not have.

More information on "issues" and "lessons learned" will be provided later, but the first lesson is that the test lab and the attacks that happen there are not necessarily what will happen in the real world. I have recommended many attacks, but the tester must customize any attack to be as "real" as possible for their local context, while conducting the attacks within a realistic test environment such as a lab. However, the lab and attacks can lack many aspects of the real world. In Figure 11.2, for example, if we think of it as the whole world, we see that the test lab is not the real world but is likely a subset of the conditions that devices may encounter. The software under test gets inputs that we provide and responds with outputs, but if these data points are not realistic compared to the real world, the testing may be flawed. Determining the degree of realism is a context-dependent situation, where there is no single method or measure possible. For this example (Figure 11.2), a tester could be doing manual testing and interacting directly with the device or software under test. For smart phones, this might be "good enough" realism, but for a heart pacemaker, direct manual test interaction might not be "good enough."

In this chapter, we will build the lab concept up from the simple to the complex. Past just having a person manually inputting all of the test information to the device or software test, I introduce a computer lab test environment, which includes supporting secondary computer(s) and automation, which can provide inputs as well as record outputs from the device, as presented in Figure 11.3.

Software runs on computers and to attack software, you need hardware. In Figure 11.3, the testing can interface to supporting hardware. This supporting hardware can range from the first layer of electrically equivalent hardware and computers, to all of the hardware of the system or device. Next, in some lab configurations, the software and computer under test are connected to a second-support computer system. This second-support system can serve the following functions:

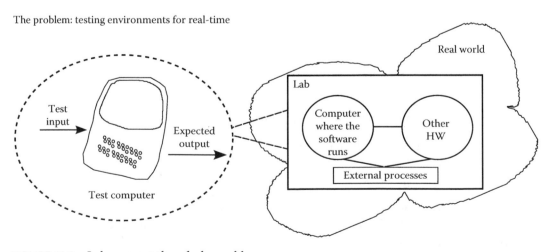

The problem: testing environments for real-time

FIGURE 11.2　Labs are not the whole world.

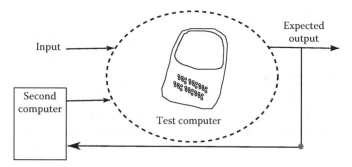

FIGURE 11.3   Simple test lab with secondary test support computer.

- Provide or monitor the data going into the device under test

- Record the outputs of the device under test

- Server to load the software under test

The faster, more complex, or critical the system under test is, the more complex the secondary "test computer" and system must be.

Key point: Do not forget to plan your test lab.

STAGED EVOLUTION OF A TEST LAB

Most test labs go through development in which the nature and use of the lab goes through stages with the development of the product. They may go from developer use, to simulation, to prototype evaluation, to full hardware in the loop testing, and finally, support to field testing. The book *Testing Embedded Software* [1] treats aspects of this. I expand the concepts for labs in the mobile and embedded attack space.

SIMULATION ENVIRONMENTS

A simulation environment supports project definition and feeds into later testing efforts. The implementation nature of this environment varies from conceptual thought exercises (perhaps using whiteboards) to tool-based modeling and simulation (such as systems modeling language [SysML] and/or unified modeling language [UML]). These factors can vary depending on the context. These environments are described in detail in *Testing Embedded Software* and are often done in the concept or proposal efforts. Pure simulated environments can provide the following:

- Early analysis or testing to help understanding for later attacks

- A jump start on modeling to support later testing

- Initial development, integration, and evolution of supporting simulation and stimulation tools

PROTOTYPE AND EARLY DEVELOPMENT LABS

Prototype or early development labs are what the book *Testing Embedded Software* calls "second stage testing lab." Many mobile or embedded systems create prototypes of the hardware and software. Prototypes can help the whole team understand the system and guide its evolution by testing the prototypes. This approach supports the ideals of "test as early as possible" and fits within the concepts of Agile development and testing. Attacks run in a prototype development lab will be quick, less formal (not looking to finalize anything), and exploratory in nature. The team will be looking to gather information to give back to developers as well as to explore risks in functionality and nonfunctional requirements (such as performance, safety, security, etc). The test team will likely also be integrating elements of hardware and/or software using threads and attacks. This feedback is a critical part of why we do prototypes and practice Agile, so the attacks should be fast, but informative.

As the prototype testing happens, the test lab, which should also be under development, is likely in a prototype stage. So now, beginning attacks can take place, although it may be with limited functionality. This prototype stage benefits both the development product and the test lab, since both will likely be evolving as the team gains understanding. Key features to have available at this stage, can include the following:

- Prototype devices

- Support computers

- Some limited modeling and simulation

- Facilities to support integration and integration attacks

- Early test tooling, hardware, and software

- Skilled lab development staff and exploratory testers to define the needed attacks

DEVELOPMENT SUPPORT TEST LABS

In Attacks 1 through 3, I defined and outlined the importance in mobile and embedded systems to have development-based testing and analysis activities as the "first line" of defense. To do this, you may need the following kinds of labs, tools, and support technologies:

- Developer support test tools (e.g., do an Internet search for "developer testing tools," searching for tools for your local project context)

- Computer instruction set simulator [2]

- Emulators [3]

- Processor boards or single-board computers with "open architectures" so attacks and debugging can "see" inside [4]

- IT support computers to run tooling, modeling, input creation, posttest checks, and analysis

- Access to development support tools such as compilers, debugs, configuration control, modeling, and others

- Static analysis tools (see Attack 1)

- Recording software and hardware data inputs and/or outputs

Developers will use the tools to create test frameworks to support execution of small parts of the software. For example, Figure 11.4 shows an example framework. It has a "driver," which supplies data, a sequence of inputs, and success criteria to check against. The driver calls the software under test, here called "unit 1." "Unit 1" will be exercised and, as needed, things such as coverage determined. If "unit 1" calls other software, the tester must create stubs or mock objects, which can be called or return data to "unit 1." This test logic is shown at the bottom with data (return) and sequencing of test information, so that a series of test cases can be run. Once the "unit 1" software under test returns results, they are recorded by the driver in return results. This is a basic framework, which many developer-level tools, commercial and open-source use. There are several variations on this concept in available tooling, although it is possible to do all of this with just basic programming. From this framework, it is also possible to understand why so many programmers dislike these kinds of attacks (Attacks 1 through 3) because developers must write two or three lines of test code for every line of executing code to create the framework.

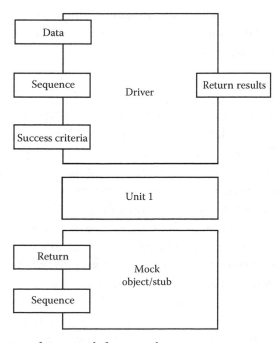

FIGURE 11.4 Developer test driver–stub framework.

Note: Any tools in use should be configured and training provided to those persons who will be using them.

INTEGRATION LABS

Building on the prototype and development labs, many mobile-embedded groups move into integration activities (see Attack 12). During this time, the lab environment will be maturing, and the real "heavy lifting" of integration testing attacks can start. The lab will be getting more hardware and newer versions of software. As the system is built up incrementally and iteratively, new attacks will become possible. The interesting part will be assessing the pieces or parts individually or as they are integrated with the attacks. The point is to try to find bugs at the interfaces, hardware, and/or software, with a focus on each new part as it is added. In integration, you do not want to add too many "parts" at once, because adding large numbers of factors or parts makes isolating the "at fault" part complex and bugs may be harder to find. This means that you can lose the clarity of isolating what is really wrong, as well as fail to recognize the factors that create the bugs. Keeping it simple and then adding things as needed to reproduce the bug are critical here. For complex systems, I recommend building up the system and lab in parallel, following the idea of integration threads and attacks. These integration activities and attacks become a key factor for many groups in larger embedded systems. This incremental build-up continues until you have a completed system–software under test and the supporting lab or environment.

PRE-PRODUCT AND PRODUCT RELEASE (FULL TEST LAB)

This is what the book *Testing Embedded Software* calls Stage 3 and is what most testers think of for testing and where most of the attacks will happen. The material later in this chapter expands on "test lab" concepts. There are as many full test lab configurations as there are mobile and embedded systems.

FIELD LABS

For mobile and embedded software, this book adds another type of test environment, since it moves out of the classic "test lab" and into the field test lab, in what some literature calls "demonstration testing." Some readers may feel this is not a lab at all, but the real world. The key here is that the product or at least a version of it, has not been released to the public but is used in the real-world testing. This environment may have support "tools" to help test the product and may be followed up by public use of the product by "unofficial testers." Such field testing is often called beta testing. But for this type of specialized "in the field" lab, we have dedicated testers, who are in the real world (or field). Testers will be doing attack testing plus whatever other types of tests might be necessary, for example, checking, verification, compliance, performance assessment, validation, or others.

I have mentioned in many attacks the idea of field testing. For many systems, it is quite important to have test staff running these attacks. I have seen this field testing type of lab used with airplanes, cars, mobile devices, and medical systems, to name just a few. In field labs, the mobile or embedded device is basically or totally the completed system. The idea is to be as "real world" as possible, in part to confirm the attacks done in the lab, but also to demonstrate the product in the real world because of the limitations a lab may have. The field test lab may have such features as the following:

- Test instrumentation, hardware, and software

- Built-in test features that can be leveraged

- Test recording devices, examples include downloads, temperature, altitude, and so forth

- Location information

- Other log files, equipment, and tools

I have seen "test vans" and mobile labs onboard ships or airplanes, which provide these test support features. They "follow" the device under test around and are used in information

SIDEBAR 11.1

Test Vehicles: I live in the mountains of Colorado. On several occasions over the last 12 years, my wife and I would see vehicles driving around covered in heavy drapes, so that no one could actually see the vehicle. At first, we wondered, "What are those and why are they covered up?" Finally, we got up close to a couple of these draped cars, only to have an armed guard chase us way. Come to find out, they were test vehicles from an "unnamed" auto manufacturer. The vehicles were being followed by a van with test equipment—and probably testers doing lab field testing. But why Colorado? Well, here are a few possible reasons: highest auto roads in the country with thin air and steep hills, aside from great scenery and nice places to stay. These field tests were done in "secret" (covered in drapes) with a mobile field lab to capture live (or real-world) data. These vehicles were tested in a real-world environment to provide the most useful information (actual data that could be used by real customers). Where are your tests done?

SIDEBAR 11.2

Testing Mobile Devices and Being Careful Where You Do It: A car or a plane is large and a little hard to lose. Many of us test small mobile devices. Teams have been given early versions and even prototypes of the mobile and embedded device that they take into the field to "test." They set them down. They forget them. Oops! Pre-release Apple or other vendor's devices have been found in bars even sold to reviewers. Was this an accident or marketing? We do not know, but when you are doing lab field testing, make sure your mobile and embedded device is where you want it to be and maybe keep it under wraps. Hold onto your product!

gathering and assessment. A lot of field testing is performed with equipment and configurations that are taken from the lab. Likewise, mobile devices may have "monitoring" apps or configurations to record what is happening in the field. The choice of configurations and tools is limited only by imagination and context (cost, schedule, and risk).

OTHER PLACES LABS CAN BE REALIZED

For years, there have been third-party companies and vendors who have provided test labs and facilities. Also, with wider acceptance of the concept of crowd testing, new options for field testing become possible.

A variety of companies and vendors have test facilities and labs, some of which are customized to a particular customer or product domain while others are, more or less, totally generic. For many projects, being able to add test facilities or lab resources makes sense. Why build your own test system when you can "rent" one for probably less cost. I do not advocate any particular third-party lab or vendor. They come and go too often and some change their names. With a little bit of Internet searching, knowing your product domain, and efforts, you will most likely find a company or vendor who can or will help your testing efforts by providing lab resources (see my website). The types of organizations and the services they offer include the following:

1. Providing equipment, tools, and other resources, which you can "rent" to be placed inside of your facilities

2. Providing labs and facilities where you bring your attacks and testers

3. Providing virtualization and cloud facilities allowing you to connect remotely to run attacks

4. Providing items 2 and/or 3, plus they provide test staff to do the testing

It is worth noting that items 2 and 3 seem to be gaining popularity in the mobile and smart app world, since the time to generate and field an app is often so short (weeks). Therefore, creating test facilities is a major burden. You may be much better off if you have access to a third-party facility or lab that is ready right now. Also these environments support concepts such as automation, running lots of attacks quickly, and addressing areas like performance or security testing. Using third-party test facilities can save large and small companies large investments of time and money.

Also, the idea of virtualization and cloud-based test environments for systems builds on the movement in testing to leverage computer resources rather than building your own. If you can run a hundred virtual attacks with some automation from a single desktop by renting some cloud time, think about how much money you can save. These approaches may not be available to every mobile and embedded project, but it is something to consider.

Finally, with the advent of companies such as uTest, Mob4hire, and other crowd testing organizations, the nature of field test labs and efforts for many companies may change.

Crowd source testing is very popular in the mobile and smart device world. Companies building apps are faced with a problem of combinations of hardware, software, service providers, and so on (see Attack 32). With the advent of Internet and crowd testing companies, now you can "ship" your testing to hundreds (or thousands) of testers and devices. The nature of this type of attack and lab is partially out of the control of the development organization. However, given the success of some of these ventures, in some product domains, the trend is likely to continue such as in the app world.

> **Key point:** There are and can be many kinds of test labs supporting different attacks at many life cycle points for different types of tests. Much planning and careful thought are needed for the simplest lab (the smart phone in your pocket) to big labs (for testing the newest super airplane with hundreds of computers and users). Know the requirements up front for whatever lab you use.

DEVELOPING LABS: A PROJECT INSIDE OF A PROJECT

Creating test facilities and labs is like a project inside of a project. I will not get into the details of management, systems, hardware, and software development and engineering as there are already many books covering these topics. It is sufficient to say that all of the problems that come with any project can haunt development labs.

PLANNING LABS

The development and planning of the lab should be done in parallel with product development. For smaller labs, the team may understand that minimal up-front work is needed, but for large and complex labs, the effort may truly be simultaneous and large even when compared to the product development.

When planning the lab, keep in mind that it needs to address such questions as

- How can you create lab interfaces for something that does not exist?

- What do you do when the hardware or software that you need to develop the lab is not available?

- What development efforts does the lab need such as in the life cycle phases?

- What general hardware, software, tools, people, and other resource do you need?

- What "special" features and equipment do you need that may be one time, temporary, and so on?

- What visibility do you need into the software or aspects of the software during testing, such as clear box versus black box?

- How much realism is needed in the environment, and once created, how close to reality is the lab?

- What is the need for simulation and modeling such as simulation and stimulation, open-loop models, high-fidelity closed–loop, real-time simulations and models (see Simulation, Stimulation, and Modeling in the Lab Test bed and Continuous Real-Time Closed Loop Simulations to Support Lab Test Environments sections)?

- When does the lab need hardware including prototypes, special versions, and/or the real thing?

- What is the overall cost, schedule, quality, and resources compared to the context of the software under development?

Lab development planning should start early and evolve as the project unfolds. To complete the planning of the lab, you may need to address the following:

- Test lab overview (use cases, operations concepts, levels of testing, etc.)

- Test lab schedule (coordinated with all organizations)

- Test lab budget (labor, material, travel, subcontractors)

- Test lab capabilities including risks and limitations (what can the lab do and or not do)

- Test lab requirements (see next section)

- Test tools, modeling, and simulation (custom and/or commercial)

It should be recognized that proper planning can reduce program costs and schedule risks. The items provided earlier may be in a written plan, but documentation is only an artifact of considering these earlier items. It is the planning and consideration of these factors that is really important.

The bigger the lab is, the more planning is needed, and the more likely the plan will change.

REQUIREMENT CONSIDERATIONS FOR LABS

This section provides requirements, which might be a consideration for a lab and facility. The areas listed are not complete, but are intended as key examples for a starting point in defining a specification of needs and desires for software test lab.

FUNCTIONAL ELEMENTS FOR A DEVELOPER SUPPORT LAB

The following are considerations and requirements for a developer support lab:

- Developer test environments: simulators, single-board computer(s) with probes, emulators, debuggers, cross compilers, tooling (mock objects), and so forth [5]

- Tools to provide measurement coverage and code structures (a short list)

  Statement

  Branch

Decision

Paths

Looping

- Developer test automation tools, which can be commercial or open source

- Integration of the software with other pieces of software and hardware

- Hardware may need to be simulated

- Profiler or analyzer (often built into compilers) (for call trees, race, and deadlocks)

- Static analysis tools

FUNCTIONAL ELEMENTS FOR A SOFTWARE TEST LAB

The following are considerations and requirements for functional support labs:

- Facility physical requirements

  Physical features (space, lighting, floor loading, and space)

  External interfaces (live interfaces, simulators, stimulators, network connections, and so on)

  Power (voltage, receptacle types, wiring)

  Grounding (locations, lug sizes, wire sizes)

  Heating, ventilation, and air conditioning (HVAC)—heating and cooling loads and air flow (a common item not often considered)

  Safety (fire or smoke detectors, static discharge, vent hoods, etc.)

  Security (locks, badge readers, hardware and software intrusion detectors, software)

  Special features (limited people access, other test equipment, customer equipment)

- Define functional requirements

  Functional: features, shall statements, needs, etc.

  Nonfunctional: quality(ies) requirements (safety, security, performance, reliability, and others)

  Software and hardware tooling to support testing (e.g., see simulation and stimulation)

  Support resources

  Input generators

  Outputs recording for the device under test

SIDEBAR 11.3

Poor Lab Planning: I once heard of an embedded software project that was well into development of the hardware and software. They were concerned because they had forgotten a major type of test facility and were wondering how to do the testing. They decided a lab was needed, and it cost millions of dollars, which was not in the original planning or budget. Management was not happy.

Other recording devices (time, temperature)

Automation tools (planning, management, execution, reporting)

• Test team (staffing) or resource use and restriction

Usability

Languages

Certifications

Standards, which the team or lab must comply with (contractual, legal, or company)

Requirements for a test lab may be captured in a document or something less formal, such as a presentation or on whiteboard. The formality of requirements and associated analysis depends on project context. Again, like planning, the key is to do the analysis and capture the information for the lab context.

TEST LAB DESIGN FACTORS

Since test labs range from single device to full hardware, software, and systems with many "features," the design will likewise vary. Design elements to consider include the following:

• Design to meet requirements and allow expansion flexibility, as needed

• Build-in testability (to test the test lab)

• Design the lab to be easily configured for testing with hardware and software (see Figure 11.5)

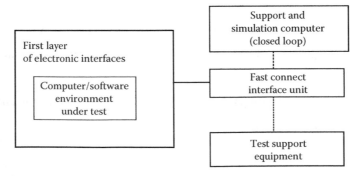

FIGURE 11.5   Design for testability reconfiguration.

- Define and allocate resources

- People to develop and/or run the lab

- Computer, hardware, and software to support lab development

- Time and schedule (both equate to cost)

- Test process and supporting management

- A design that works within context (cost, schedule, quality, functionality, and so on)

A lab test design, as shown in Figure 11.5, with a quick connect/disconnect interface unit allows hardware, test, and/or simulation computers to be configured quickly. Quick reconfigurations allow quick changes between attacks, thus saving time.

The design process of putting together a lab needs to address the facility, which would include room(s), power, and other considerations. Next, the hardware, both the embedded hardware (part of the system under test) and other equipment, which creates the test bed, needs to be designed such as computers, interfaces, support equipment, and so forth. Supporting test software must also be designed and implemented. This software would include off-the-shelf tools, customer-built software (see model of simulation and stimulation), OS, and so on. And finally, you will need to design the installation, integration, and operational concepts for these items. As shown in Figure 11.6, these design elements should be produced and documented to support later efforts, including generations of attacks.

In designing a lab, consider its use and maintenance. Just like a product, a lab that is hard to use, has bugs, or does not meet tester's needs may be useless. I have seen test labs whose design documentation was lost. I have seen labs where testers suffered finding many bugs in the lab itself (instead of the device under test), which wasted valuable test time and project money. This is a careful balancing act. You need just the right level of lab design

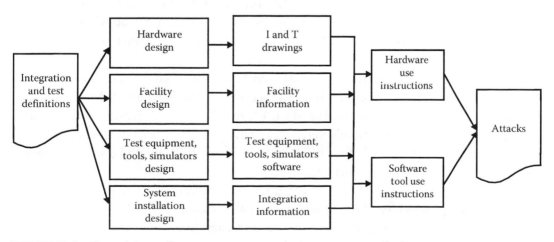

FIGURE 11.6   Comparison of answers across attacks improves error finding.

information. Good potential test lab documentation to consider producing for the design includes the following:

- Test facility drawings: facility design (normally done by facility services), unique hardware (cables, mounting hardware, adaptors), general test equipment, tools, special test equipment, simulators, and so on, system installation (into test facility)

- Software design information: commercial tools, OS, custom software, simulations, models, and other pieces of software including help files

- Test facility standard operating procedures: operational (equipment start-up, configuration, and operation), general support (safety, scheduling, security, problem reporting), special support (installation, checkout, and maintenance documentation)

- Vendor training for acquired tools and equipment

- Test lab user guide books

LAB IMPLEMENTATION

During the implementation of a lab, testers will do the classic development activities for hardware and software including ordering parts and equipment. Testers or support staff will develop hardware and software. You will integrate and populate the test lab with the hardware and software, usually following the incremental and iterative approach defined earlier. Agile concepts can be very useful in lab development too. I recommend key implementation concepts such as peer reviews, early checkout testing, and analysis. Finally, the implementation of test facilities and labs needs to be managed and controlled, just as on any project. There will be change management and error or bug reports to decipher and meetings to attend about these topics.

LAB CERTIFICATION

A "chicken and egg problem" exists in the software test lab world. Which comes first—the lab, the products to be tested, or the tests we run? And how do we know which is right (or wrong) when something unexpected happens? As the lab is built, you will need to test the test lab, but to test the lab, you will need the hardware and software under test. This is circular.

You have to start some place. Most of us who have developed mobile and embedded labs followed the prototyping approach mentioned earlier. We build a little lab. We build a few attacks. We get a little (prototype) product. We run the tests. We repeat the tests. All of this is parallel, incremental, certainly iterative, and evolves. We figure out what is wrong or right. If something is wrong, first we think the test lab or attack might be wrong. If those seem okay, maybe there is a product bug.

I always knew I was in trouble if I walked into the lab and saw a team of developers (hardware and software), testers, and lab creators all looking at a problem. It meant that nobody knew what was wrong. This is incremental and iterative development. I have almost never seen it work any other way.

So testing the test lab (some call it certification for lab use to not confuse it with "regular" testing) is a major task, which usually spans time, iterations, and versions of software, hardware, tools, tests, and people. Depending on the lab's context, the certification effort can be minor (just call it good) to a large effort with documented attacks against the lab and reports back to the stakeholders.

Certification of a test lab can sometimes be very important. If the lab is wrong and it misses a problem in a product, which is possible, then the bug escapes into the field. This is not good. If the lab is wrong, and you spend lots of developer time looking for bugs in their code that are really bugs in your lab, you have wasted project time and schedule. That also is not good. If the lab is wrong and you build bad attacks because of it, you waste tester time and schedule. This is also not good. If the lab is missing features that you need, some attacks may not be possible, and bugs may be missed, until they are seen in the field. This is even worse. Among many other skills, a tester needs to make a determination that your test facilities and resources are "good enough" to support attacks.

Lab certification efforts may need to be applied to hardware, software including models, simulations, and stimulations, and operations within the lab. Certification of test labs (and the tools) in some context can even have legal and regulatory considerations.

Key point: Do not forget to test the test lab.

OPERATIONS AND MAINTENANCE IN THE LAB

Most embedded and mobile software systems spend most of their life running in the field with changes and fixes coming back for the development of new versions in hardware, software, or both. This means more testing. The attacks may be new. The attacks may be regression (rerunning earlier tests to make sure the change has not broken something already working). The attacks may be retests of the change itself.

To support all of this, the test environment must be readily available. I have seen labs that were used for years (decades even). This means that the lab must be maintained and even upgraded. To do this, there must be management, control, and updated development of hardware and/or software, as well as tools. Again, it is like a project inside of a project, but now in maintenance mode. Factors to consider in this phase include the following:

- Configuration management of the facility

  Driven by lab bug or change tracking and closure

  Documentation

- Lab management such as scheduling, budgeting, risk

- Hardware maintenance such as spares, replacements, updates, chemicals, calibration

- Test software upgrades and fixes

- Tester support and availability such as who does the attack, training, and so forth

You must plan and develop the lab throughout a project, including lab operations and maintenance. This extends to the end of a project, and then someone must determine how to decommission the lab. Does the lab recycle to the next project? Do you scrap the facility? How does lab decommission work, who pays for the efforts, and when is the work done? In most cases, you cannot throw the lab into the trash, since this may violate environmental laws. Yet more thinking and planning may be required even when you are trying to end a project and its lab.

> **Takeaway note:** Test labs must be developed like any product following a complete development life cycle.

LAB LESSONS LEARNED

As important an asset as test facilities and labs are, it is important to realize that they are not perfect. I present some of the lessons I have learned about labs here. The lessons are not complete, but they do apply to many test lab contexts. You may have different lessons learned for your test lab(s).

First, the lab may not have all of the hardware, software, and systems seen in the real world. If you are testing a telecom device, will the lab replicate all the hardware and interfaces seen in a global telecom network? Chances are nil. If you are testing an embedded control system for a factory, will all the factory elements of every factory type be in your test lab? Probably not. So what limitations does your lab have from what it is "lacking" and how do you address this in your attack testing? These limitations can become risk items that you will need to disclose to the stakeholders.

The actions and processes of every user are basically infinite, and so attacks using the lab on the device will always be lacking some possibility. Sure models, use cases, operations concepts, and other ways to capture user actions help, but they can miss a case that the lab tests will not cover. Again, how does your lab and attack planning address your risks?

Next, the environment of the real world is likely not totally recreated in the lab. Can you create outer space in your lab? What about thousands or millions of smart phone users in your lab? What about all the different operating systems, apps, network service providers, or other "elements" of a mobile phone? In all cases, the answer is "no," meaning you probably cannot recreate every possible instance or factors in a lab. Therefore, you have yet another possible limitation of your lab and your attacks. This might be yet another lab-introduced risk.

Limitations in labs can "hide" in missed errors. If a bug happens, but is not a system failure or crash and the lab does not have the resources to flag the error, it may be overlooked. I have seen little inconsistencies, which happen in only a few attack tests, be overlooked but then be seen in the field by a customer. When we went back into the test data and looked, sure enough we could see the error, but hindsight is always 20/20. In many projects, it is better if the lab resources or posttest analysis can spot the bug before the user does, which is often easier said than done. You may have another risk to consider.

The test planning should recognize lab abilities, limitations, as well as risks that are known and unknown [6]. Disclosing these factors to the stakeholders provides information for people to make decisions about. Should you increase testing? Should you do more field testing? Should you improve the labs? These and other questions offer the chance to improve the labs and testing.

A tester must consider the resources for a lab to support testing. Each lab resource translates into money—an ultimate commodity, which is almost always in short supply. What many software program managers underestimate is the cost of developing a test program and associated lab(s). This lab development effort can be a very large percentage of the total project costs or schedule. Hence, many lab efforts are forgotten or minimized, so testers get less of a lab than what is needed for even adequate testing. Testers must always trade the need for a "right sized" or even "perfect lab" against cost, limitations, features, and risks.

> **Key point:** A test lab is a project inside of a project. A lesson learned to share is this: it will take all of the skill of a development team and the knowledge of testing to create any lab.

AUTOMATION CONCEPTS FOR TEST LABS

Associated with many of the attacks are tools and test automation concepts. Tools are incorporated within a test lab. This section builds automation ideas from the attacks, adding tool, simulations, and automation concepts within the lab development [7]. Not all automation tools, tool concepts, and tool options are presented. I do cover the concepts that I have seen to be the most useful in mobile and embedded test labs.

TOOLING TO SUPPORT LAB WORK

Most any profession is defined, in part, by its tools and the skill of those using the tools. A skilled master plumber has many tools specialized to that trade that the average person just does not know about or know how to use. It is a little sad that many mobile and embedded testers know only a few or no tools at all. In this section, I provide a list of tool categories, hardware and software, that I have either used or witnessed in labs. I have also witnessed people being called experts simply because they were master of several of the tools. Testers need skills in testing concepts, tools, software, and the product domain. It certainly takes time to learn a tool, and most organizations have many people in test labs who have mastered multiple tools. Taken as a whole, such teams of people can have enough staff resources to address all of these tool categories. A tester would do well to learn as many of these tools as possible, or at least know whom to call for help in using the tool. It is noteworthy that I have seen labs in different companies testing a variety of mobile and embedded systems, and yet still having many tool categories in common. The reoccurrence of this situation means that teams have independently identified, developed, and put into use tool concepts to solve a test problem.

This section summarizes tools and a concept presented with some of the attacks, plus expands tool concepts. I am not advocating heavy automation or use of tools into every aspect of every test attack. However, in some places, tools can be very useful. Certainly, in many mobile and embedded test labs, tools are an advantage, if not a necessity, which can lead teams into thinking that they should have lots of tools in every category. Testers would be better at testing if they understood techniques and attacks supported by a few tools rather than trying to automate everything blindly.

Large teams working on critical mobile and embedded software do have tools with high levels of automation, freeing the tester to think creatively on attacks and testing the software. This can be costly, but it may be more costly not to automate. For the small teams and projects that cannot have large teams of testers and tool support staff, you will likely need to select a few tools based on priority and context of the device(s) under test. One or two good or even free open source tools can make the single exploratory tester much more productive and efficient. In any case, you must select tools wisely and evolve with them as tools in the mobile and embedded domain are often immature.

Table 11.1 defines generalized categories of tools associated with activities in test. Each category gives example subgroupings, functionality, and potential benefits. Specific tools names and vendors are not provided, but with a little Internet research and looking in the general reference sections of this book, a tester can find specific names. A good reference book on experience in test automation is [8].

TABLE 11.1 Standard Sample Tools

Attack Activity	Example Tool Type	Function Provided	Benefit
Test data set-up	–Tools to define test case inputs –Tools to populate databases –Communications and data streams	Provides input data	–Saves tester time –Avoids tester bias –Supports random test techniques plus several other test techniques
Test case execution	Developer test tools	Supports the test during CPU execution	–Automation of testing –Repeatability –Insight into test
Product Analysis	SCA Model analysis Document analysis Security test tooling	Checks code, model, and document for certain types of bugs and issues Finds or supports security testing	Flags potential classes of problems without executing the software
Tools for the lab test results recording, reporting, and analysis	May include data storage, visualization, and reporting tools	–Recording fast incoming data –Large amounts of data –Analog and digital data	–Supports data for posttest analysis –Posttest data collection, analysis, and reporting
Performance attack tooling	–Load test tools –Test monitors	–Virtual users –Execution of scenarios –Insight into timing	Support testing that is hard to do otherwise (such as with 1000 testers)
Basic generic test and analysis support tools	Test management	–Management –Configuration –Bug reporting –Traceability –Documentation	–Central or common repository that all testers and stakeholders can use –Provides information such as measurements and status
Lab oracles and modeling tools	See Section: N-Version Testing Problems in Labs and Modeling	Success criteria to go into test cases	Support test planning and definition

Many teams, particularly Agile teams, focus up front on tools that detect programming and abstraction errors. In Attacks 1–3, we look for tools to help in "white box" or structural testing to very low levels of the computer, including a digital simulator or a hardware system such as an emulator. Tools here need to help ensure that the code implements such things as detailed software requirements (developer stories), design, configuration controls, and software standards. This testing is usually done at a module level, object, or on small segments of the code, which are executed somewhat in isolation from the rest of the system (see Attacks 2 and 3).

Most tools do not provide a way to easily determine the success criteria for the comparator function of the tool. Determining success criteria is usually human intensive and time consuming, although some use of automated comparisons based on test oracles has been achieved [9].

Other teams focus tools on the black box lab testing side of attacks. Here, the tools support attacks, which are targeted at errors of abstraction and deduction, where we are trying to find errors that deviate from things such as requirements (customer facing stories), where big bugs exist, and answer the question "did we build the right thing (hardware or software)." But there are problems with automation, determining inputs, running the attacks, and determining success criteria. Tools are not a universal solution to test challenges.

Each of these broad categories of tools is expanded in the following subsections.

> **Key point:** There are many kinds of tools and options. I present a few common examples and categories of tools to serve as starting point for you. A test team with extensive lab needs will do well to have good test automation engineers on staff since this is a specialized skill that not all testers have. A good test automation engineer will understand most aspects of testing, software development, and, in some embedded cases, hardware.

Any tool or vendor list is likely to become dated as well as being incomplete. A common question that I get is about which tools to consider. The reader should not take any tool mentioned in this book as a blanket recommendation. I believe that you should do your own studies and evaluations on tools and make up your own mind about the tools for your local context. The information provided in this chapter is intended only as a starting point for you. I suggest a good Internet search when you get ready to start any attacks or are beginning to build a lab. For some up-to-date tool listings based on my research, see my website (http://breakingembeddedsoftware.com).

TEST DATA SET-UP

Different chapters in this book as well as several attacks define tooling to help generate inputs. Table 11.2 lists some these tools.

Described with attacks, you will find a short description of how to use the tools. However, tools that generate test inputs should not be viewed as a replacement for thinking testers. Tools require active creativity to properly set up an attack using supporting tools. Test data aids the configuration of the tools to help with repetitive tasks (e.g., typing data

TABLE 11.2   Test Data Creation Tooling (Sampling Only)

Tool Type	Examples	Provides
CT tools	Advanced combinatorial testing system (ACTS), RDEXPERT (see Attack 32)	Systematic test inputs selected mathematically when dependencies are known
Database set-up	http://www.compuware.com	Inputs for many fields and monitoring
Configuration check	Do an Internet search on "mobile software configuration tools"	Finds out which software and apps are running on your device to aid in testing such as in Attack 32

into endless fields) or avoiding human data selection bias (e.g., selecting values that have already been tested or may not be very interesting).

To support Attack 32 in the generation of combinatorial test (CT) set inputs, I provide a partial list of tools. I have used most of the tools in this list, and they worked for my purposes.

- James Bach's AllPairs (Open Source) [10]—http://www.satisfice.com/tools.shtml

- Bob Jenkins' jenny (Open Source) [11]—http://burtleburtle.net/bob/math/jenny.html

- NIST ACTS [12]—http://csrc.nist.gov/groups/SNS/acts/documents/comparison-report. html

- Hexwise [13]—http://hexawise.com

- rdExpert [14]—http://www.phadkeassociates.com/rdexpert.htm

- IBM's Combinatorial Testing Service [15]—http://www.alphaworks.ibm.com/tech/cts

TEST EXECUTION: FOR DEVELOPER TESTING

This section refers to Chapter 2 on developer testing. Many tools exist to support low-level developer testing in both the commercial and open-source worlds. Current examples of tools for the developer level of mobile or embedded space include the following:

- CMock and Unity: http://cmock.sourceforge.net

- Googletest

- Grenning's tools: http://www.renaissancesoftware.net/blog

- cppUTest framework tool

- LDRA tool

- Wind River tools

- MATHWORKS tools

These tools assist in automating aspects of developer testing, including stubs, drivers, mock objects, instrumentation, reporting, set-up and initialization, framework, and other features.

Tools have very specific features. Use the reference materials or the individual tool information to understand more about tool's features and how to use the tool for specific cases.

TEST EXECUTION: GENERAL

Mobile and embedded system may need an "environment" to host the software under test. The environment is made up of hardware, connections, other pieces of software, and the operations.

For mobile and some embedded devices, you may have a user interface (UI) to deal with during test execution. Tools that automate the input of data, the monitoring of the attack or test as it progresses and records the results of the testing, can be very useful.

For many devices, there is no UI (keyboard, touch screen, display screen, etc.). In this case, the lab and tooling must provide the inputs, monitor the software, as well as record the results. There are some commercial tools to help with this situation, but many labs require the generation of customized tool sets to achieve these capabilities. Current examples of mobile and embedded test execution tools (or vendors) include the following:

- www.windriver.com
- www.testplant.com
- FoneMonkey: iOS
- Apple xCode: Mac, iPhone and IPAD
- ROBOTIUM: ANDROID
- Load Tester PRO 5.1
- Borland Silktest Mobile
- Xunit (do an Internet search for this one)
- HP_QuickTest_Professional

Note: This list of tools is subject to constant change, and even though I list tools here, you should not take that as a recommendation for any specific tool. For the latest mobile and embedded tools listings, see my website.

In both cases, devices with a UI and no UI, tooling to aid test execution can be a benefit. Execution of many mobile and embedded system tests may be fast, so the automation tooling helps.

These tools can support test scripting—keyword, capture, and/or playback. Such execution tools have existed for years in the IT or PC world with varying degrees of use and success. I expect similar situations in the mobile and embedded space. There will be new mobile and/or embedded execution support test tools from major tool vendors. This book, based in attacks, is such that many of the attacks can be used with scripted automation tools, while other attacks can be manual. Most organizations will mix the tools, automation, and manual testing.

Here are some considerations with automated execution to consider when looking at any tool for mobile or embedded testing:

- Ability to work with different device configuration (e.g., open or sealed boxes).
- Ability to update to software versions (or not) (such as software under test, other software, and operating systems).
- Stability of a device, hardware or software, which may impact testing.
- Security risks (e.g., of jail broken or rooted devices).
- Is a tool and associated processes "legal"? (Device vendor and/or regulator have rules on tooling and automation.)
- Does the tool impact test fidelity? (Works the same way in the field or will it create false results or unverified failures.)

The following is a list of potential execution tool features that a tester might want or need.

- Scripting for automation (e.g., keyword testing, data-driven testing, and others)
- Screen capture abilities

  Optical character recognition (better known as OCR)

  Object recognition

  Images
- Performance testing support (e.g., analysis, stress cases, load testing, and others)
- Ability to load, clean, and check the configuration of a system device or the software
- Tool ability to deal with firmware, gate arrays, and computer memory of various types
- Tool ability to work in limited environments (e.g., "walled garden" [16])
- Running in the cloud or in a virtualized environment
- Tool network and interface abilities such as through universal serial bus (USB), Wi-Fi, cell signal, radio wave, or others
- Tool test results ability (e.g., log files, recording, telemetry, signals, or others)
- Regression–retest test support

Many of these features are common to most software test labs, but can be very important in the mobile and embedded lab environments due to limitations of the devices, such as cross compiling, limited interfaces, networking, vendor restrictions, and others. There are limitations (such as how to load a device with software) and restrictions on what can be done to a device (sealed hardware and legal restrictions), which are used to "protect" various stakeholders, yet these limitations make testing harder [16]. For testers in the developing organization, you may often have prototype or "broken" devices to overcome. Other testers get these limitations as restrictions where you cannot break the device or load the software easily, so your test environment and tooling will need to support getting the software under test loaded onto hardware.

Each device will have instructions for loading. There are too many devices for this book to detail loading procedures for all of them, so testers will need to refer to device information for details on loading and what "walled gardens" can be played in. Many embedded and mobile devices have different (from the PC world) network interfaces. Different network interfaces introduce challenges for many execution tools, which testers at all levels must consider both in getting inputs into the software under test and many times getting results out (recorded).

PRODUCT AND SECURITY ANALYSIS TOOLS

In this book, product analysis tools include static code analysis (SCA) tools and security testing tools. There are large lists of these kinds of tools, and the lists and tools change regularly. I provide a sampling in Table 11.3 of tools that I have some familiarity with and as examples to get you started.

Longer and more up-to-date lists can be found at

- http://en.wikipedia.org/wiki/List_of_tools_for_static_code_analysis

- My website (http://www.breakingembeddedsoftware.com)

- Do an Internet search on your own

TOOLS FOR THE LAB TEST RESULTS RECORDING

In the IT and PC world, often the tester can view the results of a test or attack in real time, right on the computer display screen, and determine success or failure. Even so, recording test results in the IT and PC world is still important and often supported by tools such as loggers,

TABLE 11.3   Tools to Support Analysis, Penetration, and/or Security Checks

Category	Specific Tool Example	Notes/Usage Example
SCA tools	LDRA, POLYSPACE, KLOCKWORKS, PARASOFT, X-CODE	Conducts Attack 1 by "scanning" the software
Sniffers	WIRESHARK	Look for signals that can be hijacked
Port scans	FAST TRACK	Look for the entry and/or penetration points that the hacker might use
Password cracker	JOHN THE RIPPER	Standard passwords that can be used in a brute force hack of passwords
Software radio	Any radio wave receiver	Connect to any PC to become a radio wave receiver
Security	CHARLES, ANONYMITY, Memory Readers, ANDROID NATIVE SOFTWARE DEVELOPMENT KIT (SDK) tool	Software scans, forensic and log analysis utilities, port knocking and scanners, sniffing utilities, spoofing tools, tunneling tool, wireless scanner, packet constructor

capture–playback devices, screen views, and so forth. Likewise, many times, the results of some mobile and embedded testing can be reviewed by the testers during the test, but some systems generate so much data (volumes) and different types of data that the test team may need many different kinds of tools to record the results of testing and later analyze the tests.

In the mobile and embedded space, it is critical that test results be recorded given considerations such as the speed of devices, lack of a UI or graphical user interface (GUI), limitations of size, internal space, or memory, output data volumes, and other "features," where it is easy to miss a test result. Not all bugs announce themselves as a hard crash or obvious system–software failures. The bug can be a wrong trend line in thousands of data points. Additionally, you may have data in the form of digital and analog data to capture and later review from these subtle bugs. In mobile and embedded testing, it is often necessary to fully record the data and later judge the test results after the attack during a posttest analysis effort. These factors make comprehensive data results "recording" for mobile and embedded devices important.

The tools used on devices will often look more "hardware" based than most classic software test tools, which testers might already have experience on. The list of recording tools include devices used by the hardware world as well as software, because attacks need both hardware and software data elements to detect bugs. All of the tools can be useful in setting up, running, and understanding a test attack (see posttest data analysis section). Common questions a tester may face are the following:

- What tool(s) to use in data recording?

- How much data needs to be recorded?

- How to use the tool(s) in test set-up?

- How to use the tool(s) during the test?

- How to use the tool(s) posttest?

- Has the tool itself been tested or certified for the intended use (regulatory compliance)?

Not having the right tool or data can make mobile and embedded testing much harder or even impossible. This is not to say that every team needs every tool, but a good tester will know what kinds of tools are available, so that when faced with a mobile and embedded attack problem, the attack can be made to happen using the tool. Too often, I have heard "we did not run that test because we did not know how to deal with the data," and later a crucial bug was found. The selection of data recording and reporting tools must be made based on risk, cost, and schedule. This chapter has additional sections on dealing with the unique aspects of data in the mobile and embedded space.

PERFORMANCE ATTACK TOOLING

In Chapter 6, a series of attacks was defined to assess timing and performance concerns. The IT and embedded industries for years have had to work with test performance tools or risk missing bugs. Here, the lab tooling concepts for performance attacks are expanded.

Generically, in the mobile and embedded space, testers should look for or create tools to help perform, analyze, and monitor performance attacks. Look for tools that provide visibility into the processor usage and timing. Historically, these types of tools need customization and/or special configurations. These tools will provide the data points that the attacks identify. Tools can be software- to hardware-based solutions and include the following:

- In-circuit emulators or single-board computers that provide samples of time data
- Profiler tools that monitor central processing unit (CPU) usage, if your environment supports one
- Probe tools that monitor bus and CPU usage and can "time" events
- Low-level monitoring systems implemented in hardware directly or in software (built-in test)
- Signal input or output processing tools
- Rate Monotonic Analysis and timing analysis tools
- Performance load testing tools

In the mobile and embedded space, unlike the PC and IT space, there is not yet a large variety of load and performance testing tools for mobile smart phones, and even less in other embedded devices. This is changing, and as you read this, you should do an Internet search to see what has come onto the market after the publication of this book or check my website. Mobile testers should know about the PC and IT tools in the performance testing world to some extent, because likely the same lessons and rules of the PC and IT world will apply to the mobile space—at least for now.

Performance testing tools in the IT and PC world have been very successful in allowing the creation of hundreds or thousands of "virtual" users with different use scenarios. These IT and PC tools allow rapid attacks with load, stress, and peak performance tests. As similar tools expand in the mobile and embedded space, I recommend their use to support the attacks of Chapter 6. The use of such performance tools has been one of the "bright spots" in the test automation world.

Finally, there are features, capabilities, and tools in the mobile and embedded test lab that are needed to support timing of interrupt testing. If you are setting up for an interrupt or performance test environment in the lab, you should ask and be able to answer the following questions:

- Do we have the ability to trigger certain events (interrupts) at just the "right" time?
- Do we have the ability to capture what is happening in the system and/or inside the software during interrupt processing?
- Can we emulate, simulate, or stimulate interrupt processing?
- How realistic are each of these factors to our project in the lab?

You will need to configure the lab and analysis tools to support the test and analysis of these questions. If you can trigger interrupts, or correct triggering conditions into interrupts, testing this interrupt performance will be possible—but still difficult. If you cannot trigger the events and measure the performance numbers in the software (using the lab), you will have difficulty conducting interrupt–timing testing. As a result, this lack of testing poses a serious risk for many company's software.

BASIC AND GENERIC TEST SUPPORT TOOLS

I have mentioned numerous automated test tools associated with various attacks. I am often asked for recommendations on basic and general test support tools. This section is not a comprehensive treatment of tools or test automation; however, the references contained in this chapter can help your work.

Tool selection should also be made within project context. First, tools must fit the team's overall approach to testing. A common mistake in tool selection is picking a tool that has concepts or methods different from what the test team is doing. This results in tools not being used. (These tools are commonly called "shelfware.") Second, tools are often picked because the team knows them and because they have used them in the past. But here again, if the tool(s) does *not* fit the context, they are less likely to produce accurate analysis and results.

Test tool automation takes some specialized skills within the team. Programming skill can be very useful. Also skills from the systems or hardware domains are useful for some test automation tools, which might include modeling and even on the use of hardware. If a test team is weak in some areas, I recommend broadening tester's skills to improve the likelihood of success in configuring and using tools.

Tools must also fit the cost and schedule context. Sometimes teams select tools because they are open source and "free." But don't forget that often such tools need to be modified and customized to really work—and sometimes that can cost more in man hours (effort) than the tool itself. So the tool is not really "free." Other teams will select a commercial product, because perhaps it is better supported or easier to customize to a mobile or embedded domain. And while that may well be true, test team time and effort must still be spent learning and configuring the tools, which still costs time and money.

Finally, no single tool or tool family is likely to cover all of test automation needs. Most mobile and embedded lab environments provide examples of a variety of configurations. Due to the variety and numbers, careful planning and development of tool environments are needed. Further, I recommend that such environments start simple and evolve over time. It is better to have a couple of good key tools to start. Once these tools are working, add to them over time. My experience has shown that out of every 10 tools tried, 2 or 3 improved the project's testing, 3 or 4 were workable but did not offer much test process improvement, and the remainder should have been abandoned.

Takeaway note: To be successful, test tools should be carefully selected, developed, and configured.

There are books that address aspects of test automation including mobile and embedded automation [8,17–19]. Mobile and embedded tester knowledge would be increased by reading these references—for those serious about test automation. You must be prepared to provide resources: time, money, and people to do test automation. You must also have management and team buy-in. Lacking such things puts test automation and tooling at risk for becoming shelfware.

There can be a tendency for test teams and management to believe that "a tool" will save the day for projects in trouble. I call this irrational exuberance. Do not be so foolish. Think about which of these tool types might help your project. Areas where I suggest automation be considered include thus:

- Basic test management and planning

- Work flow management or tools that aid and guide the test process, based in

  Diagrams

  Checklists

  Web enabled

- Requirements management tools for tracing

- Configuration control and management

- Defect trackers

- Test reporting tools

> **Takeaway note:** Software test labs can contain many types of tools to make attacks more efficient and effective.

AUTOMATION: TEST ORACLES FOR THE LAB USING MODELING TOOLS

The holy grail of software testing is a test oracle. Test oracles are ways to determine if a test works or fails. There are no perfect oracles, and most testers work with only the most basic oracle—the human mind. Testers guess, make a judgment, and use requirements or other heuristics to determine if a test result is good or has found a bug. Such judgments can work great in many software systems, but in the mobile and embedded space, these factors can be lacking due to the speed of a computation (milliseconds), volume of results (megabytes or terabytes), the complexity of the answer (tens or hundreds of data values), subtleness of the bug (the bug is hard to see), or other complicating factors.

Many teams have learned to use tools, results from different levels of testing, models, or standalone pieces of software to serve as an oracle. These concepts can be better than anyone's best guess or judgment. In many situations, you may use extensive computer simulations and models to analyze systems and to serve as oracles for the actual "black box" results of the software. Oracles can be based on requirements, operational concepts, use cases, and even tester expectations or experiences. Each simulation or model is usually designed to concentrate on one error class (deductive or abstraction), smaller areas of the system (one feature), or one level of the system. Whole system oracles are unusual, unless the system is small. The oracles used before a test, during a test in real time, or after a test attack have been completed.

Examples of such tools can include the following:

- At lower levels (code or software integration), tools are based on the software design. They simulate aspects of the system but lack the overall functionality of the total system. These tools allow the assessment of software for unit of code, class, or single function of the software.

- At a more integrated level, oracles can be built on a larger functional basis. For example, a simulation may model the entire boost profile of a launch rocket with a 3–degrees of freedom model, while another simulation may model how the antilock brake subsystem of a car is required to work.

- Finally, while difficult, an oracle for a whole mobile and embedded system can be created.

This allows system evaluation starting from a microscopic level up to a "macroscopic" level. Identical start-up conditional tests on the actual hardware/software can be compared to these tools. Also, cross checks between results and between different models can be made (see Figure 11.7).

To implement such oracle tools, teams have used the following:

- Models based on state machines, sequence charts, behavior diagrams, and so forth, in modeling languages such as UML, SysML, or other specialized languages.

- Normal programming languages such as C, C++, Java, to create simulations;

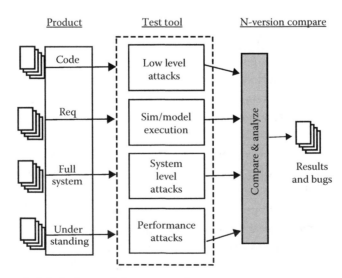

FIGURE 11.7　Example lab design and documentation process.

- Formal languages to generate test results such as vectors, Z, VDM, customized, and so on.

- Framework tools such as MATLAB® and others specialized to a domain such as area, automotive, telecom, and so on.

- Inline assertions used at code or other levels, where assertions are rules about what is true inserted into code (not part of main code's logic), then used in checking, either during the developer test or after test.
  (*Note*: There are tools and approaches to support using assertions to "check" results, normally used in developer attacks, such as Attacks 2 or 3.)

While such tools have been successfully used in many mobile and embedded environments, they come with certain issues and risks, which must also be considered. This does not mean that we should not be using these tools, just that some thought is needed when examining the issues and risks for your local context. Here are some examples of risks and limitations of modeling:

- The time and cost to develop and debug, so usage must be justified with project context.

- Many such tools require a high degree of skill such as programming, modeling, system knowledge, and so on.

- The tools or simulations/model should be created using design and implementation methods such as coding, and so the tool itself can have bugs.

- If the model and/or use case fails to represent reality, the test results may be invalid.

- These types of tools and models must be tested before attacks can be run as well as in a regulated environment where the tools may also need to be fully documented and certified.

- Test models and environments can indicate software bugs/faults when none were present, which creates a "wild goose chase" for testers and developers (see N-version testing section).

- Test tools/models can have hidden or missed problems, which can cause problems with the product under test, and a bug may be missed until it surfaces in the real world.

Each of these cases can end up costing time and money, but they can also be a benefit to mobile and embedded testing.

SIMULATION, STIMULATION, AND MODELING IN THE LAB TEST BED

Simulation, stimulation, and modeling are important concepts for many mobile and embedded testing situations. They are closely related to oracles and automation. This section expands on these concepts from early attacks, particularly Attack 17.

> **Key point:** Modeling, simulation, and stimulation in embedded and mobile systems are often harder to develop than in other software domains due to unique hardware, control problems being solved, and "mobility" issues. Using models and simulations properly is another specialized testing skill that can offer projects a competitive advantage in solving many test problems.

In the systems and software world, there are many overloaded terms that testers need to be careful of when using. Simulation, stimulation, and modeling are such terms. If an organization has systems and/or software engineering, these groups may apply analysis activities such as modeling or simulation—activities that a systems or business analyst person might perform. This section does not address the modeling and simulation an analyst or systems engineer might do. Where UML is in use by a software person, these terms might also be used. This section will not cover where UML is in use by a software person. This section is about how many test labs can use simulation, stimulation, and modeling as tools to support the attack tests and has supporting references, books, and articles on simulation, stimulation, and modeling.

Testers and analysts have successfully used these tools and modeling to do the following kinds of things in support test labs:

1. Allow automation of defined test processes in a test lab environment

    - High-fidelity closed-loop simulation

    - Simulation of hundreds of network system inputs (or users)

    - Simulation of hardware that can/does not exist, such as ordinance, safety, or hazardous events, hardware items that had not been created

    - Generation of "hard to reach" situations

2. Support generation of inputs

    - Run hundreds of random simulations of code to find input sensitivities for execution in a full test lab

    - Generation of random or statistically meaningful inputs

    - Generation of inputs to test coverage

    - Create specific data to force test conditions

3. Allow generation and/or assessment of outputs (oracles)

When testing mobile and particular embedded systems, there can be great benefit from being able to simulate aspects of the system or stimulate inputs into the software under test. Simulation and stimulation modeling tools help to do this and directly support situations such as those described in Attack 17. Simulations can be used to create test situations such as the following:

- Slow response and traffic

- Sending partial messages or other data

- Corrupted data

- Forcing error responses or

- Interrupted data

Many of these cases are the "hard to reach" software test cases and are defined more in Attack 8. You can use the simulation and stimulation to force hardware, network, or other system cases, which may be difficult to induce in the real world or hardware. There are two basic types of simulation and stimulation tools: open loop and closed loop.

In open loop, the simulation and stimulation model simply generates an input and passes that into the system under test and watches for and/or records the behavior. This is the end of a test attack. We judge if the test passed or not, and move on to another test case/attack. This is adequate for many systems. The open-loop simulation can stack up a series of inputs and pump them into the hardware–software. But many embedded devices run in real time or have inputs that are dependent on the last set of outputs. These devices are controlling something with a feedback loop. The loop may be very fast, say 10 ms. The tester cannot hope to provide a response back that fast before the next set of inputs is needed. This means that we need a closed-loop simulation or stimulation system. I will focus on closed-loop simulation modeling in the next section, since it is more complex than open-loop simulations and often very useful for critical and large mobile and embedded systems.

I further classify models into a couple of basic groups. One group is discrete in that they tend to treat time and events as a quantized element. Time can be sped up, slowed down, and so on. Time can be in "chunks" such as seconds (or smaller), minutes, hours, days, or even years. This type of simulation and stimulation modeling is normally used in analysis, oracles, and understanding of a system. Commercial tools such as MATRIX-X, and MATLAB–SIMULINK® tools fit into the discrete–executable modeling category. These simulations and stimulations allow the tester to model some aspect of the system such as inputs or outputs, or to model sequences of events and actions that the software will see. Tools like MATLAB allow engineers to calculate mathematical functions that the software will implement and serve as an oracle. While each of these tools has many other features and uses in testing and engineering, I have successfully used aspects of these tools in software test modeling including test planning, data design, algorithm checking, oracles on test results, and posttest analysis.

The other group, I call continuous simulations (and closed loop) in that they tend to function more with time as an integral element of their function. Continuous models can run in real time or run by calculating input data to the device under test based on last cycle's output at each time step of the system. Time effects, while less critical to some software, has impacts on all software to some degree, so its effect in modeling, in many cases, should not be ignored, for example, when finding bugs in interrupts or embedded devices that operate in the real world and where time is a factor. The devices are often controlling something or interacting with the world via input/output signals in real time. Testers can expand this type of continuous real-time, closed-loop simulations in the lab to provide a time-based stream of inputs, based on outputs, which replicate many aspects of the real-world and closed-loop feedback systems.

CONTINUOUS REAL-TIME, CLOSED-LOOP SIMULATIONS TO SUPPORT LAB TEST ENVIRONMENTS

Many mobile and embedded test labs have developed tooling, which we call continuous real-time, closed-loop simulations, to aid the testing and attacks. In such test support systems, we model the real world and elements external to software under test to generate many of the inputs and environments to allow the testing to be done. For example, if the computation cycle of the software under test is 10 ms and you have hundreds of inputs, how can a human hope to keep up providing new test data every 10 ms? Worse, if the software outputs need to be factored into the hundreds of inputs, you must compute each new input every 10 ms. Continuous real-time, closed-loop test environments provide a solution to this problem by having a software simulation model compute each of the 100 inputs each 10 ms cycle.

Figure 11.8 illustrates the beginnings of a closed-loop, real-time simulation showing only the most basic lab set-up, when we connect or include a second computer to the computer with the software under test. This second computer can be used to simulate aspects of or the entire external world that the test computer interacts with. If we only provide inputs and record outputs, we have an open-loop simulation, but notice the feedback line where we take the output from the system under test computer and provide that information back to the second computer, where our simulation is running. If our simulation models are fast and "smart" enough, we can compute another input back to the system under test before the software under test is expecting the inputs. This feedback loop must be faster than the timing cycle of the system under test. For example, if the system under test is running on a 10 ms cycle, the simulation must take the output from the last cycle, compute a new set of inputs, and provide them before next cycle begins, in other words in less than 10 ms. This means

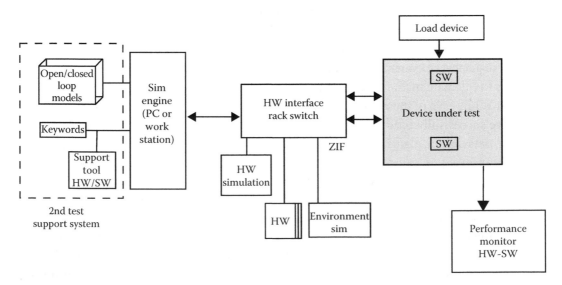

FIGURE 11.8   An example closed-loop simulation with real hardware in the loop.

that the second computer usually has a lot of computing speed and power. Variations and different implementations of this concept exist. For example, we might set up the second computer as a specialized processor, or a series of parallel processors could be used.

Additionally, the second computer and its modeling software can be used for the following:

- Generation of test set-up conditions prior to running attacks including load files and data

- A pre-execution test oracle to provide success criteria

- Loading the software under test into target computer

- A test oracle during test execution

- A posttest analysis tools including serving as a completion test oracle

These simulation models, which provide simulation and stimulation to the software under test, can provide the following device traits or behaviors:

- Input devices: such as sensors

- Output devices: such as motors

- Other systems: such as computers in the system but not part of testing, networks

- Outside world: weather, noise, humans, anything

- Other users, such as a human patient (human response, breath rate, heart rates, and others)

Since we are interested in using the lab in many different types of attacks and various levels of testing, it can be advantageous to be able to reconfigure the closed-loop, real-time simulation environment quickly. We may want to be able to run with one or more pieces of the real hardware, sensors, or output devices, then quickly reconfigure it to run different attacks with just simulations. A flexible lab configuration is needed where we can "switch" in simulations or the "real thing." An example of how to do this is shown in Figure 11.8. In this figure, we have the "under test box," which has hardware and software (and might really be a black box). The "under test box" provides outputs to the "simulation engine" (computer), where the closed-loop simulation models run.

This example figure provides a comprehensive test environment. We can run with simulation, or as the figure shows, we can "switch" to the real hardware or real world. In my experience, this switching ability is very important because it allows a flexibility to use the real hardware in one test and then use a simulated hardware feature in another attack without lots of time spent reconfiguring the lab. In the figure, I show a "HW interface rack-switch" assembly with "zero insert force" (ZIF) connectors. This is basically a hardware patch panel with the ZIF that allows quick reconfigurations between simulations and hardware

(*Note*: There may be electrical differences to consider in this patch panel.) These environments can be very expensive to create. The cost is directly dependent on the complexity and size of the system under test. Also, the interplay between simulation, hardware, real world, and the system under test tends to be tricky and complex. An actual drawing of such a test system would be very large, but such systems are often the only way to test the software in a realistic environment. Some test facilities I have seen include operational systems, all system cabling, human simulations, connections to the real world, and the ability to "switch in" a complete system configuration. However, just because we can "switch in" and run in the real world does not mean that we can do away with simulations since there are aspects of critical systems that cannot be fully duplicated in a field test environment and thus must be simulated. The ability to quickly configure labs by "switching" between the real hardware and the closed-loop, real-time simulations can be important lab features to have.

Examples of where this lab configuration approach has been used include the following:

- Space systems where many of the external sensors of the flight environment behave differently in zero gravity than on the ground.

- Mobile medical devices, where until fully tested, you cannot "risk" damage to the human patient.

- Automotive devices, where early prototype cars cannot provide all the cases needed to fully test software logic.

- Aircraft systems, where the interplay of many other computers and devices cannot fully be tested in the real world or is cost prohibitive.

- Network systems, where the test system cannot go "live" in the real world without high risk.

- Home "smart" power-related systems, where the number of combinations of hardware and providers is too large to test realistically.

- Any system where the tester needs to provide the ability to implement the "hard to reach" test cases such as system failures, error cases, tight timing, crash-and-burn logic, and so forth.

Finally, for all of their advantages in some situations, closed-loop, real-time simulation models for mobile and embedded testing come with issues and limitations. These limitations can include the following:

- Fidelity and realism of the hardware models such as does the model really reflect what will happen in the field.

- Expense and time to develop and certify.

- Complexity and realism such as how do we simulate hundreds of network system inputs (or users) in a realistic manor.

- Simulation of hardware or system, which is a "black box" to the test team such as how do we create a model of a device we do not understand.

- Simulation of real-world systems such as human behavior and conditions, weather, traffic patterns, chaos, and so on.

- The N-version problem (see section N-Version Testing Problems in Labs and Modeling).

- Simulation models and automation will only pay back if you test the system multiple times.

Takeaway note: Modeling and simulation can solve many problems in testing complex mobile and embedded devices.

KEYWORD-DRIVEN TEST MODELS AND ENVIRONMENTS

We have been talking about the use of automation, models, and simulations to drive the test environment, but what we need is a language to tell the models and/or simulations how to behave. In some cases, the model, simulation, or tool input may be simple, just some simple set of inputs to the mobile and embedded device under test. However, what if we want to vary inputs, change them over time, have automation, detect some event, and react to it differently? How do we tell the models, simulations, and/or tools how to be different? The answer is that we need a language much like any programming language to convey such instructions.

In much of the software test automation world today, the concept of keyword-based testing has come into vogue [20]. Keywords are a large topic, which I only partially consider in this book. Here, you will use words that represent the key actions of the system such as "log in," "run report," or "log out." Some keyword systems combine the keyword with data such as "log in, Jon Hagar, password," where "Jon Hagar" and "password" are data we input with the keyword. In simulation and stimulation modeling systems, we also usually create a language to program the simulation and models, including the idea of data elements. For example, imagine a keyword simulation for an antilock braking system where the road is wet and I want to apply the maximum force to the brake. We might have "Pavement value = 5 (where 0 = ice and 10 = dry road); and Brake force = max with a keyword of "Apply Brake." This keyword and the values would be fed into the test and simulation and stimulation so that the test simulates braking on a wet road, which was value 5.

The keyword-based mobile and embedded system and language could be constructed to represent such things as follows:

- Test set-up conditions

- Test execution conditions at the start of test

- Test conditions, which may be time based during the test that we would like to see changed

- Error conditions to introduce in the hardware and/or software
- Actions for the models to take
- Test end conditions
- Test reporting

The variations and nature of the keyword are governed by the simulation and stimulation modeling recreation of the real world. Testers must also account for events, time, and even what you expect the simulation and stimulation to do under failures of the system under test such as the system under test may stop, but you may not want the simulation and stimulation to stop.

The keyword and associated data form a test language. Further, the keywords can be defined to support the test lab simulation and stimulation system. I have seen rockets, airplanes, human patients, communication networks, and many other aspects of the "real world" modeled or simulated. These keyword, model, simulation lab test systems can be large and complex, often in tens or even hundreds of times the amount of code as the device under test. Keyword languages are usually unique to the system under test with tens or even hundreds of constructs with each possibly having unique values.

These keyword model/simulation-driven systems allow the programming of attacks and offer automation of much of the attacks. A tester can define an attack or even a series of attacks into a system usage scenario, program them into the keyword language with data values, and then run many tests and test cases quickly. The level of automation and human interaction with the system under test can vary. I have seen systems that, once keyword programmed, are basically automated in the attacks. I have seen lab systems that mix keywords and human interaction (mostly 50% each). And I have seen systems that have very few keywords except for a few tools or simulations and stimulations, and mostly are human-driven manual testing. There is no "one way."

Now, not every mobile or embedded device can afford (or needs) a large and complex closed-loop simulation and stimulation environment with keyword languages for the models. However, if you are risking things such as life or large dollar amounts, the cost to benefit may be realized. And as mentioned previously, different levels of keyword automation are possible, but basically most of these systems are custom-built keyword mobile and embedded systems. But vendors are now starting to offer keyword tools.

> **Key point:** Modeling, simulation, stimulation, and keywords are large topics, which we have covered only at an introductory level. Teams and testers spend years developing the knowledge, tools, and skills to create and use them.

DATA COLLECTION, ANALYSIS, AND REPORTING

In embedded and for some mobile devices (usually not smart phones), data collection, analysis, and reporting are major and important activities. Some systems will generate huge amounts of data. For example, I know of systems that will generate terabytes of data in a single system test. Here are some questions to keep in mind when dealing with large amounts of data.

1. How do we record the data?

2. What is the data?

3. Who gets the data?

4. How do we analyze this much data?

5. Can we understand and/or visualize the data?

6. Where are the bugs the data indicate?

7. How is the data transferred for network communication?

8. How is the data packaged in a message?

9. Is the data only received and not resent?

10. With what other applications will the data interact?

This part of the chapter will examine and consider some of these questions in the mobile and embedded space.

We need to have the right people look at the data, be able to analyze it, compare it when necessary, and report their findings. I have already mentioned data hiding in some of the attacks. How can you find one wrong trend line of data points in a terabyte of data? Will someone die if you miss that one wrong thing?

I can recommend the following:

- Online data repositories such as databases, web systems, downloads, and so on

- Data analysis tools such as MATLAB, PROBE, and so forth, to "strip," filter, and display interesting data

- Visual data display tools such as Excel

- Tools to pour over the data with trigger values, checks, and tolerances

Many of these tools are commercial and support much customization. Teams that spend the time to do the customization find more bugs. In larger systems, no tester can process all of the information coming from data-intensive mobile and embedded devices. PC and IT tools with a nice, user-friendly interface can or may be manually evaluated by testers. Typically, mega- and even terabytes of data will require automated post processing since humans do not deal well with large amounts of data.

Key point: The data aspect of embedded and some mobile systems introduces problems that many PC or IT systems do not see. Analog, digital, mixed, network, and time-based data may all be unique to a device and will factor ino the test process.

POSTTEST DATA ANALYSIS

Some big data issues are unique in embedded systems, so I will touch on a few of them here.

First, there is an issue on analog and continuous data streams of inputs to and in some cases outputs from the system under test. As mentioned in earlier attacks, the analog inputs must be converted, usually in the hardware, to digital data (see analog-to-digital/digital-to-analog [A2D/D2A], Attack 5). You might want to measure both analog and digital values to make sure that the conversion does not lose information that the software might need. Likewise on outputs, there is the same problem and data to record except in the reverse direction. But besides this, you have the timing issues to record such as when did the data values happen, did data get lost at any time, were there data spikes, and if so where did those happen? So, time must be recorded. Specific posttest data analysis to look for when setting up the test environments and/or tests includes the following:

- Conversion—Digital (D) to Analog (A) and A2D
- Noise
- All data inputs
- All data outputs
- Sampling (are enough bits captured/used) and
- Timing rates and information

Once the data is some place in the system, you still have risks. Systems can have data communication problems, which create or expose bugs, including:

- Instrument dropouts and "hardwired" data stream (gaps)
- Network communication problems (more gaps)
- Internal computer bus communication problems (more gaps)

With all of the sources of data problems and risks associated with data, if your software suffers from any or even some of these issues, your testing and perhaps your environments must account for them. The considerations start in the creation of your test lab.

Key result recording and analysis tools for a real-time software system can include thus:

- Data bus logger terminals
- Logiscopes and oscilloscopes
- Electronic strip charts
- Test benches
- Bus logger terminals

- Timing–performance analyzers

- Voltmeters and electrical analysis devices

- Data recorders for networks, communication lines, signal characteristics, and so forth

- Bus/CPU/computer monitor and interface/control/probe devices

- Capturing test data from the internals of the processor such as memory, CPU usage, registers, timing, and so on

- Facility lab condition monitors for

  Clocks for time

  Temperature

  Voltage

  Other lab "physical" influences

If you need to support data analysis for your local context, you may want to set up the lab considering the following features:

1. Analog data analysis

    - For recording use

      Strip charts, tape recorders, and medical device recorders

      Dedicated computer/recording systems to trap things such as signals, electrical, external world sensors, and so forth

    - Have playback ability

    - Display and analysis features:

      Oscilloscopes

      Bus analyzers

      Noise and data filtering

    - Time issues to consider:

      Sampling of the system such as what is your frequency and are you three to five times faster in your lab.

      Data spikes can get missed such as can you have "trip triggers."

      Data operating different frequencies (time correlation): can you adjust and time relate data?

Old data and lags such as do you have time tags.

Jitter (data timing variability) such as can you get enough data to identify the jitter.

2. Digital data analysis

- Have recording ability:

  Scope

  Bus logger (commercial systems exist for this)

  Computer files

  Data conversion—D2A and A2D

- Have the ability to playback the data

- Have the ability to display and analyze the data, often cross correlated to the analog data

- Time phasing of all data streams

Note: This can be less of an issue if time signals exist (time tags).

3. Mobile data analysis, special considerations:

- Do you have "access" to the inside instruments and data streams (if not do you need it)?

- Can you see and/or control battery voltage (system–software may be impacted by low battery)?

- Can you see and/or control incoming/outgoing signals such as cell, GPS, Wi-Fi, and so forth?

- Can you control "environmental" factors such as sun, temperature, humidity, which may impact the system–software?

Note: Many software testers might say "why bother" because they do not understand the impacts that environmental factors can have on some software. It can be hard to create "environmental" factors, which influence the software–device, but not testing something may let bugs out into the field.

Finally, here are the considerations for analyzing big data (or any data in this space):

- How do you deal with large volumes? Consider the following:

  Plots and over plot tools such as MATLAB, PROBE, and others

Process "arrays" of data

Scaling and views to condense or expand data views

- Can you use visual displays and graphics to understand the data? Consider the following:

Graphics

Visualization

Color coding

- If you must deal with and display mixed Data (A and/or D)? Consider the following:

Tools support

Overplot inputs, outputs, analog, and digital

Tool to refine/time synchronize data

Data analysis will be dependent on your local project and context. Simple systems will have less or none of these factors. A large complex airplane may have these and many more. Dealing with analog, digital, large, and complex data sets will take planning, analysis, and likely iterations in the lab to get things "right." The factors provided earlier give you a starting point.

Once the data is recorded, most test teams will have the problem of how to analyze the digital and/or analog communication data results. For smaller systems, humans may be able to analyze the resulting files both during and after the attack. However, data rates from some mobile–embedded systems are high, resulting in data files no human can analyze in any reasonable amount of time. Test file sizes have been seen in mega-, giga-, and even terabyte ranges. For this situation, teams should rely on tools such as MATLAB, PROBE, or others, which can analyze and/or graphically display data communications and results. Here again, the test support infrastructure must be designed ahead of time, keeping these features in mind. I have seen test teams scrambling at the end of a project to digest huge data files, only to run out of time and miss errors until those errors escaped into the field.

POSTTEST DATA REPORTING

Along with analyzing the data from a test attack to understand it and hopefully find bugs, teams must preserve, be able to display, and report the data, in the near term and long term. Testers and analysts need to be able to look back at the test data. This means that the data must be organized and saved in ways and places so that it can be found. I have seen rooms filled with paper in boxes to the ceiling. It was a little hard to find an individual attack. Better systems exist now including database systems, web data storage, and test management systems, all of which are online with file storage systems based on the attacking. Each such system should be tuned to the local needs and must work for the individual team. Also, these systems should include reporting to development, management, and customers.

The customer and management may want less detail and more summary information. This information can include bug information, what worked or not, visualization of the test data, metrics, and so on. The ability to communicate between the test team and the other stakeholders is important, and each stakeholder has their own perspective. Some just want "it works, we can ship it." Some want more data and visualization of complex test results, but in an easy-to-understand graphic. Tools and systems exist to do this, but a tester needs to know they are needed and how to set them up. Yet others may want all of the test data and details so that an in-depth posttest analysis can be done. Depending on context, you display and storage system may need to support such varied levels of reporting.

The "information" of software test attacks brings up the issue of test documentation, which can be a source of records from the testing. ISO29119, part 3 [21], defines many different types of test documentation. It covers the whole test life cycle with documents being produced for each set of efforts. This can include test planning, strategy, design, procedures, bug reports, final reports, and many others. This book is not about test documentation, but test documents may be one of the "information" sources that stakeholders expect and pay for. The documentation needs are part of the project context and so will be defined for the test team. I have worked for Agile teams where the final report was a "GO" statement. I have worked on other teams where we produced full and long test reports because customers valued them (i.e., paid for them). I recommend that the test team understand which documents and pieces of information their stakeholders need, put great value in, and want. Certainly, bug reports are tops on my list, but I am not a stakeholder for your testing.

Besides the outside interest of test stakeholders, the test team should be interested in having access to the test data and reports. I built this book based on a taxonomy of public, mobile, and embedded bugs. I would have preferred more data points, but it can be hard for companies to release all of their bug data (which I understand). I do recommend to test teams that when the attacks are complete, look back at the test results and data. This can be done in retrospectives, bug fests, and post mortems looking for lessons learned. Efforts can be done to create locally very specialized bug taxonomies so that test teams and individual testers can improve their testing and attacks. Improving testing and attacks can be done by data mining the bug database and test attacks. Finally, summary data and reporting make good materials for publishing papers and presentations to the public. We can all learn from each other as testers. Again, to do any posttest analysis, the data must be preserved, accessible, and able to be understood such as in reports and visualizations.

Finally, in many areas of the mobile and embedded space, maybe even more so than other software industry segments, there is often a need to keep attack/test recorders for legal and regulatory reasons. Mobile and embedded market segments such as medical, financial, aircraft, space, automotive, power, and other groups work under regulation, legal constraint, safety, or risk factors, which mandate that the testing that is done has a high level of record keeping. Some projects must keep records and reports for 5 years, while others keep data until the end of a product life cycle plus some number of years. There have been lawsuits where the test records were required in support of legal actions on both sides. U.S. and international governments have strict rules on testing, reporting, and record keeping. Some of the rules are harder to comply with than others. Some require much more testing than others. Basically, test groups and the projects they work for must understand the legal implications of the testing, as much and as well as the regulatory factors. Forgetting or ignoring these factors has resulted in project termination, lost lawsuits, and heavy financial burdens to companies (tens to hundreds of millions of dollars could be at stake). Attack-based testing in the mobile and embedded space can be an activity to show auditors, regulatory agencies, lawyers, and other interested parties. Attack-based testing provides necessary documentation and test data as proof rather than a company or project relying on the "trust me" approach.

WRAP UP: N-VERSION TESTING PROBLEMS IN LABS AND MODELING

In this chapter, I have covered labs and tooling with such things as test languages, modeling, simulation, and test oracles. The last few sections of this chapter contain some final thoughts and problems areas of which mobile and embedded testers should be aware.

In the lab, when testers use models, simulations, or software-based tools that "replicate" aspects of the software under test, this scenario is similar to a form of multiple version (or N-version) programming, but I call it N-version tooling to support testing [22]. In test labs, there can be multiple versions of code, oracles, models, and simulations and partial implementations of the system. These N-Version tooling results provide data that directly aid testers and are generated by the test team, independent analysts, or others independent of the implementation. However, even though this independence is established, common mistakes (bugs) still happen.

Figure 11.7 illustrates N-version tooling and associated efforts at a variety of levels. In this example, the lowest level is the code/developer level where the N-version tooling judges what is right or wrong at the code implementation level during tests, such as Attacks 2 or 3. Further, these models or oracles can support a comparison of results across level boundaries as in Figure 11.7. Additionally, the results of levels can be compared to each other. These comparisons and different implementations (models, simulations, or oracles) form a conceptual problem similar to an N-version programming system. This problem is because these tools are compared at different points to each other and developed independently of the product under test. In one project area that I studied, we had a minimum $N = 3$, and a maximum $N = 8$ during all life cycle stages. Thus, we compared and checked points for these various "Ns."

> **Key point:** N-version testing and comparisons offer test teams ways to solve the oracle, success criteria, and test input problems, but this takes time and effort to set up.

However, N-version programs have been criticized for not significantly aiding the reliability of a software system, according to Knight and Leveson [23], when using two (or more) versions of the program to support actual operations. Knight and Leveson's data indicate that different teams tend to make the same mistakes in "independent" versions of software, yet the underlying assumption of N-version is that independently coded programs fail independently. Knight and Leveson's information indicates, in the general case, that independent implementations contain the same bugs in high percentages of cases and so will fail in the same ways. This issue has been, and continues to be, a concern to anyone trying to implement systems with N-version programming. However, using N-version tools in testing has successfully detected errors. And I judged the value of tests concepts and tools on their ability to assist in detecting errors.

Testers, modelers, toolmakers, and lab builders must understand the issues of N-version programming. It is quite possible to have an error in the lab, oracles, or simulation N-version test software mask or miss a bug in the product under testing, because of the N-version problems. The N-version risks must be weighed against various factors of cost, schedule, quality of the test environment, and impact of bugs escaping into the field. Here again, there is no one right answer for every context. Some simple mobile apps may not have any test environments, simulation, and stimulation, or other such tools. Their risk in this area is low with an "N" of zero or one, but a mobile medical device used to help make life-and-death decisions may benefit from a few more "N" versions than a billion dollar satellite.

In the development of such labs and tools, teams using N-version tooling to support testing have found that it is worthwhile in the right contexts [22]. The team must use supporting methods and techniques to ensure that the problems of N-version are partially mitigated. Use of good engineering practices and numerous cross compares is advised. Additionally, tools or simulations should be developed and tested in controlled software development.

Test teams must recognize the benefits as well as the limits of N-version-based testing. It is not a complete answer in and of itself. However, it is an additional approach where mobile and embedded testing can become successful in error detection.

FINAL THOUGHTS: INDEPENDENCE, BLIND SPOTS, AND TEST LAB STAFFING

For years, there has been a debate within the test world about the level of independence needed within the test staff. Everyone has blind spots (the bug is right in front of you and you cannot see it). Hence, many have advocated large amounts or even total tester independence. However, in many of the attacks in this book, a team of hardware, software, and systems engineers is identified, including having development staff involved. This may seem in conflict with independence. I have found that such mixed teams are advantageous by having the experience and knowledge of developers, but this approach of intimate teamwork may have a price of lessened independence.

An advantage of mixed teams is that the team engages engineers who are responsible for defining testable requirements and code in the first place and testers continue to gain knowledge of the architecture, which is necessary in embedded and mobile software for system integration testing. A requirement, story, or piece of code that is testable is better than one that is not. This mixed approach is common on Agile teams. Such engineers also understand what the system should be doing and so can define testing and stress testing quicker that an engineer with no history on the system. However, all teams should be aware of prejudice and blind spots, which can be a real issue for all teams of engineers.

To compensate for blind spots, testing staff in the lab can do the following things (tricks, see [24]):

- Establish mixed teams with staff who can provide the same or complementary skills as those of development staff.

- Attack test in multiple passes, switching your attention to different parts of the screen, parts of the system, or different qualities of it. For example, look for performance checks on one pass and correct answers on the next.

- Diversify your testing vision by using paired testing, where one person looks for the answers and the other looks for bugs.

- Be aware of the risk of any testing that relies exclusively on narrow oracles such as N-version tooling, blindly following written test procedures, or total automation, where a human never looks at the results.

- Use posttest analysis to highlight questionable results differently such as with color or sound.

- Convince yourself and the team that blind spots are very real such that you are always looking for them (e.g., avoiding the complacent attitudes of people).

- Have teams with a mix of experience, junior testers and other people from many disciplines (arts, math, user history, science, and others).

- Look at the device or software differently such as from the side, upside down, in black or white, and so on.

- Use multiple tools, techniques, and attacks when their cost can be justified.

- Have independent test groups on critical projects such as man-rated projects, high-dollar or high-visibility projects, and so forth.

Consider these lab staffing requirements when building a test team. This combination of "tricks" and different engineers enables a comprehensive independent testing effort. The effort combines the experience of people, their knowledge of attacks, and test environments to show compliance of code to expectations such as requirements, but more importantly, to identify any anomalies in the software–system. However, highest levels of independence (such as independent verification and validation (IV&V) do not come without great cost and so are uncommon. IV&V is reserved for government contracting or for life-critical systems such as medical or nuclear industries. Most teams will just use the "mixed" team and "tricks" approach.

Further, in the lab, many successful projects attack and test a series of objectives through a series of phases to offset issues of independence, including the following:

- Bug finding (really what this book is about)

- Developer testing (see Chapter 2)

- Early prototype testing and integration

- Total independence from development team "blind spots"

- Requirements testing such as checking and verification (did we build the system right)

- System testing by other teams such as validation (did we build the right system) and attack-based testing (much of this book)

Teams could use a mix-and-match approach from those listed here since many mixes are possible. In bug finding, developer testing, and integration attacks, we are trying to provide information to the development staff. In verification, we test to check compliance of the code to design, design to requirements, and even a binary executable configuration to its source files. Some verification can even be done mechanically.

In attacking, we try to find cases that show bugs where there should be none.

Good test teams (in the lab) use most or all of these approaches to independently provide information about the qualities of the device. Really good test teams figure out how to do this quickly and efficiently (cost and schedule). It can take a lifetime to get great at these things.

> **Key point:** Testing software is a very human-intensive problem. It is a challenging problem that takes heuristics and deep thought.

EXERCISES (ANSWERS ARE ON MY WEBSITE)

1. You are field testing a car. List the environmental factors that you might try to find in the field to "stress" the car and how your mobile lab would account for these factors?

2. You are working for a company testing new smart phones with 4-inch touch display screens. List the support items (hardware and/or software) you might want to have in your test lab environment to conduct your tests.

3. You are testing an embedded pacemaker for humans, which has upload and download communication channels (wireless). Define the models and simulations you might want to have in your lab to test this software device.

4. You are testing an embedded process to control an airplane. List the kinds of tools for test automation that you might want to consider and why.

5. You are being asked to define the test lab life cycle for a new first-of-a-kind "smart" electric car. List the stages your lab might evolve through and why.

REFERENCES

1. Broekman, B. and Notenboom, E. 2003. *Testing Embedded Software*, Addison-Wesley, Boston, MA.
2. Instruction set simulator. http://en.wikipedia.org/wiki/Instruction_Set_Simulator (last accessed April 11, 2013).
3. Emulator. http://en.wikipedia.org/wiki/Emulator (last accessed April 11, 2013).

4. Single board computer. http://en.wikipedia.org/wiki/Single-board_computer (last accessed April 11, 2013).
5. Grenning, J. 2011. *Test Driven Development for Embedded C*, The Pragmatic Bookshelf, Raleigh, NC.
6. Green, G. and Hagar, J. 1997. *Testing Critical Software: Practical Experiences, Critical System Conference*, Denver, CO.
7. Metodi, T. 2010. *Space Vehicle Testbeds and Simulations Taxonomy and Development Guide*, Aerospace Report No. TOR-2010-8591-16, EL Segundo, CA.
8. Graham, D. and Fewster, M. 2012. *Experiences of Test Automation*, Addison-Wesley, Boston, MA.
9. Hagar, J. and Bieman, J. 1995. *Adding Formal Specifications to a Proven V&V Process for System*, IEEE Computer Society, Washington, DC.
10. James Bach's AllPairs (Open Source) tool. http://www.satisfice.com/tools.shtml (last accessed April 11, 2013).
11. Bob Jenkins' jenny (Open Source) tool. http://burtleburtle.net/bob/math/jenny.html (last accessed April 11, 2013).
12. NIST ACTS tool. http://csrc.nist.gov/groups/SNS/acts/documents/comparison-report.html (last accessed April 11, 2013).
13. Hexwise tool. http://hexawise.com/ (last accessed April 11, 2013).
14. rdExpert tool. http://www.phadkeassociates.com/rdexpert.htm (last accessed April 11, 2013).
15. IBM's combinatorial testing service. http://www.alphaworks.ibm.com/tech/cts (last accessed April 11, 2013).
16. Closed platform testing or walled garden testing. http://en.wikipedia.org/wiki/Walled_garden_ (technology) (last accessed April 11, 2013).
17. Harty, J. 2010. *A Practical Guide to Testing Wireless Smartphone Applications*, Morgan & Claypool Publishers, San Rafael, CA.
18. Pries, K. and Quigley, J. 2011. *Testing Complex and Embedded Systems*, CRC Press, Boca Raton, FL.
19. Gardiner, S. (Ed.) 1999. *Testing Safety–Related Software, a Practical Handbook*, Springer-Verlag, London, U.K.
20. Keyword-driven testing. http://en.wikipedia.org/wiki/Keyword-driven_testing (last accessed April 11, 2013).
21. ISO29119 software test standard, to be published in late 2013. http://www.iso.org/iso/catalogue_detail.htm?csnumber=45142 (last accessed April 11, 2013).
22. Hagar, J. 1998. Industrial experiences in establishing laboratories and software models to effectively execute software test, *STAREAST Conference*, Orlando, FL.
23. Knight, J. and Leveson, N. 1986. An experimental evaluation of the assumption of independence in multi–version programming, *IEEE Transactions on Software Engineering*, 12(1), 96–109.
24. James Bach's Web site http://www.satisfice.com/ (last accessed April 11, 2013).

CHAPTER 12

Some Parting Advice

HERE ARE SOME PARTING THOUGHTS on attacks and testing for your consideration. First, keep in mind that testing is intractable. Books on testing are only a starting point. Readers should *beware* of any tool, class, or author who claims to know or do it all. The authors cited in this book have spent their careers learning, as have I. This book represents one set of mobile and embedded software attacks, but there are many more. Start here and then grow your own attacks. Seek other testers, developers, engineers, and thinkers.

ARE WE THERE YET?

As I stated at the beginning, defining mobile and embedded software is not clear. What constitutes time or real time is even less clear. Testing becomes even harder depending on the factors involved. When coming into a new situation, ask, listen, learn, and think. Do not be afraid to discuss factors, topics, and other testing topics with other people. Yes, we have opinions, but wise people learn many attack styles and ask even more questions. The test world is often divided into "schools," certifications or no certifications, camps, tool fanatics, and so forth. Embedded test engineers need a broad knowledge base and in-depth understanding of many different topics (system, hardware, operations, software, test, sciences, math, etc.), as well as good critical thinking skills [1]. All of this takes effort. All of this takes time. All of this knowledge is valuable.

WILL YOU GET STARTED TODAY?

I hope so. When will you reach the end of test learning? Probably never. It is not the destination that matters anyway. The interesting part is the journey and that never ends.

ADVICE FOR THE "NEVER EVER" TESTER

I hope that some of you will try testing for the first time using this book and concepts within—maybe on your mobile smart phone as a beginning. We all live in a time when people will change jobs and careers many times over their lifetime. We also have young students looking to "play" with testing. For these people, whom I call "never-evers," I offer some parting advice on getting started in the world of testing.

Review Wikipedia on software testing [2]. Next, go to this website [3] and take the classes there: http://www.guru99.com/

After these two basic steps, consider your next move with the following steps.

1. Buy more books like this one (I have over 40 books on software testing and hundreds more on high-tech subjects).

2. Read more on testing on the Internet (Google can be your best friend for concise searches).

3. Consider taking classes at conferences, college, or elsewhere on engineering, software, software testing, and other related topics. Here is another great place [4] and great mentor for pertinent classes.

4. Practice your testing. Then, practice some more.

5. Consider certifications and degrees (there are pros and cons to support each side as I know great testers who have neither and others who have both).

6. Learn. Learn. Learn.

7. Have fun!

> **Key point:** I think that many people with some training in software testing and some skills in other knowledge areas (a few are named in this book) could earn a good living in software testing.

BUG DATABASE, TAXONOMIES, AND LEARNING FROM YOUR HISTORY

I have based the attacks in this book on bugs of the past, as shown in Appendix A. Two things to keep in mind are learn from history or repeat it. These histories are generic and, hopefully, common to many embedded domains. It is likely that as you gather data, the bug trends and clusters will change—or at least one can hope they will. Your local project may have more or less bugs in any given area, meaning an attack will be more or less effective or new attacks should be created. It is surprising how few testers have looked into bug histories, but I feel that more testers definitely should.

One approach when arriving on a new project or beginning work with a new company where there are historic bug or data lists is to mine the database. When doing this on a project, learning different things is more likely. If the project has history, one approach is to ask the senior people and management what their expectations are. Then, analyze the bugs or data using a taxonomy to confirm or disprove their opinions using the trends in the bugs—both may happen at the same time. If the project is new, but the organization has history, seek the organization's history and database. Organizations repeat patterns of behavior, which yield the same errors over and over. Programmers have blind spots and make the same mistakes. (They are only human.) Historical bug databases and taxonomies prove this. Look for the trends and patterns. If the organization and project are new (a start-up), the history will unfold in front of you. So, you must pay attention and look at embedded industry history, hopefully, captured in part in this book. Still, I advise you to create your own taxonomy.

LESSONS LEARNED AND RETROSPECTIVES

Closely related to bug histories is the team's history. One of the great concepts in Agile and predating Agile are retrospectives, bug fests, and/or lessons learned meetings. These are meetings or sessions where "tribal knowledge" is collected and shared. Software and software testing are very human-intensive intellectual activities. We do not learn or improve by just going to class or reading books. The attacks in this book and others must be practiced and their lessons learned. The retrospective is a place where we can reflect and learn for the next time. Many cultures have no written history, and in fact, much of written history is new (over the last few thousand years) compared to how long humans have been learning. Often written histories are interpreted in the present tense by the writer or altered by biased individuals. Humans have, for tens of thousands of years, learned by doing and through camp fire stories, as well as through listening to elder's stories passed down to them. Test teams can also learn from storytelling. Good sources for how to do this exist. Review the books and other references by the authors cited in this book. Testers on the team can and should facilitate lessons learned or retrospectives. Try to get everyone on the team involved and to share by seeking the diversity of opinions, viewpoints, and knowledge. Consider a session during lunch—or over a beer, coffee, or tea. As this type of meeting happens, capture the stories and then mine them for new attacks, written histories, and social media.

IMPLEMENTING SOFTWARE ATTACK PLANNING

Planning attacks can follow many patterns. I present a flow and a mind map as example starting points for planning. The keys are to capture this information, repeat it regularly, think about it, and understand it.

1. Select a test planning approach such as exploratory testing, touring, design of experiments, script based, a set of attacks (from this book), and so on.

2. Review implementing project contexts such as software, language, tools, methods, and so on, for any "weaknesses" that should be tested given the nature and functions of the system, as well as "mandatory" activities such as standards, contract statements, regulations, legal items, and so on.

3. Conduct a technical risk exercise with a first pass of interesting attacks to consider.

4. Draft a rough test strategy and plan.

5. Take this plan, in whatever form (document, yellow sticky notes, whiteboard, on line, etc.), review it with owning "experts," systems engineering, or the "customer" to define uses or operational scenarios that are likely to occur.

6. Repeat the test planning for "off-normal" and stress attack cases.

   a. Look for function(s) and select data and/or conditions that result in computation, time, and/or data "overflows" and interactions.

   b. For any projects using modeling and/or simulations or with large data sets, consider running random-based (Monte Carlo) or combinatorial (CT) data selections over the function's input or data space and look specifically for interesting cases, singularities, under- or overflows, illegal outputs, and so on. These can be run in the test lab also.

7. Conduct the checks, testing, and attacks.

8. As attacks are run and errors found, revise your lessons learned and feed those lessons into your test planning, test design cases, and test execution. Constantly evolve the plan as you learn.

9. Learn.

10. Repeat within cost and schedule.

An example and simple mind map of test attacks defined for a test plan on a mobile or embedded device might look like that of Figure 12.1.

The mind map contains developer level attacks, an integration attack cycle, basic features and functional testing, and finally, followed by specialized attack that the team deems "important." Your plan will look a bit different. A basic equation for assigning these might be the following:

```
Total tests/efforts = 20% (developer attacks 1,2,3) + 30%
(integration and exploratory tests before a final product is
reached attacks 6,9,10,12,13) + 30% (basic functional and control
attacks 4,16,17,29,33) + 20% (specialty attacks 28,29,30,32)
```

This equation would be adjusted based on context, risk, and project histories. It is only a starting point. Plus, some attacks might be used in different subareas depending on how they are defined.

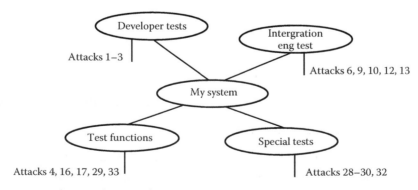

FIGURE 12.1 Example test plan mind map.

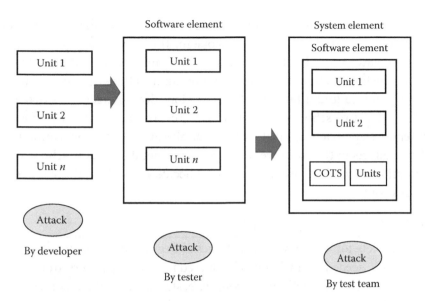

FIGURE 12.2   Example series of stages in a test plan.

Software testers who plan to test applications on mobile devices need to go beyond thinking "if I tap on the icon, does the application start" or "if I press the cancel button, does the operation or function get cancelled." A variety of tests, attacks, and levels is appropriate for most test plans. The application is dependent on the entire system.

Test plans on a project typically address a series of levels or stages of testing conducted by different testers. Figure 12.2 depicts such a set of stages (as shown here). In stage 1, developers are doing the attacks of Chapter 2. This is followed by integration tests for multiple units in the software element, where this is conducted by "independent" testers. Finally, at the system element level where we have both hardware and software, the test plan defines another series of attacks at this stage. This is only an example of test planning and incrementally testing the software and system.

As already mentioned, there is much more to test planning defined in the testing literature, but these ideas will give your test team a good start. It is important to do planning, and the total planning of this section can be done in 10–20 min (see planning game exercise), which is practiced by some Agile teams.

REGRESSION AND RETEST

Plans are put in place, then tests are executed. Bugs are found. Code is fixed. Then people will ask: "Did the fix work and then did the fix break anything that we thought was already working?" This is the regression test problem. There are several schools of thought on attacking regression bugs. One school just ignores regression and runs new attacks or tests. Certainly this approach, if well thought out with sufficient attacks, can find regression bugs or other bugs. Another school of thought wants to repeat all attacks and testing. If a project is automated, this may be possible, but if manual testing has been done, a project will likely not have the time to redo all of the tests. A middle school of thought wants to select regression tests, retests (repeated old attacks), and new tests as a small but doable subset of tests. There are a variety of approaches to select a reasonable subset of tests, but a full discussion of how to do this is out of scope for this book. A real concern of the first school about the latter two schools is that testing time may be wasted on tests that do not provide much new information. If we have already run the attack, and it did not find a bug, it is likely not going to find a bug on later execution. Boris Beizer calls this a "pesticide paradox." The other two schools worry that if regression bugs are not properly targeted by the first school's approaches, a feature that was already found to be working will break, because of side effects in the software. I do not recommend a specific school and I do not define a regression attack, because the *team* must decide what, if anything is needed, based on the project's context.

WHERE DO YOU GO FROM HERE?

Books are incomplete tomes. They provide topics as beginning points for critical thinking. Critical thinking is needed past what can be offered here. Many testers and managers look for "the answer," be it a tool, method, person, class, standard, book, or something else. The motto of many testers is "it depends." There is no "best" or "right" answer—unless it is bound in context and specifics. College students and new testing people find this disconcerting. In grade school and even college, we are almost always given absolute answers. The real world is full of gray areas and answers such as "it all depends." Good testing is very real world.

Happy bug hunting!

EXERCISES (ANSWERS ARE ON MY WEBSITE)

1. Attack planning—define a starting list of potential attacks for

    a. Your project's device

    b. A travel website app for cell phone

2. Test planning game

    a. Define characteristics you would test "Angry Birds" for (after 10 min stop).

    b. Define attack tests for each "Angry Birds" characteristic (and how you evaluate these) in 10 min cycles for each characteristic.

REFERENCES

1. Critical thinking. http://en.wikipedia.org/wiki/Critical_thinking (last accessed April 11, 2013).
2. Software testing. http://en.wikipedia.org/wiki/Software_testing (last accessed April 11, 2013).
3. Guru99 best testing learning resource. http://www.guru99.com/ (last accessed April 11, 2013).
4. Kaner, C. http://www.testingeducation.org/BBST/foundations/ (accessed September, 2012).

Appendix A: Mobile and Embedded Error Taxonomy: A Software Error Taxonomy (for Testers)*

INTRODUCTION

"To defeat an enemy, one must know that enemy" (*The Art of War* Sun Tzu). But how does one *know* a software bug? As a tester, errors or bugs are your enemy. As part of the development team, testers are obliged to bring to the team as much detailed information about those enemies as possible.

Papers published on errors, case studies of errors, error classification systems, and error taxonomies exist, but there are few. Performing error analysis in the mobile and embedded systems domain is still related to debugging software. The public sees Mars rovers stopping, hears of problems at automobile manufacturers, may experience problems with pacemakers, fears airplane failures, and many others that all have a cause or fix in mobile and embedded software. So apparently, in the mobile and embedded software world, many testers do not know their enemy well enough.

Most mobile and embedded software testers gain their ability to know their enemy—bugs—by hands-on experience. Some testers learn that "to test" means to use methods and techniques in error analysis (whenever time is taken to perform this). Unfortunately, some testers just pound out tests blindly simply to demonstrate that a system works, finding only a few of the enemies as they go or until they run out of time or money.

As a mobile and embedded software tester, I would like to see something different. I would like management to value what I identify in the world of bugs and, therefore, to value my talents.

Software patterns have emerged as a major area of interest in software design [1]. Likewise for testers, software errors form patterns of bugs, which can be called a "taxonomy," which is one way to understand the enemy. Taxonomies and patterns of bugs lead us to attacks on

---

* Based on Hagar, J.D. September 2010, Testing embedded software using an error taxonomy, *Conference STARWEST*, http://www.stickyminds.com/s.asp?F=S16886_CP_2.

the enemy [2–4]. This is true of software in general, as well as specialized domain contexts such as mobile and embedded software systems, but in this modern progressive world of graphical user interfaces (GUIs), what are mobile and embedded software systems?

The exact definition of an embedded software system is open and has many views. I give an extensive definition in Chapter 1.

CHARACTERISTICS OF THE ENEMY: A TAXONOMY

From Wikipedia [5]:

> **Taxonomy** is the practice and science of classification. The word finds its roots in the Greek τάξις, taxis (meaning "order," "arrangement") and νόμος, nomos ("law" or "science"). Taxonomy uses taxonomic units, known as **taxa** (singular taxon). In addition, the word is also used as a count noun: **a taxonomy,** or taxonomic scheme, is a particular classification ("the taxonomy of …"), arranged in a hierarchical structure.

Developers have and use patterns in their code and design to help them. There are even books and tracks on patterns at conferences. As testers, a taxonomy of software errors presents bug patterns or schemes of occurrences that happen over and over and over, which testers should be able to recognize. Many of our tools are built on, and techniques are based on our ability to find certain kinds of errors. Table A.1 illustrates a top level taxonomy that was researched over a period of years. How many error categories do you recognize for your mobile and embedded software system? How many do you wish you did *not* recognize?

SUMMARY OF THE TAXONOMY TABLE

In Whittaker's *How to Break Software* book, he lists four fundamental capabilities that all software possesses:

1. Software accepts inputs from its environment.

2. Software produces output and transmits it to its environment.

3. Software stores data internally in one or more data structures.

4. Software performs computations using input or stored data.

I expand and refine this list based on mobile and/or mobile and embedded contexts for

- Time and

- Hardware (a sub of items 1 and 2)

The implications of the table are summarized here, although the full meaning of these is a work in progress.

First, I see patterns of errors related to time: basic time, clock time, time sequences, interrupts, time boundaries, and even dates. Time, the fourth dimension, causes many

TABLE A.1   Super Mobile and Embedded Error Taxonomy (Percentages)

Super Category (Percentage)	Aero and Space	Medical Systems	General Areas	Mobile/ Smart
Time—general (time boundary not met)	3	2	3	—
Interrupt processing—saturation (over time)	5.5	—	—	—
Time boundary—failure resulting from incompatible system time formats or values	0.5	—	2	—
Time—race conditions	3	—	—	—
Time—long-run usage problems	4	—	1	20
Interrupt—timing or priority inversions	0.7	3	—	—
Date(s)	0.5	—	—	—
Clocks	4	—	3	—
Computation—flow	6	23		19
Computation—with data	4	1	3	1
Data	4	5.00	—	—
Data—initialization	6	2.00	3	5
Data—pointers	8	2.00	23	5
Logic and/or control law ordering	8	43	3	5
Loop control—recursion	1	—	—	—
Decision point	0.5	1	—	—
Logically impossible and dead code	0.7	—	—	—
Operating system (OS)—fault tolerance (lack of fault tolerance)	1.5	2	3	—
Software—hardware interfaces	16		13	—
Software—software interface	5	2.00	—	—
Software—bad command	3	—	—	—
UI—user/operator interface	4	5.00	13	10
UI—bad Alarm	0.5	—	3	—
UI—gaming	0	0	10	10
UI—training—system fault resulting from improper training	—	—	3	—
Security	2	4	3	24
Others	8.6	5.00	8	1
Percentage	100	100	100	100

problems, so the taxonomy has several category parts for it. Time complexity, interdependencies, and vagueness become difficult for the hardware and software people to fully understand, so errors in these situations slip in and by them as test cases are missed. Differences in the category appearing in the taxonomy come from differences in the domains and contexts. Different attacks and testing are implied. Can we find these with testing and attacks? Yes, but not with single attacks or approaches. Can we find some of these with analysis or reviews or other kinds of attacks? Yes. Some of them can even be found using nontest methods. Why should we as testers limit ourselves to just running tests? Years ago, I learned about analysis, simulation, scheduling, modeling, rate monatomic analysis, as well as critical thinking while looking at the code, testing, and/or dealing with time. I use many tools and continue to learn more to expose my enemy.

Computation bugs include ordering of the computation, the wrong equation or computation, and data such as using the wrong data in computations. For example, how to compute the speed of a vehicle or object is in most first year physics books. But get something wrong in the logic of the software, and you have a bug in that software. Bugs include logic ordering, missing equations, bad guards, or incorrect operands (+, −, /) and numeric representations. In some contexts, from 1/4 to 1/5 of bugs fit into these categories. Computation here was considered inside of the module or object boundary. Given these high percentages, testers should target these enemies using numerous test techniques and attacks.

Related to computation is pure data: data values, pointers, and initialization or lack thereof. These are internal data values such as the values of Pi have been found to be wrong, a number stored in an integer when a floating point number was needed, storing to a wrong variable, or just simply using a data element that had not been initialized. Computers are quite dumb and they operate on data. If the data is wrong or a wrong data element is used, bad things happen, which equals a bug. A special mention here goes to pointers. Many mobile and embedded projects have sought to ban pointers, yet pointers live on in code. Some mobile and embedded groups use them as commonly as information technology (IT) projects do. One report indicated as many as 70% of errors were associated with C pointers. Hmmm. Attack!

Now it is your turn; how would you go after these bugs, and will you keep a taxonomy of your own?

The biggest bug category that many products have is "logic and/or control law ordering." Here, the laws or algorithms controlling the hardware or system across several modules, objects, boundaries, and even hardware contain error(s), which resulted in unexpected or unwanted outputs. Like computation errors, the bug is math or logic based but spread here across a larger part of the system or involving interactions and complexity, making the error harder for the coder/programmer to find. So, testers really need to focus on these high-percentage areas.

Testers have known that developers make mistakes with "loops" and "decision points," resulting in mistakes or logically impossible "dead code." These are classic mistakes where testers can assist developers (coders or programmers) to use test techniques such as boundary value analysis, decision tables, output forcing, and equivalence classes. When the code takes a true path, when it should not, loops extra times or not enough, or does some other "illogical" path, a bug lives to be hunted.

An interesting area in mobile and embedded software systems is the logic built to handle the "bad day" or abnormal situation. The so-called 80/20 rule estimates that fault tolerance and/or recovery code is as high as 80% in some mobile and embedded software! Fault tolerance is the portion of the code that deals with hardware issues, noise, unusual use cases or scenarios, software bugs, and other situations. The problem is that since mobile and embedded software systems function in the "real" world, the number of cases is practically infinite, meaning there is greater likelihood that a case will be missed. Fault tolerance or lack of it is a good area to attack. It is surprising that more errors do not center here, but the classification process may have put a bug into one of the other categories when it was also a problem with faulting. More study is needed here.

Many testers feel that interface and integration activities are often overlooked by developers and other testers in the mobile and embedded world. Generic IT personal computers (PCs) have this to some degree but often have nice standards (simple object access protocol [SOAP], service oriented architecture [SOA], Internet [NET], web or W3C [World Wide Web Consortium], etc.) that help. Some mobile and embedded systems have standards, but often the interfaces are unique. The taxonomy identifies "software interfaces," meaning software to software, and hardware interfaces, meaning software-to-hardware or hardware-to-software interfaces, which when they do not function as expected have a bug. For example, there have been signals that were not converted with enough resolution for the software to act on properly. These interface and integration issues become major system level targets that a tester should verify.

Next, on the output side of the bug world, software has been known to send "bad commands" to hardware, other software, or the user. These may be a problem with the command directly or come from secondary bugs making their way to the command. Testers should look to exercise all commands and even combinations of commands in sequence and time.

And almost last is the human part of the system. The taxonomy trends identified bugs in the often limited user or operator interface, including bad alarms, poor or wrong user interfaces (UIs), and even training and reference materials that mislead the user. These bugs cause improper system behavior because the user, operator, or human makes a wrong decision based on the information the software or documentation provides to them, often making a bad situation even worse.

Toward the bottom, there is the fastest rising category, security. Several decades back, no one really worried much about security. In the last few years, security has become the focus of conferences, books, and much effort. The centering of data on security issues can be seen, but one wonders, *is the focus on security driving the data*? In some ways, it does not matter, since security will remain a top attack area for the future, particularly in the medical and mobile world.

And finally there is the infamous "Others" category, which has categories with percentages that just did not make the "bug hit list."

SOME DETAILS AND IMPLICATIONS

How many detailed categories were there in the raw taxonomy? There were over 80.

These classified the enemy from industry data on the web, news sites, books, and others that I list in the reference section of this book. The data points, while interesting, represent limited sampling. Data represents x studies, reports, and articles. As more data becomes available, modification, expansion, and evolution of the table is likely. Limited error reports and failures remain a problem in the mobile and embedded industry. As Mark Twain [6] observed, "there are lies, damn lies and statistics."

The taxonomy is based on what data could be found publicly over many months of searching. Further, many current errors cluster in areas where fairly simple techniques or attacks would have an effect. This observation concerns me because it implies that testers and developers are not doing a very good job and that as better testing is applied, the defect patterns will change.

Could you have different classifications and percentages locally? Yes. I recommend that mobile and embedded testers take their local bug or error databases and rip into them using this taxonomy and any others you find as a starting point. Testers should create their own versions of this taxonomy. I fully expect different classifications, percentages, and results based on local context and evolving data. This could drive different tests, techniques, tools, and plans. Context matters and "it depends" should be familiar concepts to testers.

What should you do locally with this taxonomy data? Many conference attendees and students look for magic bullets or tools that solve all their problems with fast fixes. We know that engineering, software, and mobile and embedded test techniques are all based on heuristics. There are no simple fixes or magic potions. Thirty years of learning about testing and I still see that most test teams apply only limited approaches to mobile and embedded testing.

Can you change that where you work? Perhaps. However, I see many projects and testers living within their comfort zones and staying there. You should try your own taxonomy and new tools and techniques. Testers need creative attacks.

WHERE DID THE DATA COME FROM?

The web constitutes a great tool, but we should remain suspicious of using web data as well as published articles verbatim. The taxonomy of Table A.1 was built from public data found via search engines and literature searches:

- News reports
- Articles
- Websites
- Conference proceedings

The major references and sites are presented in the reference section at the end of this appendix. Mobile and embedded software systems testing remains my passion. And I am a mobile and embedded software Jedi tester in practice. There are many enemies on board (the central processing unit [CPU]).

WHAT DOES THE TAXONOMY INDICATE?

The data seem to indicate that many errors can be found by developer activities and/or static code analysis (SCA). The taxonomy indicates that we have not done a good job in the mobile and embedded world with these. Testers should look at what developers are doing. If these areas are lacking, figure out how to improve them. This can mean having a separate individual or group doing unit or developer testing. This has worked in organizations, although it is not as efficient as having the developer run unit testing. SCA tool vendors offer services that will do the first pass analysis for projects, and many companies have "test groups" doing the static analysis and feeding the information back to the developers. There are many ways to improve, so a tester should advocate SCA as one of those ways.

The taxonomy indicates areas in which testers should know the enemy better and devise "new" attacks. The specific attacks where testers need to focus are in the general areas of timing, computation, logic, input–output interfaces, and hardware.

We exist in a field and an industry where few have published data on the enemy but many talk about how to do test planning, which techniques work, which tools to use, and so forth. I think we should be talking about the puzzle of the bug, error trends, as well as taxonomy maps that exist in our heads.

Do you test based on risk and error taxonomies? If you do, that's good, but where do your risks come from? In part, they should come from a taxonomy. If not mine, then build your own.

CHANGING THE FACE OF MOBILE AND EMBEDDED SOFTWARE

Once upon a time, mobile and embedded software was small, contained within some unique hardware, had no UI, and had simple functions such as "turn the lights on or off." That was a long time ago. Now, mobile and embedded systems include millions of lines of code, UIs, and GUIs, which may have network connections and offer all of the problems of older more generic (IT) computer systems, plus many new ones, which could kill your family, cost big money to fix, and so on. Robots, cell phones, and devices with mobile and embedded software in them are everywhere. Technologists seem to enjoy putting a processor into everything. Automobiles now have tens to hundreds of processors—even my washer and dryer have software! All of this presents more opportunities for testers to increase their taxonomies as well as their testing skills. Mobile and embedded is really a continuum of products, development, and bugs, sharing things with the IT world and having uniqueness that mobile and mobile and embedded testers must keep in mind.

SUMMARY

A mobile and embedded error taxonomy was created from a compendium of industry sources to capture "how mobile and embedded software systems fail" and give a view of the tester's enemy (i.e., bugs). Mobile and embedded testers, as with all testers, need to know their enemy. The taxonomy presented is not an end but a crude beginning, which testers can use to know their enemy. The mobile and embedded bug or enemy has many faces, which change over time and projects, so testers should consider evolving this taxonomy into something specific to their needs, since not every mobile and embedded context pattern will be reflected in this taxonomy. Feel free to use this taxonomy (as a beginning) to create attacks, tests, and tools to find an enemy. That is future work. Testers should learn from this taxonomy and go "kill some mobile and embedded bugs."

REFERENCES

1. Beck, K. 2003. *Test-Driven Development*, Addison-Wesley, Boston, MA.
2. Whittaker, J. 2003. *How to Break Software*, Pearson Addison-Wesley, Boston, MA.
3. Whittaker, J. and Thompson, H. 2004. *How to Break Software Security*, Addison-Wesley, Boston, MA.
4. Andrews, M. and Whittaker, J. 2006. *How to Break Web Software*, Addison-Wesley, Boston, MA.
5. Taxonomy. http://en.wikipedia.org/wiki/Taxonomy (last accessed April 9, 2013).

6. Twain, M. (1906-09-07). Chapters from my autobiography, North American Review Project Gutenberg.

Notes: The following references were used to create the taxonomy:

- 2010 *CWE/SANS Top 25 Most Dangerous Programming Errors*, http://cwe.mitre.org/top25/, 2010 (updated annually).

- ACM SIGSOFT Software Engineering Notes, Vol. 6, no. 2.

- ACM SIGSOFT Software Engineering Notes, Vol. 9, no. 5.

- ACM SIGSOFT Software Engineering Notes, Vol. 23, no. 1.

- "Airbus," *Aviation Week*, February 13, 1989.

- Avižienis, A., Laprie, J.-C., and Randell, B., Dependability and its threats, a taxonomy, Web Article.

- Bailey, G.M. and Piziali, A. 2007. *ESL Design and Verification: A Prescription for Electronic System-Level Methodology*, Morgan Kaufmann Publishers, Burlington, MA.

- Beatty, W., March 2002, *Finding Firmware Defects*, Class T-18, High Impact Services, Inc., Noblesville, IN. http://test.techwell.com/sites/default/files/articles/XDD3619filelistfilename 1_0.pdf

- Encontre, V. November 24, 2003. Testing: Do you have the guts for it, *IBM Developer Works*. http://www.ibm.com/developerworks/rational/library/459.html (last accessed April 11, 2103).

- Freimut, B. 2011. Developing and using defect classification schemes. Fraunhofer tech report. Kaiserslautern, 2001, VII, 44 pp. : Ill., Lit. IESE-Report, 072.01/E Report no.: 072.01/E, English language version.

- Halfhill, T.R. March 1995. The truth behind the pentium bug, *Byte*.

- Jetley R.P. and Anderson, P. April 2008. Using static analysis to evaluate in medical devices software, *Embedded Systems Design*.

- Kandler, J. Automated testing of embedded software, lessons learned from a successful implementation, Web Article.

- Kuhn, D.R., Wallace, R., and Gallo, A.M.Jr. June 2004. Software fault interactions and implications for software testing, *IEEE Transactions on Software Engineering* 30(6).

- Kuhn, R. April 1997. Sources of failure in the public switched telephone network, *IEEE Software*, 31–36.

- Leveson, N.G. July–August 2004. Role of software in spacecraft accidents, *Journal of Spacecraft and Rockets*, 41(4).

- Maji, A.K., Hao, K., Sultana, S., and Bagchi, S. 2010. Characterizing failures in mobile OSes: A case study with android and symbian, *Conference ISSRE10*, San Jose, CA.

- McCracken, H. Bizarre bugs: 9 of the strangest software glitches ever, *PC World*, June 17, 2009.

- Neumann, P. 1995. *Computer Related Risks*, ACM Press, a Division of Addison-Wesley, Boston, MA.

- Noergaard, T. 2005. *Embedded Systems Architecture: A Comprehensive Guide for Engineers and Programmers*, Newnes, Burlington, MA.

- Peng, W. and Dolores, W. 1993. *Software Error Analysis*, NIST Special Publication 500-209. U.S. Department of Commerce Technology Administration, National Institute of Standards and Technology, Computer Systems Laboratory, Gaithersburg, MD.

- Sridhar, T. 2003. *Designing Embedded Communications Software*, CMP Books, San Francisco, CA.

- Van der Spuy, R. 2009. *Foundation Game Design with Flash*, Friends of Ed, New York.

- Vijayaraghavan, G. and Kaner, C. May 2003. Bug taxonomies: Use them to generate better tests, Presented at the *STAR EAST 2003*, Orlando, FL.

- Wallace, D.R. and Kuhn, D.R. 2001. Failure modes in medical device software: An analysis of 15 years of recall data, *International Journal of Reliability, Quality and Safety Engineering* 8(4):351–371.

- Wescott, T. 2006. *Applied Control Theory for Embedded Systems*, Newnes, Burlington, MA.

Industry Reports

- ARIANE 5 Flight 501 Failure Report by the Inquiry Board, Paris, France, 1996.

- Report on Project Management in NASA by the Mars Climate Orbiter Mishap Investigation Board, NASA, March 13, 2000.

- Report on the Loss of the Mars Polar Lander and Deep Space 2 Missions JPL Special Review Board, NASA, 22 March 2000.

- Titan IV B-32 mission Failure Review Board Report, U.S. Air Force, 2000.

Appendix B: Mobile and Embedded Coding Rules

THIS APPENDIX, LIKE CHAPTER 2, MAY SEEM OUT OF PLACE TO SOME TRADITIONAL TESTERS, but the testing world is evolving. Testers need to support the team and all stakeholders—including developers. This is true even if you are not on an agile team. But, for those working on agile teams, knowledge of code, coding rules, and attacks, such as those found in Chapter 2, makes you a much more valued tester. Testing may not be dead, but it is evolving. This appendix presents coding rules and includes pointers to other resources.

Yes, Attack 1 can find "code rule problems," since many of the tools covered are really designed to check for "rule violation." But why wait for a rule to be violated? It can be better to have developers write clean code to a set of coding standards. Team members can save projects time and money by preventing bugs in the first place. This appendix can get you started down a path of prevention.

> **Key Concept:** It can be easier to prevent a bug than to find a bug. Coding rules can help with prevention and testers can help define the "rules."

Coding rules will need modification, additions from the other references, and some strong critical thinking to be made applicable to your project's software and context. You should *not* treat these rules as absolutes either in implementing them or when the team is writing code. A better way to think of coding rules is to call them guidelines. For example, I have data that suggest the use of pointers causes 70% of the bugs in C/C++ programs. That data point might tell us to just ban pointers, right? Well, some rocket programs have done just that. But pointers are a necessity for most applications. With some care, checking, and testing, the use of pointers may be acceptable, and their use makes many data structures much easier to implement. However, use caution with absolutes.

CODING GUIDANCE (TO BE FOLLOWED) AND THE EVIL OF CODE (TO BE AVOIDED)

Standards or guidelines can reduce the errors in the software by providing "rules" that the project must follow. Standard rules can be applied to the code, development processes, models, requirements, design, documents, and even the testing. Prevention is really more

TABLE B.1 Examples of Code Structures to Avoid in Many Cases

Embedded Code	Rational	Activities to "Help"
Complex code	Bugs hide in complex code. Keep code as simple as possible to accomplish a task or function. Complexity can be in size or numbers of paths.	Do peer checks, test-driven development (TDD), and user tests (UTs). Code restrictions to consider: *goto*, jumps, large amounts of code, etc.
Minimize dynamic memory usage	"Dynamic" means it may be hard to predict how much memory gets used, and you may exceed memory size (which is bad). These constructs have great power, but with great power comes equally greater risk and responsibility.	Restrict compilation and/or run static analysis checks. For example, on one system, we had a large stack space, initialized it to a hex value of "DEAD," and then, at the end of the test, checked how far stack penetration happened by checking where the value "DEAD" started or stopped.
Restrict or forbid pointers	Pointers, common in most languages, are a special "construct" of dynamic memory, which can lead to memory leaks and time issues.	Peer reviews and static code analysis (SCA) or lock out their use in the compilation.
No recursion	Engineers learn to use recursion instead of other iteration structures, but recursion can "use" memory unpredictability.	Peer reviews and SCA or lock out their use in the compilation
Limit memory indexes and stacks	Check indexing and stack use as it can be unpredictable.	Consider using a fixed value for indexing and stack usage checks. Testers can check stack usage by having a bit pattern such as the Hex pattern of "DEAD" in a data stack. Then, at the end of the test, check how far stack penetration happens by checking where the value "DEAD" starts or stops.
Global data memory	Global data memory is fast (an advantage if the software is time constrained) but increases coupling (equaling more side effects), which is bad.	Check using peer reviews, memory access analysis, or set/use frequency checks. Also, consider limiting global data memory. Attack 1 may be useful here too.
Function/object calls with large amounts of passed data	Passing data takes time (see global data memory), which can create real-time performance issues.	Check in and out lists during reviews and maybe in attacks of Chapter 2.
Loop termination and/or limit size	Loops add time if they do not terminate or when they run too long. Predictability is hard and "intermittent" problems can be created.	Static analysis, Attacks 2 and 3, peer checks, and compiler optimization
Exception processing or assertion checks	Results can be unpredictable, and it is hard to test or verify. It is kind of like an unexpected interrupt and then a jump.	Turn off exception processing or do conditional compiles and then turn off exception processing. If exceptions are kept, test and peer review them carefully.
Simple single entry points and exits in modules or functions	This is a violation of the "keep it simple" rule.	Do peer checks and run Attack 1.

TABLE B.1 (continued)   Examples of Code Structures to Avoid in Many Cases

Embedded Code	Rational	Activities to "Help"
Register usage (internal memory)	Registers all "look" alike and can be confused easily, resulting in a bug.	Run attacks of Chapter 2 and do peer reviews.
Timer usage (time out, watch dog, clocks)	Embedded systems are often time critical, so timers can be wrong.	Run attacks of Chapters 2 and 6.
Data input storage and conversions (A2D)	Data coming in from the "real world" can be converted or stored from the analog world incorrectly.	Check data and run attacks from Chapter 4.
Data output storage and conversions (D2A)	Data going out to the "real world" can be converted or stored incorrectly.	Check data and run attacks from Chapter 4.
Interrupt processing	Many embedded systems are "driven" by interrupts, and a bug in the process will likely be critical.	Run Attack 18.
Uninitialized variables	All variables should be explicitly set before use as undefined variables may have unpredictable results between systems or when dependent paths are taken.	Run Attacks 1, 2, or 3.
Meaningless variable names	All variables in the code should have names that aid in reviews, testing, and maintenance and reflect their usage. For example, bad names are X, Y, Z, variable 1 or names that do not convey information.	Run Attack 1 and do peer reviews.
Short, out of date, or unclear comments including headers	In the review of bugs for the error taxonomy, it was observed that code developed by programmers who wrote good in-line comments and maintained them tended to have fewer errors in the code. A tester can use this information in attacks when they have access to the code to determine a module or areas of code that might have more or less bugs. This applies to segments of code to focus more attacks on.	Run Attack 1 and do peer reviews.

Source: Anderson, P. 2008. The use and implementation of coding standards for high-confidence embedded systems, *Aerospace Conference*, 2008 IEEE, Big Sky, MT; DO-178B or C (industry standard) (use an Internet search for this one); Editorial, Guidelines for mobile web development, *Smashing Magazine*, July 2, 2011; JPL Institutional Coding Standard for the C Programming Language, NASA JPL standard, version 1, 2009; MISRA C coding standards (industry standard); NCST-D-MI051, *Microwave Imager/ Sounder (MIS) Program Field-Programmable Gate Array (FPGA) Development Process Guidelines*, 2011 (use an Internet search for this one).

efficient and desirable than detection and removal. Standards and guidelines can avoid many common issues, which mobile and embedded software have including memory limitation, timing performance, resource usage (e.g., batteries), and other limitations. Table B.1 shows examples of code issues that might be encoded in standards.

These are simple examples for embedded systems guidelines. Projects can tailor this set to their local context by defining their own project coding standards and then adopting them project wide. However, no one should follow concepts or standards blindly. By encouraging developers/coders/programmers to follow standards, your project and product can avoid many bugs. Keep in mind that when rules are violated, the violation needs to be analyzed and fixed, if needed.

WEBSITE

http://www.smashingmagazine.com/guidelines-for-mobile-web-development/ (last accessed April 11, 2013)

Appendix C: Quality First: "Defending the Source Code So That Attacks Are Not So Easy"

INTRODUCTION

Why Would Anyone Want to Make Testing Harder?

Most of this book has been about "attacks" made by testers, be they a "development tester" or a "tester tester" against the software, once it is coded and ready to be executed. However, there are many ways to find bugs and, in some cases, prevent the bugs from being coded and going into test. This appendix details some of these. The reason a tester might want to make their job more challenging is that most, if not all, test teams have experienced the "crunch" before a product's release. Programs will always have bugs, so projects should seek to remove them early on, reducing the last minute rush and leaving the hard bugs for thoughtful attacks.

As a tester, push the project from the day you come on board to create better software. Many testers view their job as only developing and running tests to find bugs or demonstrate that things work. An often overused phrase is "quality is everyone's job." Testers need to advocate quality in the development. This may mean ensuring that developer tests happen by helping to set up tools for developers to use, by supporting peer reviews, by helping to create coding standards, and by supporting other engineering analysis of the embedded system. This helps to make better software and can aide in understanding the software so that better testing happens. We want our embedded testing jobs to be a challenge as well as to bring value to any project—not to just run any attack or test for no reason.

We now depart slightly from the earlier format of this book with this appendix.

Testers should support peer reviews and have input into coding standards on any project. By the way, do not worry; this will not do away with your job. Instead, it should bring out the value of your contribution to the project.

Who Should Do This?
This is a team effort. Management, developers, analysts, hardware, support staff, and testers all participate. The tester's role is to be an advocate for the ideals of this section and "testing." If a tester knows about tools, standards, code inspections, and analysis techniques, then they can assist the team in considering what fits the local context. Often team members are isolated in their engineering world, thinking "it is the other guy's job." A highly competent tester has skills that span knowledge areas as well as the mobile and embedded domain, seeking every concept to make the product better. For testers, it is advisable to help set up tools, establish, and help to write documentation and practices in this area since, often during the engineering efforts, the developers are swamped.

Pull Quote: *A good team leverages everyone's knowledge to be successful.*

When to Apply This Attack
This effort should start during the initial analysis and/or proposal processes, continuing throughout the life cycle of the project. However, in many cases, testers are brought in later (during production) only to find these activities have not been done. Here, the proactive experienced tester will become an advocate, targeting one or two high payoff items such as developer testing attacks or static analysis tools. With the success and improvement of product quality, other doors will begin opening, but everything takes time.

How to Conduct This Attack
This effort can be set up using one or more of the following concepts.

Policy and governance—these are the written "rules" the project agrees to follow covering everything from operational concepts and requirements to coding and test, including useful documentation. These rules must be supported by management, since programmers are notoriously "free spirits" not liking rules. But here is the rub; *no rule is absolute*, and sometimes rules can be bent or broken. Having someone on project who knows the rules, knows why the rules exist, understands how and where to apply them, but also knows when and how to break the rules can be important

Coding standards (see Appendix B)—these are detailed rules about everything on how the code is created. Rules can address allowed constructs, logic, naming, comments, size, complexity, and other items. Many standards and rules have been proposed over the years for embedded software. There is no universal "right" standard or set of rules, but having them first, tailored for your project and enforcing them second, can avoid a lot of initial bugs, which can save money in the long run for every project.

Static code analysis (SCA) (tools)—there are classes of static analysis tools, like compilers on steroids. They can check rules and/or policy, look for classes of errors (pointer problems), and be run by developers (see Attack 1). After code peer reviews and developer testing, maybe a next logical step is SCA tools.

Peer reviews—the literature is full of why and how to perform and use peer reviews. They come in many forms, and their use is hard to doubt. The scary fact is that many embedded projects choose to pay lip service to reviews (do them, but in name only and not in real detail). I advise you to be familiar with the literature, and perform the peer reviews. Enough said.

Developer testing—pointers to developer attacks (Attacks 2 and 3).

STANDARDS AND COMMONALITY: THE DARK SIDE OF COMMERCIAL OR THIRD-PARTY SOFTWARE, CODE REUSE, AND OPERATING SYSTEMS

The rest of the software world, personal computers (PCs), workstations, mainframes, and so on use software that is not developed as part of their applications. When was the last time in the PC or information technology (IT) world that someone created their own version of an operating system (OS)?

Increasingly, the embedded world uses software not developed locally. This is certainly true in the smart- or cell phone world. However, surprisingly, many embedded systems still have their own OS and limits in trusting other software. Here are some of their reasons:

- Memory and time leaks

- Conflicts, which are hard to control

- Unused functions

- Memory space allocation issues

- Size

- Errors in the commercial software products, which are hard or too costly to fix

More and more, I see embedded systems using third-party software based on standards, using off-the-shelf (used as is), subcontracted (semi- or fully customized), open source (watch out for legal implications), or reused components. Reused and third-party software saves time and money and can even improve quality, but embedded testers need to know the dark side. There is no free lunch here. This moves embedded testing to the problems of the generic computer world and standardization.

OFF-THE-SHELF AND SOFTWARE REUSE

Reuse has made the news and has user attention everywhere. Some program managers believe that reuse is a panacea for software development. For example, Arianne, Airbus 318 (per pilot interviews), and Mars Rover *Spirit* all had bugs associated with off-the-shelf or reused software. These negatives mean that many embedded projects react with "Do not use…" statements. Expect that reuse of nonproject developed software will only increase. The following questions should be asked of any reused or off-the-shelf software:

- How much use has this software seen?

- In what environments has this software been used?

- What is the trustworthiness or measured reliability of this software?

- What are the risks of this software?

- How much testing has this software seen (and how "good" are the tests)?

- What are the required functions of this software (critical, medium, low)?

- What information do we have on this software (bugs, release notes, documents, code, tests, etc.)?

- How easy is this software to change?

- Can we protect our software from this software?

The answer to these questions should influence the attacks to be run. I recommend that the project conduct acceptance testing of the reused or off-the-shelf software functions as they come in the door, particularly testing functions, which the software under development will be dependent upon. This is to ensure that the reused or off-the-shelf software elements meet the current project's needs. Off-the-shelf software or reuse attacks can be done very early before the "traditional" test attack cycles. This is, in fact, an advantage to using this type of software construction.

EVERYONE HATES DOING DOCUMENTATION BUT SURELY MISSES IT WHEN THEY NEED IT

Standards should include documentation: comments, design documents, requirements, and so on. Many programmers have written code one day, only to look at it the next day and not remember what logic they were thinking about when they wrote it! Studies have shown that good coders/programmers tend to write better comments. However, this is not a "more is better" situation. Successful projects have "just right" levels of documentation such as on agile projects. The trick here is understanding what "just right" and "meaningful" documentation actually is. This often means peer reviews and testers as second eyes to determine if the documentation adds value. This may be nontraditional for many teams. Meaningful documentation can be found in the following:

- Requirements (with traceability and data dictionaries)

- Architecture

- Design and model information

- Pictures

- Comments

- Operation concepts

- Help files

- Training material

- Test cases and

- Other project files available to testers

Some have advocated that all a project really needs is code with comments and test cases. This is probably the smallest set of "documentation." Most embedded projects will have more documentation than this, but what is "right" is very context and project unique. Think about documentation carefully.

Testers should find, use, mine, and evaluate any documentation they can get their hands on. However, care must be exercised, since documentation often does not match the product and code. Other attacks may be needed. The code is the software. The documentation is added information. When the documentation and code do not match, a bug may be in the making.

IS TESTING DEAD?

Some testing approaches, such as Agile in the early days and cloud systems, have advocated that traditional testing is dead. Well, it is not dead, but traditional end of life cycle verification and validation testing may be dying, and good testing may look really different.

Traditional testers started efforts at the end of the product's life cycle when all products were available, used requirements, created big long scripts (maybe with attacks), ran tests for hours (days or weeks), beat on the software until there was no time or money left (and management was upset), and then declared that the software was "good." This approach still exists, but it may be on life support in many domains, including on some embedded projects.

What we (testers) are seeing is the use of parallel test efforts, integrated and incremental tests as the software evolves, more automation, very short "final validation" test cycles (a few hours or a day), and daily product releases to the cloud. This means that more errors escape into the field. But if a project can hear a report of a bug, find it, fix it, and test it all in a day, will their user or customer be that unhappy?

Now some embedded projects will not be able to follow such a life cycle model due to the long process of hardware development, for example. But in some domains, particularly in mobile apps, such short and fast test cycles might be of great benefit. Consider this story. I get my new app published to the cloud fast after only a little attack testing. The app needs to be just "good enough," so I have hit it hard—in parallel with a mix of automated attacks and exploration. The app has integrated reporting features, which allow the project to understand the bugs as they are reported. My testing is set up to attack the "new bug" and to replicate it based on the user report. When I get the fix, I attack it. Then, for good measure, I attack around the bug and the fix (in the case of regression). I do this in just a few minutes, and then we can push the code to the cloud for users to access. This all takes place in a day or two—or in an hour or two. Do the rewards and risks of my project justify this? Well, that is a local call, but testers need to keep in mind that "testing is not dead; it is just different." *When in Rome...you know?*

SUMMARY: THE P WORD: PROCESSES AND STANDARDS

This section is in part about process, since standards apply and may be trying to enforce a common way of doing engineering. The software and test industry has preached process for years. Some managers want to believe, "follow the process and all will be well no matter the staff." Many others such as Agile followers abhor process and think that "people" are the key to software. But processes and standards are not bad words. Processes, tools, measurements/metrics, and standards have their place, particularly in the embedded world. However, processes are not a substitute for critical thinking as well as having an experienced staff.

In summary, many errors can be avoided when teams follow or at least pay attention to standards and common processes (call them *patterns of engineering*, if you wish), standards, and the concepts found in this appendix. These represent knowledge and lessons learned from past projects, which when applied to new efforts can avoid many errors—but not all errors. Testing and attacks will still be needed. If the "easy" bugs can be avoided, the interesting or hard testing will remain.

BIBLIOGRAPHY

Gilb, T. and Graham, D. 2002. *Software Inspection*, Addison-Wesley, Wokingham, England, U.K.
Jones, C. 2008. *Applied Software Measurement: Global Analysis of Productivity and Quality*, 3rd edn., McGraw-Hill, New York.
Wiegers, K. 2001. *Peer Reviews in Software: A Practical Guide*, Addison-Wesley, Boston, MA.

Appendix D: Basic Timing Concepts

THIS APPENDIX PROVIDES ADDED BACKGROUND MATERIAL ON TIMING CONCEPTS. Time and performance are complex subjects in embedded and mobile systems. This appendix is not complete so for those wishing to further expand their understanding of time, you must use the reference material cited in this book.

Mobile and embedded real-time performance computing is about satisfying time constraints acceptably well. Here, acceptable includes all of the following timing factors:

- Predictable

- Meets application and situation-specific data criteria

- Provides functionality given the system–software context circumstances

BASIC TIMING CONCEPTS AND TERMS

The following outlines a few basic timing terms, which a tester may encounter when dealing with performance concepts and testing.

Time performance constraints are typically considered in two name forms: hard real time and software real time (see the Glossary and the factors that follow). There are two fundamental factors involved in both of these:

1. Time constraints on individual activities and

2. Criteria for using those time constraints to achieve system performance that is acceptably optimal and predictable

The most familiar example of a timing constraint is called a *deadline* by many. A deadline can be defined as a time:

1. Until the completion (answer) of software processing can be used by the system and

2. After which the completion has less use to the system

Deadlines have forms, typically, hard and soft. A *hard deadline* has two situations where the utility (functionality of the software system) is all or nothing:

1. Deadline is met and the utility is achieved, as in a shutdown message is sent on time to the engine controller or

2. Deadline is not met and no utility is achieved, as in a message is sent to the controller but after the time needed to shut the engine down causing missed accuracy.

First, we have "hard deadlines," where a utility value can be determined and measured in absolute terms of being met or not:

- Positive utility value, for example, if the missile is destroyed miles before it hits a ship that it targeted.

- Zero utility value, as in a message is sent from the receiver station after a mobile device is out of range, meaning the message cannot be received. (The message might as well not have been sent.)

- Negative utility value, as in an incoming missile is destroyed but only 20 ft from the ship it was aimed at, meaning the ship may be damaged.

The next term you may hear is "soft deadline." A *soft deadline* is a general case, where utility of the software in time performance is measured in terms of the deadline being

- Met, so that use and/or functionality is achieved, or

- Missed, where a difference of completion time minus deadline is calculated, with tardiness being the difference being positive lateness or late and earliness being negative lateness or too soon.

Larger values of lateness or tardiness usually represent lower utility (not good), and consequently larger positive values of earliness may represent greater utility (better), since this time is available for other processing or activities, although this may not always be the case in timing. Examples of soft deadlines met might be the time it takes to update a smartphone web display, which may vary from 0.1 to 2.0 s. The longer times might be acceptable since most users expect some "delay." An example of large values of lateness might be where the display on the smartphone takes "too long," which may be a number more than 5.0 s. Studies have shown that 5.0 s is as long as a typically tolerant user will wait. The more time past 5.0 s, the worse performance gets, and the unhappier a user may be. However, if we finish the screen update early, say, less than 0.1 s, the smartphone can do other processing, which equates to a happy user. We may also want the longer times of performance to include a display that says something like "working" or shows a status bar or a clock spinning indicating that the system is processing, so the user is willing to "wait" as long as they do not have to wait too long (in the minute range).

Deadlines are context unique and often misapplied. Missing a deadline is a bug or issue, which may be either

- Relatively insignificant such as a sensor data sample is missed but an old value can be used and the reading can be repeated on the next cycle (user is not aware of a miss) or

- Relatively significant such as a sensor data sample is missed and then the target is missed.

Further, any given deadline value may be either

- Relatively insignificant such as a smartphone screen performance update for a game is missed or

- Relatively significant such as a screen performance update on a mobile device is missed, and because of that, someone missed a big stock sale.

Finally, we should note that performance deadlines and times are not intrinsically short or long. They can be

- Short, microseconds such as an engine stop within 100 ns

- Longer, 5 s to load a page on the smartphone using a 4G network, which has 70% traffic usage while the device has 10 active apps running

- Really long, megaseconds such as a fire in a spacecraft engine 300 days into a flight to Mars

IMPACTS OF TIMING

Many mobile and even embedded systems may have all of these types of deadlines and performance considerations. The timing concepts presented in this appendix are only a beginning. Testers need to understand these concepts and more. Beyond deadlines are more general time constraints and measures of performance.

We often create written performance specification requirements of a feature's time utility, which can be tested against, but many projects do not have the "written" numbers to test to. In either case, timing attacks are possible. Time-based measures of performance for noncritical (no deadline) process can be a little "looser," in other words not written down. For example, "How long does a person on a cell phone wait for a message or call before getting through? What happens to a game if excessive wait time happens? How many apps can a user have running before their smartphone becomes stupid (i.e., freezes)?" These can all be measured and considered with or without written performance numbers. Next, what performance data does the team really need? *There is no one right answer.* An embedded controller for a rocket or airplane will likely really need a lot of analysis, written numbers, and testing, but the mobile human-dominated world should not be short changed on this attack just because there are limited written or recorded performance numbers.

Appendix E: Detailed Mapping of Attacks

I AM OFTEN ASKED WHICH ATTACKS APPLY IN SOME CONTEXT. There are many context areas, and everyone will have different attacks to consider. While every test team should plan and consider a variety of test concepts, strategies, plans, and supporting attacks, it is useful for many of you who are new to the idea of attack-based testing to see a mapping of attacks to a context, which might be close to their respective test problem.

Table E.1 is provided as a sample starting point for attacks that you should consider. Many people may read the table and a context and say "why that attack and not another?" This mapping is based on my experience and a lot of years of teaching classes, where people from different contexts considered how testing might be done. However, many variations and options are possible, so local context and critical thought is needed. There is no single absolute "right."

Note: Each chapter also has a mapping of attacks to general mobile and/or embedded contexts.

TABLE E.1   Detailed Attacks to Consider plus Some Optional Attacks

Context Area	Attacks to Consider	Optional or Possible Attacks
Telecom (general)	1, 2, 3, 6, 8, 9, 10, 11, 12, 13, 14, 15, 16, 17, 18, 19, 20, 21, 22, 23, 24, 28, 33	29, 30, 31, 32
Switch	1, 2, 3, 4, 5, 6, 7, 8, 9, 10, 12, 13, 14, 15, 16, 17, 18, 19, 20, 21, 28, 33	11, 22, 23, 24, 29, 30, 31, 32
Cell (network side)	1, 2, 3, 4, 5, 6, 7, 8, 9, 10, 12, 13, 14, 15, 16, 17, 18, 19, 20, 21, 28, 33	11, 22, 23, 24, 29, 30, 31, 32
Handheld	1, 2, 3, 4, 5, 6, 8, 9, 10, 12, 14, 15, 16, 18, 19, 20, 21, 22, 23, 24, 25, 26, 27, 28, 29, 30, 32, 33	13, 17
Medical (info system)	1, 2, 3, 4, 5, 6, 7, 8, 9, 10, 12, 13, 14, 15, 16, 17, 18, 19, 20, 21, 22, 23, 24, 25, 26, 27, 28, 29, 30, 32, 33	11
General info support	1, 2, 3, 4, 5, 6, 7, 8, 9, 10, 11, 12, 13, 14, 15, 16, 17, 18, 19, 20, 21, 22, 23, 24, 25, 27, 28, 29, 30, 33	32
Smart device	1, 2, 3, 6, 8, 10, 13, 14, 15, 16, 18, 19, 20, 21, 22, 23, 24, 25, 26, 27, 28, 29, 30, 31, 33	4, 5, 7, 12, 17, 32
Cell device (simple)	1, 2, 3, 6, 8, 12, 14, 15, 18, 19, 20, 21, 22, 23, 24, 25, 26, 33	4, 5, 7, 12, 17, 28, 30, 32
Phone (smart)	1, 2, 3, 4, 5, 6, 7, 8, 9, 10, 12, 13, 14, 15, 16, 17, 18, 19, 20, 21, 22, 23, 24, 25, 26, 27, 28, 29, 30, 32, 33	
Tablet	1, 2, 3, 6, 9, 10, 12, 14, 15, 16, 19, 20, 21, 22, 23, 2, 4, 25, 26, 27, 28, 29, 30, 32, 33	
Medical (life critical)	1, 2, 3, 4, 5, 6, 7, 8, 9, 10, 12, 13, 14, 15, 16, 17, 19, 20, 21, 22, 23, 2, 4, 25, 26, 27, 28, 29, 30, 33	32
Patient support	1, 2, 3, 4, 5, 6, 7, 8, 9, 10, 12, 13, 14, 15, 16, 17, 18, 19, 20, 21, 22, 23, 24, 25, 26, 27, 28, 29, 30, 33	11, 32
Inserted in humans	1, 2, 3, 4, 5, 6, 7, 8, 9, 10, 12, 13, 14, 15, 16, 17, 18, 19, 20, 21, 22, 23, 24, 25, 26, 27, 28, 29, 30, 33	11, 32
Robotics	1, 2, 3, 4, 5, 6, 7, 8, 9, 10, 11, 12, 13, 14, 15, 16, 17, 18, 19, 20, 21, 22, 23, 24, 25, 26, 27, 28, 29, 30, 31, 32, 33	
Agile	1, 2, 3, 4, 5, 6, 7, 8, 9, 10, 11, 12, 13, 14, 15, 16, 17, 18, 19, 20, 21, 22, 23, 24, 25, 26, 27, 28, 29, 30, 32, 33	31
Life critical	1, 2, 3, 4, 5, 6, 7, 8, 9, 10, 11, 12, 13, 14, 15 ,16, 17, 18, 19, 20, 21, 22, 23, 24, 27, 28, 29, 30, 32, 33	31
Mission critical	1, 2, 3, 4, 5, 6, 7, 8, 9, 10, 11, 12, 13, 14, 15, 16, 17, 18, 19, 20, 21, 22, 23, 24, 27, 28, 29, 30, 31, 32, 33	
Avionics/control	1, 2, 3, 4, 5, 6, 7, 8, 9, 10, 11, 12, 13, 14, 15, 16, 17, 18, 19, 20, 21, 22, 23, 24, 27, 28, 29, 30, 32, 33	
Automotive (gen/info)	1, 2, 3, 4, 5, 6, 7, 8, 9, 10, 11, 12, 13, 14, 15, 16, 17, 18, 19, 20, 21, 22, 23, 24, 25, 26, 27, 28, 29, 30, 32, 33	
Automotive (control)	1, 2, 3, 4, 5, 6, 7, 8, 9, 10, 11, 12, 13, 14, 15, 16, 17, 18, 19, 20, 21, 22, 23, 24, 27, 28, 30, 32, 33	
Rockets	1, 2, 3, 4, 5, 6, 7, 8, 9, 10, 11, 12, 13, 14, 15, 16, 17, 18, 19, 20, 21, 27, 28, 30, 32, 33	
Spacecraft	1, 2, 3, 4, 5, 6, 7, 8, 9, 10, 11, 12, 13, 14, 15, 16, 17, 18, 19, 20, 21, 22, 23, 24, 27, 28, 30, 32, 33	
Aircraft	1, 2, 3, 4, 5, 6, 7, 8, 9, 10, 11, 12, 13, 14, 15, 16, 17, 18, 19, 20, 21, 22, 23, 24, 25, 26, 27, 28, 29, 30, 32	

TABLE E.1 (continued)   Detailed Attacks to Consider plus Some Optional Attacks

Context Area	Attacks to Consider	Optional or Possible Attacks
Industrial/buildings	1, 2, 3, 4, 5, 6, 7, 8, 9, 10, 11, 12, 13, 14, 15, 16, 17, 18, 19, 20, 21, 22, 23, 24, 27, 28, 29, 30, 31, 32, 33	
Machines/milling	1, 2, 3, 4, 5, 6, 7, 8, 9, 10, 11, 12, 13, 14, 15, 16, 17, 18, 19, 20, 21, 22, 23, 24, 27, 28, 29, 30, 31, 32, 33	
Lights	1, 2, 3, 4, 5, 6, 7, 8, 9, 10, 11, 12, 13, 14, 15, 16, 17, 18, 19, 20, 21, 27, 28, 29, 30, 31, 32, 33	22, 23, 24
Heating, ventilation, and air conditioning (HVAC)	1, 2, 3, 4, 5, 6, 7, 8, 9, 10, 11, 12, 13, 14, 15, 16, 17, 18, 19, 20, 21, 25, 26, 27, 28, 29, 30, 31, 32, 33	22, 23, 24
Building control	1, 2, 3, 4, 5, 6, 7, 8, 9, 10, 11, 12, 13, 14, 15, 16, 17, 18, 19, 20, 21, 22, 23, 24, 27, 28, 29, 30, 32, 33	31
Home/office/entertain	1, 2, 3, 4, 5, 6, 7, 8, 9, 10, 11, 12, 13, 14, 15, 16, 17, 18, 19, 20, 21, 22, 23, 24, 25, 26, 27, 28, 29, 30, 32, 33	
Control	1, 2, 3, 4, 5, 6, 7, 8, 9, 10, 11, 12, 13, 14, 15, 16, 17, 18, 19, 20, 21, 22, 23, 24, 25, 26, 27, 28, 29, 30, 32, 33	
Gaming (dedicated)	1, 2, 3, 6, 11, 12, 21, 22, 23, 24, 25, 26, 27, 28, 29, 30, 32, 33	18, 19, 20
Transportation (railroad, traffic control)	1, 2, 3, 4, 5, 6, 7, 8, 9, 10, 11, 12, 13, 14, 15, 16, 17, 18, 19, 20, 21, 22, 23, 24, 27, 28, 30, 31, 32, 33	
Utilities (power, water, sewer, others)	1, 2, 3, 4, 5, 6, 7, 8, 9, 10, 11, 12, 13, 14, 15, 16, 17, 18, 19, 20, 21, 22, 23, 24, 27, 28, 30, 31, 32, 33	

Appendix F: UI/GUI and Game Evaluation Checklist

THIS APPENDIX MOVES MORE INTO THE CONCEPTS OF USABILITY, which is a wide and complex topic. I will treat it lightly here, but added research may be needed [1–5]. As a start in usability, you will need to consider the following general factors, which are expanded on in the checklist:

- The user (age, ability, disability, language/culture, experience, and others)

- The app (what is it and what does it do)

- The project context (cost, schedule, risk, and others)

- The mobile or embedded device (physical features such as screen, connectivity, central processing unit [CPU], memory, and others mentioned in the attacks of this book)

- The environment the device will be used in (factory, weather, sea, air, space, and the universe)

- Level of mobility of the device (vendors, networks, location, and others)

These can all vary and change over time, even for the same app. When considering aspects of any attack where there is a user interface (UI) or graphical user interface (GUI), usability may be a consideration, and so these factors to one degree or another can come into play. I offer this checklist as a simple starting point that can be used with games, UI, and GUI evaluations.

This evaluation, as most attacks, is iterative and incremental over the project stages, involves the whole team (hardware, system, software, test, management, and other stakeholders), and must reflect the UI/GUI factors. There is no one "right" combination of factors, so the checklist evaluation conducted by testers with any attacks can provide useful information throughout the life cycle.

The checklist provides a very simple "model" that the evaluation can be run against. There are many models that might also be used, including experience, other products, industry standard reference models (when they exist), top performance example products, performance data (see Chapter 6), and user surveys. These are out of the scope of this appendix and book.

There are a number of ways in which you assess these models, but here I use this simple checklist and the testers' "expert" judgment heuristic (do you like it or not, maybe with some scale rankings 1 to 5). If your team has several testers, you can get a better sampling of judgment by using the whole test team. If your project wants real usability testing, this is outside of the scope of this book and checklist. There are publicly available usability surveys such as Software Usability Measurement Inventory (SUMI) and Website Analysis and Measurement Inventory (WAMMI) (search the Internet for the latest versions). Such heuristic evaluations have been useful in many settings, including information technology (IT) and the web world, so applying them to the mobile and embedded domains makes sense.

This checklist does have specializations for Attack 26. In the game app mobile device world, there are a variety of things to check, which can easily be forgotten, so I provide them in the checklist. Also, consider using this checklist with or in Attacks 33, 26, and 25.

Finally, I recommend that you customize your checklist to the specific app, with such questions as "Is it a game? Is it an informational app? Is it life critical? Is it to be fun?" in the overall context of the software. I would expect the list items to change over time. Finally, the tester should have these items in a spreadsheet or other kinds of list, so that the items are actually "checked" off as you go, so you will make sure not to forget an item.

CHECKLIST OF USABILITY FACTORS AND CRITERIA

Check off these for each level or page of the app:

1. Buy/install/uninstall interfaces: Download from the app store/load site works

    a. Interesting and correct download promo info

    b. Interesting graphics on promo

    c. Key load restrictions are correct (accept, legal, age info, fees, etc.) and function

    d. Program actually installs and launches (on different devices)

    e. Start-up page works per story/use cases

    f. Confirm an uninstall works (app removed, clean up old files, system still works after)

2. Start-up app landing splash page is correct and informative (company, networking, welcome, graphics, fees, authors, start features, copyrights, trademarks, date, version, disclaimer, and other information)

3. Menu function UIs for each display page (test each of these with Attack 33 and track each page)

    a. Identify a mapping of what functional items exist

    b. Coverage of each function, display screen

    c. Force outputs (as needed of the game, app, server (Attack 8))

    d. User expectation functions work (Attack 22)

    e. Check that online/written help features are correct (Attack 24)

4. User inputs

    a. Touch screen (tap, double tap, corners, objects, zoom, long, short, slow, etc.)

    b. Keyboards (size, position, correct label, use, hard or soft keys)

    c. Voice (male, female, accent, child, soft, loud, etc.)

    d. Optional "plug" devices (readers, sensors, trackballs, mice, etc.)

    e. Device "orientation" and status (on network, off, flat, rotated, etc.)

    f. Ease of using inputs (1 to 5 scale)

5. Graphic/display rendering—check (if they exist)

    a. Fits in screen size (different sizes and devices)

    b. Screen orientation (try all combinations)

    c. Static—art, terrain, maps, figures, objects, texture, color, location, check for missing parts, and look at different screens and/or under different lighting conditions

    d. Dynamic—movement, look and feel, realism, update frame rate, maps, special features, character/object movement interactions, and look at on different screens and/or under different lighting conditions

    e. Text—correct display, location, visible on screen is the meaning clear, spelling, and reader level

    f. For games or graphic programs with camera view check: zoom, rotation, movement, replay, quality, and ease of use

    g. Check the whole display environment (including any hidden parts)

    h. Performance is sufficient—see Attack 31

    i. For gaming, does the end of the game work (the user either quits, saves, or really finishes the game)?

    j. Control display work in different device environments (is it device dependent)

6. Sound

    a. Sync to graphics

    b. Volume controls

    c. Clarity (on different mobile systems, local speaker, headphone jack) of music, voice, sound effects, and modes (on/off/vibrate/multi)

    d. Correctness (alarms, an alarm, not soft music)

7. For games, does game flow and logic (is it fun and playable)?

    a. Story and/or storyboard (design) adherence or consistency and flow of the game (does it make sense)

    b. Levels and difficulty (is it too hard or easy)

    c. Options (at each level)

    d. Correctness of levels: Actions at a level; characters at a level; abilities at a level; resources at a level; are the conditions to move levels met when a level is moved

    e. Events, triggers, what is needed to play

8. For games, scoring

    a. Pregame and postgame conditions

    b. User

    c. High score

    d. Score at a level

    e. Current score

    f. Reset states to initial scoring when selected (if an option)

9. For games, player logic

    a. Role to story line

    b. Movement

    c. Capabilities

    d. Positioning

    e. Multiplay working

    f. Rules match storyboard and reasonable expectations

    g. Game interface controls (working with pads, pointers, device movement, sensors, touch screen, and others)

10. "Fun" factors (apps and games)

    a. Is it hard but not too hard (for level or age)?

    b. Is it easy but not too easy?

    c. Does it "meet" advertised ideals?

    d. Does it compare to other apps in this segment of the industry?

    e. Does the user get bored, frustrated, "get it" (quickly with minimal "help"), excited ("I want that because it is new and cool")?
    Note: Consider creating a ranking of fun/usefulness factor (1 = worst and 10 = best) for each of these.

    f. Does the app have localization issues (one country vs. another country or language)?

11. Breaking the app and system (from many attacks in this book)

    a. Stress all inputs valid and invalid

    b. Device physical limitations (CPU, memory, network, secondary storage)

    c. Power management (drains on battery, sleep mode works, etc.)

    d. Bandwidth connect impacts (network, cell, Wi-Fi, radio link, fast, slow, dropped, etc.)

    e. 2G, 3G, 4G, airplane mode, or other protocol systems

    f. Security working (if any, see Chapter 9: lock/unlock, penetration, fraud, etc.)

    g. Performance test (Chapter 6)

    h. Long-duration test (Attack 6)

    i. Stress all interfaces

    j. Network—fast, slow, in, and out

    k. Combinations of interfaces (see Attack 32)

    l. Shock/vibration of hardware

    m. Lighting in environment

    n. Noise in the environment

    o. Bake or smoke test of hardware (hot environments)

    p. Reset mode, if any, works

    q. App in low-power setting or situation

    r. Test mobile in different environmental modes (weather, analog, digital, on- or off-line)

    s. Different hardware configurations (Attack 32)

    t. Save data and app information—recall default, persistence of unwanted data, resets

TABLE F.1   Example of a Simple, Multilevel Rating System When Needed for Checklist or Survey

Rating	Description
1	This feature is "usable."
2	Cosmetic concern on usability and fixed if time is available on project.
3	Minor usability problem with a low-priority fix (only after higher priority).
4	Major usability problem with a high-priority fix (do soon).
5	Catastrophic usability problem (fix before any further testing or before release of product).

u. App messaging/output forcing—force all errors, warnings, status, or other messages

v. Run on different hardware platform configurations (see Attack 32)

w. Run on different software configurations (run Attack 32 and monitor what is or can be running at same time, other apps, and so on)

Some of the aforementioned checks recommend a rating system, and many of the criteria could be expanded to have a rating system applied. Here is simple rating system example (Table F.1).

REFERENCES

1. Jeffrey, E. 1994. *Handbook of Usability Testing: How to Plan, Design, and Conduct Effective Tests*, John Wiley & Sons, New York.
2. J. Kohl Web site http://www.kohl.ca/ (last accessed April 9, 2013).
3. Rogers, R. 2011. *Learning Android Game Programming: A Hands–On Guide to Building Your First Android Game*, Addison-Wesley, Boston, MA.
4. Schultz, C.P., Bryant, R., and Langdell, T. 2005. *Game Testing All in One*, Course Technology PTR. http://tech-lib.net/Course%20Technology%20-%20Game%20Testing%20All%20in%20One/Game%20Testing%20All%20in%20One (last accessed April 9, 2013).
5. Milano, D.T. 2011. *Android Application Testing Guide*, Packt Publishing, Birmingham, U.K.

Appendix G: Risk Analysis, FMEA, and Brainstorming

THROUGHOUT MUCH OF THIS book, I have talked about risks, risk identification, and using risk in testing. This appendix gives some basic risk analysis concepts. Risk analysis is a big subject with many books and classes [1,2]. I wanted to give you enough points and information to get you started, but you must understand risk associated with your product and then adapt risk analysis for each attack.

The software and systems engineering vocabulary (SEVOCAB) [3] defines risk as

> (1) an uncertain event or condition that, if it occurs, has a positive or negative effect on a project's objectives (A Guide to the Project Management Body of Knowledge (PMBOK(R) Guide)—Fourth Edition) (2) combination of the probability of an abnormal event or failure and the consequence(s) of that event or failure to a system's components, operators, users, or environment. (IEEE 829-2008 IEEE Standard for Software and System Test Documentation, 3.1.30)

In this book for simple risk assessment, a tester starts risk analysis with information collection, which can be done using interviews, history assessment, review of taxonomies, independent checks, workshops, checklists, organized brainstorming, and/or customer interviews. The knowledge coming from these efforts should be captured and recorded using note cards, spreadsheets, tables, text files, tools, or even on white boards with pictures. There are risk tools and systems to help in doing formal risk analysis, but here I will keep it simple by using a list in a table of risk statements.

A risk statement defines a clear description of the risk, which can be understood, so that it can be tested. Classically, risk descriptions capture a single condition followed by details of the potential consequence (potential problem or risk). One way to do this is with a statement structure of

```
<if condition>, then <consequence(s)> + <time factor>
```

If condition—a single phrase citing a single key circumstance or situation that can cause the concern, doubt, case, or uncertainty. The "if" should not have an "and" since this can indicate multiple risks.

Consequence—a single phrase or sentence describing the key, negative outcome of a condition. A consequence can have "and" statements.

Time factor—a single phrase or sentence that captures a time factor or implication of a time factor of when a risk can or may occur. Time factors are optional.

Examples:

> If the hardware has a noise factor of 10 in the software, then software may hit the trip level for shutdown, resulting in engine shutdown prematurely and causing the crane to drop within 20 s.
>
> If the display screen on the smartphone is not bright enough on adjustment, then the user may not be able to see fine screen details resulting in decreased usability (unhappy user).

In risk analysis, to create the risk statements, I recommend an organized meeting and/or brainstorming session be held. It is okay to bring ideas and cases to the meeting since the purpose of this kind of meeting is to promote communication among team members as well to bring forward information useful to the entire team. The team can be composed of testers, systems and/or hardware/software personnel, users, and other interested parties. Everyone should have a voice and be heard at these meetings.

Testers can also generate the own risk statement list working by themselves. While not as effective as a group meeting or brainstorming session, testers can use this book, taxonomies, histories, their experience, industry data, and general "fear" to identify risk statements.

Throughout the book, I have presented summaries of the Appendix A taxonomy and general list of risk factors associated with attacks. These are repeated from Chapter 3.

Testers can define attack cases based on these common areas of risk:

- Safety—when the well-being of humans is threatened

- Security—data or information can be exposed to the bad guys

- Hazard—damage to equipment or the environment is possible

- Business concerns—bad computations generate wrong information and ends up impacting profits

- Regulations—the product could result in lawsuits or be at odds with government regulations

- Environment and input factors—hardware inputs for devices and electronics, which are susceptible to influence (noise) either systematic or random, including outside communication lines and characteristics, the "real world" (weather or conditions in the real world such as wet roads or rain), and even human operations

TABLE G.1 Example Risk List Table

Risk Statement	Impact	Likelihood	Test/Note
If the hardware has a noise factor of 10 in the software, then software may hit the trip level for shutdown, resulting in engine shutdown prematurely and causing the crane to drop within 20 s.	High	Medium	Attack 20—hit this one hard
If the display screen on the smartphone is not bright enough on adjustment, then the user may not be able to see fine screen details resulting in decreased usability (unhappy user).	Medium	Medium	No—test later if time allows

- Output noise—outputs to devices and electronics that are susceptible to noise influences

- Complexity—the size of the system or some aspect of the system makes missed cases likely

Testers should use these lists as starting points for risk statements.

For attacks and risk-based testing, a tester should focus on technical risk (i.e., a potential problem in the product) associated with the mobile/embedded device. Generally speaking, in defining risks for attacks, you should be less concerned over management types of risk, which mostly concern cost, schedule, and project performance. Testing does not directly address these types of risks. The resulting risk statements get captured in a table such as Table G.1.

Once risk statements have been defined, they should be prioritized. Setting priority for cases and risks can be hard. The literature talks about many methods, but maybe the easiest is just to decide how many tests you have the budget and time for during this attack effort and then sort the cases into two buckets, "test" and "don't test," making sure the test bucket does not exceed how many attacks you can do. If the "test" bucket gets too full, this is useful information to provide back to the team, management, and other parties. It is possible to make the case for more budget and/or time for this attack. It may be more likely to reduce the items in the "test" bucket. In Table G.1, I rated the tests by Impact and Likelihood with categories of High, Medium, and Low. For risks that were Impact = High and Likelihood = High, I assigned a test for others I just made notes, in case I found myself with more time to do attack testing.

The team (management, development, hardware, customers, users, and others) should review the table and classifications. This may shift things. However, in the agile concept of team involvement, getting consensus and agreement is a good thing. We will test the important things first.

Assigning an attack type is only basic planning. Later, the risks will help refine attack details and design. This process of defining risk, prioritization, and planning will be interactive and iterative.

Risk analysis can create the need for many attacks. Good testers can define many more risks than they can test in any one attack or set of test plans. Additionally, as testing and attacks continue, you will think of new or refined risks. The risk list should grow, new tests and attacks will be needed, prioritization will change, and these efforts will likely continue until you run out time and/or money. This is why you should use the risks to prioritize the attacks, attacking the "scariest" things first.

Risk-based testing is advocated by many standards such as in ISO29119 [4] and in concepts such as exploratory testing. Risk-based analysis and testing are mentioned in many attacks, but its use can be applied to almost all attacks, test approaches, and techniques. Good testers think risk and run scared.

FMEA/FMECA

A close concept to risk analysis is failure mode and effects (FMEA) [5], which SEVOCAB defines as

> [Technique] an analytical procedure in which each potential failure mode in every component of a product is analyzed to determine its effect on the reliability of that component and, by itself or in combination with other possible failure modes, on the reliability of the product or system and on the required function of the component; or the examination of a product (at the system and/or lower levels) for all ways that a failure may occur. For each potential failure, an estimate is made of its effect on the total system and of its impact. In addition, a review is undertaken of the action planned to minimize the probability of failure and to minimize its effects. (*A Guide to the Project Management Body of Knowledge (PMBOK(R) Guide)*, 4th edn.)

I suggest that you read more about FMEA [5] to understand how it compares to risk analysis. FMEA is more detailed, gathers more information, has longer tables, and can include "criticality analysis" [8,9] (similar to setting risk likelihood and impact). For those organizations dealing with highly critical software such as software with safety or high dollar impact, the added rigor and information generated by an FMEA may be justified (even legally required).

In several attacks, I recommended considering FMEA when it is justified. When I worked with safety critical systems, we used FMEA. By following up with the references, most testers can get a good start on FMEA. For the rest of the testing world, risk analysis is usually good enough.

BRAINSTORMING (ORGANIZED)

Another excellent supporting technique I keep mentioning is brainstorming. Brainstorming is usually a team effort to get better ideas by using everyone's knowledge and skills. It can be used in risk analysis, planning, defining which attacks to run, attack design, as well as test data and case selection—just to name a few. Although I have outlined some basics of brainstorming, by reviewing the reference materials [6], you will gain a more complete understanding of how brainstorming can help your project and how it can change your thinking.

A basic flow for brainstorming the cases for this attack could proceed as follows:

1. Define the attack area and approach, which can include subtechniques such as risk analysis, FMEA, state charts, and cause and effect [5–9].

2. Identify someone to record each idea.

3. Each person takes a turn to present a single idea, and then the next person goes in a round robin format.

4. When a person does not have an idea, they can say pass to the next person, but when it comes to their turn again, they are free to input an idea.

5. The session continues until everyone has a chance to speak multiple times or the team calls time out.

6. There should be minimal discussion about the ideas, no belittling of an idea and no idea should be considered "bad" as the activities progress.
   (*Note*: The person taking notes can ask clarifying questions.)

After the brainstorming session ends, review the ideas; organize them into a matrix or table. This may result in more ideas, which should also be captured. Some ideas can be modified and even removed but only after all ideas have been brought up. This approach should be repeated over time and test efforts. This exercise has the advantage of engaging the full team, leverages ideas from everyone, does not allow strong people to dominate, encourages teamwork, captures the ideas, and supports a variety of other methods.

REFERENCES

1. Risk analysis. http://en.wikipedia.org/wiki/Risk_analysis_(engineering) (last accessed April 9, 2013).
2. Harvard Center for Risk Analysis. http://www.hcra.harvard.edu/ (last accessed April 11, 2013).
3. Software Engineering Vocabulary. http://pascal.computer.org/sev_display/index.action (last accessed April 11, 2013).
4. ISO29119, part 4 to be published in late 2013. http://www.iso.org/iso/catalogue_detail.htm?csnumber=45142 (last accessed April 11, 2013).
5. Failure mode and effects analysis. http://en.wikipedia.org/wiki/Failure_mode_and_effects_analysis (last accessed April 11, 2013).
6. Brainstorming. http://en.wikipedia.org/wiki/Brainstorming (last accessed April 9, 2013).
7. Myers, G. 1979. *The Art of Software Testing*, Wiley, New York.
8. Leveson, N. 1995. *Safeware: System Safety and Computers*, Addison-Wesley, Boston, MA.
9. Failure mode, effects, and criticality analysis. http://en.wikipedia.org/wiki/Failure_mode,_effects,_and_criticality_analysis (last accessed April 11, 2013).

References

REFERENCES AND ADDITIONAL READING

I provide this section to help share the knowledge that I have gained through many of the books, articles, websites, links, forums, and other sources. The references represent what I use, have on my bookshelf, or have bookmarked in my browser, but none of these are comprehensive. Some of the references in this appendix were cited in the main chapters, but some are new. These references are intended to (1) give credit and (2) provide places for testers to continue learning their craft.

It should be noted that the software testing field is still young. Many of the first and second generations of testers are still alive in 2012. Some data points and concepts are the subject of disagreement between reasonable parties (authors and testers). The discussion and research should continue. Readers should be aware of this, be well read, and analyze their own testing for the local context. I hope that the information provided will help in your personal journey.

Books on My Bookshelf and Many Used as References

Andrews, M. and Whittaker, J. 2006. *How to Break Web Software*, Pearson Addison-Wesley, Boston, MA.

Beck, K. 2003. *Test Driven Development*, Addison-Wesley, Boston, MA.

Beizer, B. 1995. *Black Box Testing. Techniques for Functional Testing of Software and Systems*, John Wiley & Sons Inc., New York.

Broekman, B. and Notenboom, E. 2003, *Testing Embedded Software*, Addison-Wesley, London, U.K.

Cheng, Albert M. K. 2002. *Real Time Systems: Scheduling, Analysis, and Verification*.

Copeland, L. 2003. *A Practitioner's Guide to Software Test Design*, Artech House Publishers, Boston, MA.

Craig, R. and Jaskiel, S. 2002. *Systematic Software Testing*, Artech House Publishers, Norwood, MA.

Crispin, L. and Gregory, J. 2008. *Agile Testing: A Practical Guide for Testers and Agile Teams*, Addison-Wesley Professional, Boston MA.

Enough Software Team. 2012, Don't Panic, *Mobile Developers Guide to the Galaxy*, Enough Software (free).

Gardiner, S. ed. 1999. *Testing Safety-Related Software, A Practical Handbook*, Springer, London, U.K.

Graham, D. and Fewster, M. 2012. *Experiences of Test Automation*, Addison-Wesley, Boston, MA.

Grenning, J. 2011. *Test Driven Development for Embedded C*, The Pragmatic Bookshelf, Raleigh, North Carolina.

Gilb, T. and Graham, D. 2002. *Software Inspections*, Addison-Wesley, Reading, MA.

Harty, J. 2010. *A Practical Guide to Testing Wireless Smartphone Applications*, Morgan & Claypool publishers, San Rafael, CA.

Jones, C. 2008. *Applied Software Measurement: Global Analysis of Productivity and Quality*, 3rd edn., McGraw-Hill, New York.

Kaner, C., Bach, J., and Pettichord, B. 2002. *Lessons Learned in Software Testing*, John Wiley & Sons, New York.

Kaner, C., Falk, J., and Nguyen, H. 1993. *Testing Computer Software*, 2nd edn., Van Nostrand Reinhold, New York.

Labrosse, J.J., Ganssle, J.G., Oshana, R., Walls, C, Curtis, K.E., Andrews, J., Katz, D.J., Gentile, R., Hyder, K., and Perrin, B. 2007. *Embedded Software*, Newnes, Burlington, MA.

Leveson, N. 1995. *Safeware: System Safety and Computers*, Addison-Wesley, Reading, MA, Appendix A.

Marcus, E. and Stern, H. 2000. *Blueprints for High Availability: Designing Resilient Distributed System*, Wiley, New York.

Marick, B. 1995. *The Craft of Software Testing*, Prentice Hall, Englewood Cliffs, NJ.

Milano, D.T. 2011, *Android Application Testing Guide*, Packt Publishing.

Myers, G. 1979. *The Art of Software Testing*, Wiley, New York.

Neumann, P. 1995. *Computer Related Risks*, ACM press, a Division of Addison-Wesley Publishing, Reading, MA.

Pries, K. and Quigley, J. 2011. *Testing Complex and Embedded Systems*, CRC Press, Boca Raton, FL.

Rorabaugh, B. 2010. *Notes on Digital Signal Processing: Practical Recipes for Design, Analysis and Implementation*, Prentice Hall, Hoboken, NJ.

Roychoudhury, A. 2009. *Embedded System and Software Validation*, Morgan Kaufmann, Burlington, MA.

Rubin, J. 1994 *Handbook of Usability Testing: How to Plan, Design, and Conduct Effective Tests*, John Wiley & Sons, New York.

Scambray, J., and McClure, S., and Kurtz, G. 2001. *Hacking Exposed: Network Security Secrets and Solutions*, 2nd edn., Osborne/McGraw-Hill, Berkeley, CA.

Utting, M. and Legeard, B. 2006. *Practical Model-Based Testing*, Morgan Kaufmann, Burlington, MA.

Whittaker, J. 2003. *How to Break Software*, Pearson Addison Wesley, Upper Saddle River, NJ.

Whittaker, J. 2009. *Exploratory Software Testing*, Addison-Wesley Professional, Upper Saddle River, NJ.

Whittaker, J. and Thompson, H. 2004. *How to Break Software Security*, Pearson Addison-Wesley, Boston, MA.

Wiegers, K. 2001. *Peer Reviews in Software: A Practical Guide*, Addison-Wesley, Boston, MA.

Sampling of Standards That Testers Should Be Familiar With (Used in the Production of This Book)

IEC 60300-3-9:1995, Risk analysis of technological systems

IEEE, P11073-10417_D02 2009, Draft Health informatics—Personal health device communication Part 10417: Device specialization—Glucose meter; IEEE

IEEE Standard 610.12–1990, IEEE Standard Glossary of Software Engineering Terminology™

IEEE Standard 829–2008, IEEE Standard for Software and System Test Documentation™

IEEE Standard 1008–1987, IEEE Standard for Software Unit Testing™

IEEE Standard 1012–2012, IEEE Standard for Software Verification and Validation™

IEEE Standard 1028–2008, IEEE Standard for Software Reviews and Audits™

ISO 29119 covers all of software testing (scheduled release 2013)—4 parts: general, process, documentation, and techniques

ISO/IEC 12207:2008, Systems and software engineering—Software life cycle processes

ISO/IEC 15026:1998, Information technology—System and software integrity levels

ISO/IEC 16085:2006, Systems and software engineering—Life cycle processes—Risk management

ISO/IEC 25000:2005, Software Engineering—Software product Quality Requirements and Evaluation (SQuaRE)—Guide to SQuaRE

ISO/IEC 25010:2011, Systems and software engineering—Systems and software Quality Requirements and Evaluation (SQuaRE)—System and software quality models

ISO/IEC 25051:2006, Software engineering—Software product Quality Requirements and Evaluation (SQuaRE)—Requirements for quality of Commercial Off-The-Shelf (COTS) software product and instructions for testing

ISO/IEC/IEEE 24765:2010, Systems and software engineering—Vocabulary

IEEE 730—Software Quality Assurance Plans

IEEE 829—Software Test Documentation (being retired for ISO29119)

IEEE 830—Recommended Practice for software requirements

IEEE 12119—Quality requirements and testing

NASA-STD 8719.13 Software Safety, Revision Level: B with Change 1, Date: 7/8/2004

OMG UML Testing Profile (UTP)

Tool Information (To Name Only a Few)

http://experitest.com/

http://ready.mobi/launch.jsp?locale=en_EN

http://www.deviceanywhere.com

http://www.kozio.com—VTOS™Suite1.0. Kozio's Verification

http://www.ldra.com/

http://www.mathworks.com/verification-validation/embedded-software-test-verification.html

http://www.neweagle.net/support/wiki/index.php?title = MotoHawk_Control_Solutions

http://www.parasoft.com/jsp/resources/embedded_software_development

http://www.perfectomobile.com

http://www.qatestingtools.com/mobile-testing-tools/list (test pointer list)

http://www.testplant.com/ (eggplant)

http://www.windriver.com/ (leader in embedded space, e.g. OS, COTS, tooling)

MOBILE EMBEDDED NEWS AND WEBSITES

Testers should follow these sites that cover many topics (chapters) in this book including mobile, embedded, testing, security, tools, education, and others:

http://en.wikipedia.org/wiki/List_of_mobile_network_operators

http://en.wikipedia.org/wiki/Motor_Industry_Software_Reliability_Association

http://karennicolejohnson.com/2011/07/sites-for-mobile-testing-news-etc/ (good site on mobile testing)

http://mashable.com/mobile/

http://mobithinking.com/mobile-marketing-tools/latest-mobile-stats

http://tapcellphone.com/

http://techcrunch.com/mobile/

http://testingeducation.org/wordpress/ (training in testing)

http://usa.kaspersky.com/ (security info)

http://validator.w3.org/mobile/ (quick test assessments)

http://webtrends.about.com/od/mobileweb20/tp/list_of_mobile_web_browsers.htm

http://www.associationforsoftwaretesting.org/ (a professional society for software testers)

http://www.developsense.com/ (Michael Bolton)

http://www.eetimes.com/

http://www.eetimes.com/design/embedded/
http://www.embedded.com/
http://www.fiercewireless.com/
http://www.forbes.com/
http://www.gomez.com/mobile-readiness-instant-test/ (available quick test assessments)
http://www.google.com/ (look for mobile/embedded groups to subscribe to)
http://www.istqb.org/(tester certification)
http://www.kohl.ca/ (another good site on mobile testing)
http://www.linkedin.com/ (look for mobile/embedded groups to subscribe to)
http://www.mobilecommercedaily.com
http://www.mobileforum.com/
http://www.mobileqazone.com/ (mobile tester information)
http://www.openmobilealliance.org/
http://www.qualitytree.com/classes/ (test training in the Agile world)
http://www.satisfice.com/tools.shtml (James Bach)
http://www.softwaretestpro.com/
http://www.stickyminds.com/ (general one stop shop for testers)
http://www.wirelessweek.com/

Industry Communication Standards and Standard Location Websites (To Name Only a Few)

http://en.wikipedia.org/wiki/Communications_server
http://en.wikipedia.org/wiki/List_of_mobile_phone_standards
http://standards.sae.org/electrical-electronics-avionics/embedded-software/standards/
http://www.mobilein.com/mobile_basics.htm
http://www.rfidc.com/docs/introductiontowireless_standards.htm

Specific examples:

IEEE 802.16m—Advanced Mobile Broadband Wireless Standard
IEEE 1394—Firewire
MIL-STD-1553
OMG (object modeling group) Distribution Service for Real-Time Systems (DDS)

Glossary

O NE OF THE BIGGEST PROBLEMS in any industry is the lack of good communication due to poorly defined terminology or misuse of terms. It has been said that "a *profession* is one in which the practitioners have common meanings and usage of concepts." Readers should be familiar with standard industry terms. In addition and for quick reference, I have provided some definitions in the following table that are critical to understanding the material found in this book.

In this book, I have followed—as much as possible—common usage and definitions from the following sources:

- SEvoc—http://pascal.computer.org/sev_display/index.action

- IEEE—ISO/IEC/IEEE24765:2010Systemsandsoftwareengineering (vocabulary)

- Definitions in the *How to Break* book series by James Whittaker

If you do not find a term in this list, refer to one of sources listed here or one of the references given throughout the book, or you can do an Internet search for the term. (Google can be your best friend in helping you to find things.)

Term   Definition and/or Reference

A2D: Analog to digital.

ACTS: Advanced combinatorial testing system.

app: A software module or software application that makes a device function or compliments another app (application). This is a common term for applications for mobile and embedded devices. See http://en.wikipedia.org/wiki/Mobile_app.

As built: The configuration of a product as it will be delivered and used. This is as opposed to special product configurations for testing, prototypes, etc.

ASIC: Application-specific integrated circuits.

Biometrics: Confirmation or check of identity of humans by their characteristics or traits; see http://en.wikipedia.org/wiki/Biometrics.

BIT: Built in test-building functionality into hardware and/or software to facilitate testing, for example, test circuits, test logic code, data input ports, and data output values, all of which have a primary goal of making testing easier.

Brainstorming: A group effort where "out of the box" thinking is encouraged (http://en.wikipedia.org/wiki/Brainstorming).

Bug: See error.

Checking: The basic activities of confirming that software meets its requirements (see verification).

Coder: Also known as a programmer or developer. Someone who writes code in any software or computing language.

Concept of operations: How a system is to be used, usually over a series of activities; see http://en.wikipedia.org/wiki/Concept_of_operations.

COTS: Commercial off the shelf (can be hardware or software); throughout this book, I have used "off-the-shelf."

CPU: Central processing unit.

Critical thinking: A type of reasonable, deep, and reflective human mental activity that is aimed at deciding what to do(here, during testing); see http://en.wikipedia.org/wiki/Critical_thinking.

CT: Combinatorial testing.

CTT: Code then test.

D2A: Digital to analog.

Debug: The process of diagnosing the precise cause of a known error and then correcting the error. A developer activity that is performed before and after testing.

Developer test: Testing done at a structural or "white box" at the statement or code level, also known as unit testing.

DOE: Design of experiments.

EMI: Electromagnetic interference.

Error: A human action that produces an incorrect result, which could be in software, process, documentation, system, and so on.

Exploratory testing: Software testing that simultaneously learns, designs tests, and executes them; see http://en.wikipedia.org/wiki/Exploratory_test.

Failure: Termination of the ability of a product to perform a required function or its inability to perform within previously specified limits.

Fault: When an error in software manifests itself.

Field testing: Full-system test done at an operational site or in the real world.

FMEA/FMECA: Failure mode and effects analysis and/or failure mode, effects, and criticality analysis—see http://en.wikipedia.org/wiki/Failure_mode_and_effects_analysis and http://en.wikipedia.org/wiki/Failure_mode,_effects,_and_criticality_analysis.

FPGA: Field programmable gate array.

Functional test: Testing done to show that the features (requirements, customer needs, etc.) of the software are present.

GPS: Global positioning system.

GUI: Graphical user Interface.

Hard deadline: A deadline that must be met exactly for software functions to be provided to a customer or user.

Heuristic: Concepts that can solve a problem but cannot guarantee a solution in every case.

HIPPA: Health Insurance Portability and Accountability Act.

HMI: Human–machine interface.

HVAC: Heating, ventilation, and air conditioning.

ICD: Interface control document.

ICS: Industrial control system.

ID: Identity.

IEC: International Electrotechnical Commission.

IEEE: Institute of Electrical and Electronics Engineers

Implementation: How software is coded using models, languages, constructs and others.

Implementation testing: Also known as developer testing.

Interrupts: Hardware-based signal generated to the software for action (http://en.wikipedia.org/wiki/Interrupt).

Invalid: Data that is not expected as input into the system but may be received anyway.

IP: Internet protocol.

ISO: International Standards Organization.

IT: Information technology.

IV&V: Independent verification and validation.

Jail breaking: The process of removing restrictions imposed by vendor(s) on devices running various operating systems through the use of hardware/software exploits to gain root access and circumvent vendor "safeguard" features, for example, what you can load or do. For example, http://en.wikipedia.org/wiki/Jailbreak_(iPhone_OS).

LIB: Least significant bit.

Load (test): Testing, which puts the software under conditions where you can determine how much processing a computer performs, for example, usage of CPU, memory, time, network bandwidth. or others.

LOC: Lines of code.

Malware: Short for malicious software, which is code constructed to do "harm," for example, virus; see http://en.wikipedia.org/wiki/Malware.

Mind map: A method, usually a diagram, that captures a human's understanding to visually outlined information; see http://en.wikipedia.org/wiki/Mind_map (Last accessed April 9, 2013).

Model: A representation of a real-world process, device, software, or concept, which can be logical, physical, and/or mental.

Mutation testing analysis: A test technique in which variations of data or code are created and then used in the test activities; see http://en.wikipedia.org/wiki/Mutation_testing.

NET: Internet.

NIST: National Institute of Standards and Technology.

Noise: In the physics world and analog electronics, noise is mostly an unwanted random addition to a signal picked up by sensors or electronics, which can impact software processing.

Normal: Typical usage.

OCR: Optical character recognition using a system and special software.

Off normal: Nontypical usage.

OO: Object-oriented.

Oracle: Any approach to defining or judging results generated by test. Oracles can include tester judgment, mental models, secondary software programs, and formal models., and others

OS: Operating system.

PC: Personal computer.

PDA: Personal digital assistant.

Performance test: Testing focused on requirements and issues related to system execution in areas of speed, load, response, etc. There are numerous techniques and tools that support performance testing.

Pesticide paradox: A concept in software testing where if the exact same test is used over and over; the likelihood of it finding errors decreases with each use.

PIF: Product improvement file.

PLC: Programmable logic controller, a digital computer used for automation of industrial processes, such as machinery control in factories.

Power down: Turn off a system.

Power up: Turn on a system.

Priority inversion: Priority inversion is a scheduling problem that happens when a low-priority task grabs a resource that a higher-priority task needs and so the high-priority task is forced to wait for it, but then another priority task runs, preventing the low-priority task from finishing with the resource and releasing it, which prevents the high-priority task from ever running. This can create deadlock and system failures. This problem is often associated with interrupt-driven software systems.

Programmer: Also known as a developer or coder. Someone who writes code in any software or computing language.

Race (conditions): See priority and inversion and http://en.wikipedia.org/wiki/Race_condition.

Regression testing: Regression testing involves retesting portions of software items after modification of associated software products. Modifications that may influence previous testing can include changes to code, patches, data, requirements, interfaces, operational uses, and hardware.

Reliability: The probability that software will not cause the failure of a system for a specified time under specified conditions. This probability is a function of the inputs and use of the system, as well as a function of the existence of faults in the software. The inputs to the system determine whether existing faults are encountered.

Risk analysis: See http://en.wikipedia.org/wiki/Risk_analysis_(engineering). (Last accessed April 9, 2013)

RMA: Rate monotonic analysis.

ROM: Read only memory.

Safing: Logic or hardware constructs that place a device into a "safe state" after a negative event such as a fault, a failure, hardware breakage, and network communication problems and others.

SCA: Static code analysis.

SCADA: Supervisory control and data acquisition—a type of ICS.

Scripted: Testing in which written or automated information is generated before the test to determine the "course" (or execution sequence) of the test.

SD: Secure digital.

SDK: Software development kit.

SEVOCAB: Software and systems engineering vocabulary.

Side effect: A situation where code is changed or a bug occurs in one location in the software logic, but another area of code is impacted. This is associated with the concepts of coupling and cohesion in software.

SIL: System integration lab.

SIM: Subscriber identification module.

Smart device: Any device that exhibits some processing capability (either computer or integrated circuit, FPGA, or other). These range from smart light switches to handheld systems (phones and tablets).

SOA: Service oriented architecture.

SOAP: Simple object access protocol.

Social engineering: http://en.wikipedia.org/wiki/Social_engineering_(security).

Soft deadline: A deadline that must be met for functionality to be provided but has some degree of time flexibility.

SOX: Sarbanes–Oxley Act.

SQL: Structured query language.

Stress (test): Tests with emphasis on robustness, availability, and error handling of the software under some load. These cases can be valid or invalid test cases.

Structural testing: Also known as white-box testing.

Success criteria: The information (data) that defines when and how a particular test case is satisfied. This is specified before the test is run.

SUMI: Software usability measurement inventory.

SysML: Systems modeling language.

TDD: Test-driven development.

Test: An activity in which a system or component is executed under specified conditions, the results are observed or recorded, and an evaluation is made on some aspect of the software system or component providing information on these to interested parties.

Test case: A single set of data inputs that result in one set of test outputs for any given test environment. (A test attack may have one or more test cases.)

Test like you fly: The test environment is as close to a production, field, or operational environment as possible. Environment includes hardware, connections, data, communications, and operations. There may be practical limitations to testing, so while this is a good idea, it is often not possible to achieve, which leads to field testing.

Test strategy: The set of ideas (i.e., methods and objectives) that guide test design and execution.

Test technique: Test method; a heuristic or algorithm for designing and/or executing a test.

Test tools: Hardware and/or software aids that help to automate some aspect of testing. There are varying levels of test tool automation.

Testability: The ability of an item to be tested in a reasonable manner.

Testing: Questioning a product in order to evaluate it (Bach version); technical investigation of a product, on behalf of stakeholders, with the objective of exposing quality-related information of the kind they seek (Kaner version).

Time box: A time management and scheduling approach where limits of time (start and stop) are placed on an activity; see http://en.wikipedia.org/wiki/Timeboxing.

Time lines: An order sequence of time (linear).

Timers: A clock that measures time that can be absolute or relative.

Tours: A logically ordered sequence of test activities. For example, stories, techniques, or attacks, which are centered around a theme or concept. For example, a world tour, an error tour, or a hacking tour.

UI: User interface.

UML: Unified modeling language.

Unscripted: Testing in which there is no (or minimal) written or automated information generated before the test to determine the "course" (execution sequence) of the test.

Unverified failure: A bug or error that cannot be repeated or confirmed (that it is a bug) and fixed (you cannot fix what you cannot repeat or find). This is a problem for testing because if we see a potential problem (say the system crashes), but cannot make it happen again, you know there is some kind of bug, but not how to repeat it, find it, nor fix it.

USB: Universal serial bus.

UT: User test.

UTC: Universal time code.

Valid: Data or test cases that are within the "expected" usage of the system software.

Virtualization: Created environments that are "not the real thing" (not actual), such as a hardware platform, OS, storage device, real world, or network resources; see http://en.wikipedia.org/wiki/Virtualization.

VPN: Virtual private network.

W3C: World Wide Web Consortium.

Walled garden: The area where a service provider limits applications, content, and/or media to set platforms and/or restrictions on content. For example, on a wireless network, to an app in a s tore, or other vendor control aspects, see http://en.wikipedia.org/wiki/Walled_garden_(technology). This concept can make testing and testing with some devices difficult (see rooting and jail breaking in Wikipedia).

WAMMI: Website analysis and measurement inventory.

White-box testing: Also known as structural testing.

ZIF: Zero insert force.

Index